Universitext

Universitext is a series of textbooks that presents material from a wide variety of mathematical disciplines at master's level and beyond. The books, often well class-tested by their author, may have an informal, personal, or even experimental approach to their subject matter. Some of the most successful and established books in the series have evolved through several editions, always following the evolution of teaching curricula, into very polished texts.

Thus as research topics trickle down into graduate-level teaching, first textbooks written for new, cutting-edge courses may find their way into *Universitext*.

Ana Agore

A First Course in Category Theory

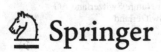 Springer

Ana Agore
Institute of Mathematics
Romanian Academy
Bucharest, Romania

Vrije Universiteit Brussel
Brussels, Belgium

ISSN 0172-5939   ISSN 2191-6675   (electronic)
Universitext
ISBN 978-3-031-42898-2   ISBN 978-3-031-42899-9   (eBook)
https://doi.org/10.1007/978-3-031-42899-9

Mathematics Subject Classification: 18-01

This work was supported by Fonds Wetenschappelijk Onderzoek, Belgium and Romanian Ministry of
Education and Research, CNCS/CCCDI-UEFISCDI.

This Springer imprint is published by the registered company Springer Nature Switzerland AG
The registered company address is: Gewerbestrasse 11, 6330 Cham, Switzerland

Paper in this product is recyclable.

To my parents

Preface

Categories were first considered in 1945 in a paper by S. Eilenberg and S. Mac Lane [21] with the purpose of formalizing the concept of "natural transformation", which was informally used at that time in many papers from various fields, especially in algebraic topology. The initial theory introduced in [21] developed rapidly, allowing for several new mathematical disciplines to arise, as was the case, for example, with homological algebra. Category theory is based on the idea that many mathematical properties can be described using diagrams of arrows of different types. Working in this very general setting allows for a better understanding of the common constructions and patterns in mathematics and leads to a unified treatment of similar concepts across different mathematical structures. An early and notable example can be found in [19], where group cohomology, Lie algebra cohomology and associative algebra cohomology are recast as derived functors in a suitable module category.

Over the years, category theory has become a universal language allowing mathematicians to achieve important advancements by exchanging ideas and techniques between seemingly unrelated domains. Using very abstract definitions that capture the idea behind a certain concept in universal terms rather than its isolated properties, purely categorical techniques have found their way into most mathematical areas.

Nowadays category theory is an indispensable tool for doing research not only in various areas of pure mathematics such as algebraic topology, homological algebra, algebraic geometry and functional analysis, but also in theoretical computer science (e.g., the development of algorithms, automata theory), physics (e.g., electrical circuits), chemistry (e.g., chemical interactions), biology (e.g., biological systems) and medicine (e.g., genetics). We refer to [52] for an approach to category theory in the spirit of the applied sciences and to [29] for applications to the cognitive sciences.

The purpose of this book is to provide students with no prior exposure to categorical reasoning with an accessible source from which to learn the basic material. The fundamentals of category theory are clearly and thoroughly covered with the aim of leaving the reader able to confidently use categorical techniques as well as to easily explore and understand more advanced topics.

The book is based on my lecture notes from the graduate course on category theory that I have taught at Vrije Universiteit Brussel. Additional fully worked examples and complete proofs have been added with the purpose of making the material suitable for self-study. Although the reader is expected to be at the advanced undergraduate level, some background and full references are provided throughout the book. The prerequisites include familiarity with group theory, rings, modules and topological spaces, as well as a basic understanding of set theory. As opposed to the standard category theory monographs by S. Mac Lane [35] and F. Borceux [8–10], and the more recent ones [5, 34, 47, 48], which are more encyclopedic in nature and oriented toward researchers rather than students, the present book serves as a first introduction to the field. The excellent monographs [1, 2, 5, 8–10, 12, 13, 22, 23, 25, 30, 34, 35, 46–48] have been used when preparing these notes and have influenced the approach and the development of certain topics.

The first chapter introduces the fundamental concepts needed in the sequel. Important notions such as (sub)categories, functors, natural transformations, representable functors, which form the backbone of category theory, are well illustrated by many familiar examples. A concise description of the duality principle, a crucial reasoning process in category theory, is also presented. The first important result we present is Yoneda's lemma, which allows us to embed any (locally small) category into a category of functors on that category. This generalizes the well-known group theory result called Cayley's theorem, stating that any group is isomorphic to a subgroup of a symmetric group.

The second chapter treats the general theory of limits and colimits. Both are very general concepts which arise in various forms in all fields of mathematics. We introduce them gradually, starting with some special cases which might be familiar to the reader such as: (co)products, (co)equalizers, pullbacks and pushouts. A variety of detailed examples are included to illustrate the newly introduced concepts. (Co)products and (co)equalizers are not only important special cases of (co)limits but also generic in the sense that all (co)limits can be constructed out of these two special cases. Certain types of functors are considered in connection to the existence of (co)limits. The existence of (co)limits in several important categories such as functor categories or comma categories is investigated in detail as well.

The third chapter deals with one of the most important notions in category theory: adjoint functors. Several descriptions of adjoint functors are presented and the theory is illustrated by a wide range of examples from various areas of mathematics. Many important constructions in mathematics are shown to be part of an adjunction, including for instance the classical free constructions present in algebra, localizations in ring theory or Stone–Čech compactifications of topological spaces. Important related concepts such as equivalence of categories, (co)reflective subcategories or localization of categories are also investigated and well illustrated by a plethora of detailed examples. Deeper connections with the concepts introduced in the previous chapters are emphasized. For instance, (co)limits and representable functors are equivalently described by means of adjoint functors. Going beyond what is usually covered by an introductory text in category theory, the book ends with a more advanced topic, the adjoint functor theorem. More precisely, two variations of this celebrated theorem, namely Freyd's Adjoint Functor Theorem and the Special Adjoint Functor Theorem, are considered. They provide different kinds of necessary and sufficient conditions for a functor to admit a left or a right adjoint.

I would like to take this opportunity to thank my coauthors as well as my colleagues and students from Vrije Universiteit Brussel for everything I have learned from them. My warmest thanks to Gigel Militaru for teaching me category theory when I was a student and for the many wise suggestions he made after reading a first draft of this book as well as to Alexandru Chirvasitu for the countless illuminating discussions. I am very grateful to the Springer editors, especially to Rémi Lodh, and to the anonymous referees for their comments and advice which greatly improved the book's presentation. During the preparation of this manuscript, my work was supported at different stages by FWO (Fonds voor Wetenschappelijk Onderzoek—Flanders) and a grant of Ministry of Research, Innovation and Digitization, CNCS/CCCDI–UEFISCDI, project number PN-III-P4-ID-PCE-2020-0458.

Bucharest, Romania   Ana Agore
Brussels, Belgium
June 2023

Contents

1   Categories and Functors 1
   1.1   Set Theory 1
   1.2   Categories: Definition and First Examples 2
   1.3   Special Objects and Morphisms in a Category 6
   1.4   Some Constructions of Categories 16
   1.5   Functors 24
   1.6   Isomorphisms of Categories 36
   1.7   Natural Transformations: Representable Functors 41
   1.8   The Duality Principle 52
   1.9   Functor Categories 62
   1.10 Yoneda's Lemma 70
   1.11 Exercises 77

2   Limits and Colimits 83
   2.1   (Co)products, (Co)equalizers, Pullbacks and Pushouts 83
   2.2   (Co)limit of a Functor. (Co)complete Categories 105
   2.3   (Co)limit as a Representing Pair 119
   2.4   (Co)limits by (Co)equalizers and (Co)products 121
   2.5   (Co)limit Preserving Functors 127
   2.6   (Co)limits in Comma Categories 134
   2.7   (Co)limits in Functor Categories 140
   2.8   Exercises 148

3   Adjoint Functors 153
   3.1   Definition and Generic Examples 153
   3.2   Adjoints Via Free Objects 154
   3.3   Galois Connections 157
   3.4   More Examples and Properties of Adjoint Functors 159
   3.5   The Unit and Counit of an Adjunction 170
   3.6   Another Characterisation of Adjoint Functors 183
   3.7   (Co)reflective Subcategories 194
   3.8   Equivalence of Categories 199

3.9   Localization 214
3.10   (Co)limits as Adjoint Functors 221
3.11   Freyd's Adjoint Functor Theorem 227
3.12   Special Adjoint Functor Theorem 236
3.13   Representable Functors Revisited.................................... 245
3.14   Exercises 248

4   **Solutions to Selected Exercises** 251
4.1   Chapter 1 251
4.2   Chapter 2 259
4.3   Chapter 3 268

References.. 277

Index.. 281

Frequently Used Notations

$\emptyset$	empty set
$\mathbb{N}$	set of natural numbers
$\mathbb{Z}$	ring of integers
$\mathbb{Q}$	field of rational numbers
$\mathbb{R}$	field of real numbers
$A \setminus B$	set-theoretic difference
$\aleph_0$	cardinality of $\mathbb{N}$ (and $\mathbb{Z}$)
$|X|$	cardinality of the set X
$\subseteq$	inclusion
$f_{|X}$	restriction of a map f to a subset X of its domain
$\otimes$	tensor product
$H \trianglelefteq G$	H is a normal subgroup of G
G_{ab}	abelianization of the group G
$\mathcal{G}(M)$	universal enveloping group of the monoid M
$\ker f$	kernel of the morphism f
$\operatorname{Im} f$	image of the morphism f
$S^{-1}X$	localization of the ring (module) X by the multiplicative set S
C_X	discrete category on the set X
$\mathrm{PO}(X, \leqslant)$	category associated to the pre-ordered set $(X, \leqslant)$
Set	category of sets
FinSet	category of finite sets
Grp	category of groups
SiGrp	category of simple groups
Ab	category of abelian groups
Div	category of divisible groups
Rng	category of rings
Ring	category of unitary rings
Ring^c	category of commutative unitary rings
Field	category of fields
$_R\mathcal{M}$	category of left R-modules
$\mathcal{M}_R$	category of right R-modules

Top	category of topological spaces
Haus	category of Hausdorff topological spaces
KHaus	category of compact Hausdorff topological spaces
PreOrd	category of pre-ordered sets and order preserving maps
Poset	category of partially ordered sets and order preserving maps
$\mathrm{Hom}_C(A, B)$	hom set in the category C between A and B
Hom_C	hom bifunctor
$\mathrm{Hom}_C(C, -)$	covariant hom functor
$\mathrm{Hom}_C(-, C)$	contravariant hom functor
Cat	category of small categories
$\mathrm{Fun}(I, C)$	category of functors from the small category I to the category C
$(F \downarrow G)$	comma category of the functors $F: \mathcal{A} \to C, G: \mathcal{B} \to C$
$\mathrm{Cone}(F, C)$	set of cones on the functor F with vertex C
$\mathrm{Cocone}(F, C)$	set of cocones on the functor F with vertex C
$\mathrm{Cone}(F, -)$	cone functor
$\mathrm{Cocone}(F, -)$	cocone functor
$\lim F$	limit of the functor F
$\lim$	limit functor
$\mathrm{colim}\, F$	colimit of the functor F
colim	colimit functor
$F \cong G$	naturally isomorphic functors
$\mathrm{Nat}(F, G)$	class of all natural transformations between the functors F and G
$\alpha * \beta$	Godement product of the natural transformations α and β
$F \dashv G$	F is left adjoint to G (or G is right adjoint to F)

Chapter 1
Categories and Functors

1.1 Set Theory

We start by setting very briefly the set theory model that will be assumed to hold throughout. The main issue that arises is that most categories of interest have as objects all sets, all groups, all topological spaces, etc. Therefore, a proper definition of a category which includes the examples mentioned above is not possible in the classical Zermelo–Fraenkel set theory. One way to get around this issue is by using the von Neumann–Bernays–Gödel (NBG) set theory which introduces, in addition to sets, the notion of a *class* to play the role of these "*big sets*" consisting of all sets, all groups, etc. More precisely, the connection between sets and classes is given by the so-called *limitation of size axiom*:

A class is a set if and only if it is not bijective with the class of all sets.

To conclude, we can use the word class to designate any collection of mathematical objects; all sets are obviously classes.

The NBG axioms are in fact a conservative extension of the ZFC axioms. Therefore, all statements about sets which can be proved in NBG hold in ZFC as well. An important axiom included in NBG which will be used in many places in the sequel is the following form of the *axiom of choice*:

We can choose an element from each of any class of nonempty sets.

Moreover, the following consequence of the NBG axioms will be intensively used throughout: if A is a set and B is a subclass of A then B is a set. A detailed account of the NBG set theory axioms can be found, for instance, in [51] or in the Appendix of [46].

© The Author(s), under exclusive license to Springer Nature Switzerland AG 2023
A. Agore, *A First Course in Category Theory*, Universitext,
https://doi.org/10.1007/978-3-031-42899-9_1

1.2   Categories: Definition and First Examples

We start by providing the definition of a category and many examples which may already be familiar to the reader.

Definition 1.2.1   A *category* C consists of the following data:

(1) a class Ob C whose elements $A, B, C, \ldots$ are called *objects*;
(2) for every pair of objects A, B, a (possibly empty) set $\mathrm{Hom}_C(A, B)$, whose elements are called *morphisms* from A to B. An element $f \in \mathrm{Hom}_C(A, B)$ will be denoted by $f : A \to B$; A and B are called the *domain* and the *codomain* of f, respectively;
(3) for every triple of objects A, B, C, a *composition law*

$$\mathrm{Hom}_C(A, B) \times \mathrm{Hom}_C(B, C) \to \mathrm{Hom}_C(A, C)$$

$$(f, g) \mapsto g \circ f;$$

(4) for every object A, a morphism $1_A \in \mathrm{Hom}_C(A, A)$, called the *identity* on A

such that the following axioms hold:

(i) *associative law*: given morphisms $f \in \mathrm{Hom}_C(A, B)$, $g \in \mathrm{Hom}_C(B, C)$, $h \in \mathrm{Hom}_C(C, D)$ the following equality holds:

$$h \circ (g \circ f) = (h \circ g) \circ f;$$

(ii) *identity law*: given a morphism $f \in \mathrm{Hom}_C(A, B)$ the following identities hold:

$$1_B \circ f = f = f \circ 1_A.$$

A category C whose class of objects Ob C is a set is called a *small category*. Furthermore, a category will be called *finite* if it contains only finitely many morphisms.

Note that, in certain references such as [35], the hom-sets are allowed to be classes. In that framework, categories as in Definition 1.2.1 are called *locally small*.

Examples 1.2.2

(1) Any set X can be made into a small category, called the *discrete category on* X and denoted by C_X, as follows:

$$\mathrm{Ob}\, C_X = X$$

$$\mathrm{Hom}_{C_X}(x, y) = \begin{cases} \varnothing & \text{if } x \neq y \\ \{1_x\} & \text{if } x = y \end{cases}, \text{ for every } x, y \in X.$$

The only possible compositions of morphisms are $1_x \circ 1_x = 1_x$ for all $x \in X$. Throughout, we denote by **n** the discrete category on a set with $n \in \mathbb{N}$ elements. For $n = 0$ we obtain the *empty category* with no objects and no morphisms.

(2) More generally, any pre-ordered set[1] $(X, \leqslant)$ defines a small category PO$(X, \leqslant)$ as follows:

$$\text{Ob PO}(X, \leqslant) = X$$

$$\text{Hom}_{\text{PO}(X, \leqslant)}(x, y) = \begin{cases} \emptyset \text{ if } x \not\leqslant y \\ \{u_{x, y}\} \text{ if } x \leqslant y \end{cases}, \text{ for every } x, y \in X,$$

where $u_{x, y}$ denotes the unique morphism from x to y. The composition of morphisms is given by the rule $u_{y, z} \circ u_{x, y} = u_{x, z}$, while the identity on any $x \in X$ is $u_{x, x}$.

Moreover, following the same idea, we can further generalize this example to the level of classes. For instance, consider the class of all sets pre-ordered by inclusion; we obtain a category denoted by **Set**$(\subseteq)$ which has the class of all sets as objects and for all sets A, B we have

$$\text{Hom}_{\textbf{Set}(\subseteq)}(A, B) = \begin{cases} \emptyset \text{ if } A \not\subseteq B \\ \{u_{A, B}\} \text{ if } A \subseteq B \end{cases},$$

where $u_{A, B}$ denotes the unique morphism from A to B. Composition of morphisms is defined as in the case of pre-ordered sets.

(3) A monoid $(M, \cdot)$ can be seen as a small category $\mathcal{M}$ with a single object denoted by $*$ and the set of morphisms $\text{Hom}_{\mathcal{M}}(*, *) = M$. The composition of morphisms in $\mathcal{M}$ is given by the multiplication of M and the identity on $*$ is just the unit 1_M. In particular, using the same idea, any group can be made into a category.

(4) The category **Set** of sets has the class of all sets as objects while $\text{Hom}_{\textbf{Set}}(A, B)$ is the set of all functions from A to B. Composition is given by the usual composition of functions and the identity on any set A is the identity map 1_A. **Set** is not a small category.

(5) **FinSet** is the category whose objects are finite sets, and $\text{Hom}_{\textbf{FinSet}}(A, B)$ is just the set of all functions between the two finite sets A and B. **FinSet** is also not a small category; this can be easily seen by noticing that $\{X\}$ is a singleton and therefore a finite set for every set X. However, as opposed to **Set**, the category **FinSet** satisfies the following property: by choosing[2] exactly one object from each isomorphism class of finite sets together with all functions

---

[1] A set X is called *pre-ordered* if is endowed with a binary relation $\leqslant$ which is reflexive and transitive.

[2] Recall that the axiom of choice is assumed to hold.

between them we obtain a category, called a *skeleton* of **FinSet**, which is small. Categories such as **FinSet** which admit a small skeleton are called *essentially small* and will be treated in more detail in Sect. 3.8. **Set** is not an essentially small category as we will see in Example 3.8.15, (2).

(6) Consider **RelSet** to be the category defined as follows:

Ob **RelSet** = Ob **Set**

$\text{Hom}_{\textbf{RelSet}}(A, B) = \mathcal{P}(A \times B) = \{f \mid f \subseteq A \times B\}$, for every $A, B \in$ Ob **Set**.

The composition of morphisms in **RelSet** is defined as follows: given $f \subseteq A \times B$ and $g \subseteq B \times C$ we consider

$$g \circ f = \{(a, c) \in A \times C \mid \exists b \in B \text{ such that } (a, b) \in f \text{ and } (b, c) \in g\}.$$

Finally, the identity is defined as $1_A = \{(a, a) \mid a \in A\}$. **RelSet** is called the category of relations.

(7) **Grp** is the category of groups, where Ob **Grp** is the class of all groups while $\text{Hom}_{\textbf{Grp}}(A, B)$ is the set of all group homomorphisms from A to B. Similarly, **Mon** denotes the category of monoids with monoid homomorphisms between them.

(8) **SiGrp** denotes the category of simple[3] groups with group homomorphisms between them.

(9) **Ab** is the category of abelian groups with group homomorphisms between them. Throughout, we use multiplicative notation for the group structure on an arbitrary group and additive notation for abelian groups.

(10) **Div** is the category of divisible[4] groups with group homomorphisms between them.

(11) **Rng** is the category of rings with ring homomorphisms between them.

(12) **Ring** (resp. **Ring**^{c}) is the category of (resp. commutative) unitary rings with unit preserving ring homomorphisms between them.

(13) **Field** is the category of fields[5] with field homomorphisms between them.

(14) For a ring with unity R, we denote by $_R\mathcal{M}$ the category of left R-modules with morphisms between two R-modules given by R-linear functions. The category of right R-modules $\mathcal{M}_R$ can be defined analogously. In particular, if K is a field then $_K\mathcal{M}$ (resp. $_K\mathcal{M}^{fd}$) denotes the category of vector spaces (resp. finite-dimensional vector spaces) over K.

---

[3] A non-trivial group is called *simple* if its only normal subgroups are the trivial group and the group itself.

[4] An abelian group $(G, +)$ is called *divisible* if for every positive integer n and every $g \in G$, there exists an $h \in G$ such that $nh = g$.

[5] Throughout, a *field* means a commutative division ring.

Furthermore, if R is commutative and $S \subset R$ is a multiplicative subset of R,[6] then $_R\mathcal{M}^{S-\text{aut}}$ stands for the category of left R-modules on which S acts as an automorphism, i.e., for all $s \in S$ and $M \in \text{Ob}_R\mathcal{M}^{S-\text{aut}}$ the multiplication map $\mu_s : M \to M$, $\mu_s(m) = sm$ is invertible; the morphisms between two such objects are given by R-linear functions.

(15) For a field K, we denote by $\mathbf{Alg}_K$ the category of unital and associative K-algebras together with algebra homomorphisms between them. Similarly, $\mathbf{Alg}_K^c$ stands for the category of unital, associative and commutative K-algebras.

(16) $\mathbf{Top}$ is the category of topological spaces where Ob $\mathbf{Top}$ is the class of all topological spaces while $\text{Hom}_{\mathbf{Top}}(A, B)$ is the set of continuous functions between A and B. $\mathbf{Top}_*$ stands for the category of pointed topological spaces, that is, the objects are pairs (A, a_0) where A is a topological space and $a_0 \in A$ while the morphisms between two such pairs (A, a_0) and (B, b_0) are just continuous functions $f : A \to B$ such that $f(a_0) = b_0$.

(17) $\mathbf{Haus}$ is the category of Hausdorff topological spaces where Ob $\mathbf{Haus}$ is the class of all Hausdorff topological spaces while $\text{Hom}_{\mathbf{Haus}}(A, B)$ is the set of continuous functions between A and B. Similarly, $\mathbf{KHaus}$ denotes the category of compact Hausdorff topological spaces.

(18) $\mathbf{PreOrd}$ is the category whose objects are pre-ordered sets and the morphisms between two pre-ordered sets are order preserving maps.[7] Similarly, $\mathbf{Poset}$ is the category whose objects are partially ordered sets[8] and the morphisms between two partially ordered sets are order preserving maps.

(19) For a field K, we denote by $\mathbf{Mat}_K$ the category whose object class is the set of natural numbers $\mathbb{N}$. The morphisms in $\mathbf{Mat}_K$ between two objects $m, n \in \mathbb{N}$ are all $n \times m$ matrices with entries in K and the composition of morphisms is given by matrix multiplication:

$$\text{Hom}_{\mathbf{Mat}_K}(m, n) \times \text{Hom}_{\mathbf{Mat}_K}(n, p) \to \text{Hom}_{\mathbf{Mat}_K}(m, p)$$

$$(A, B) \mapsto BA.$$

The identity morphism on any $n \in \mathbb{N}$ is given by the $n \times n$ identity matrix, where the 0×0 identity matrix is by definition the zero matrix. Furthermore, by convention, if either m or n is zero, we have a unique $n \times m$ matrix called a null matrix. $\qquad \square$

Remark 1.2.3 Notice that although we sometimes work with categories whose objects are sets, morphisms in the sense of Definition 1.2.1 need not be functions. This situation is best illustrated in Example 1.2.2, (6).

---

[6] S is called a *multiplicative subset of the ring* R if $1_R \in S$ and for all $s, s' \in S$ we have $ss' \in S$.

[7] Given two pre-ordered sets $(X, \leqslant_X)$ and $(Y, \leqslant_Y)$, a map $f : X \to Y$ is called *order preserving* if $x \leqslant_X y$ implies $f(x) \leqslant_Y f(y)$.

[8] A set X is called *partially ordered* if is endowed with a binary relation $\leqslant$ which is reflexive, antisymmetric and transitive.

Definition 1.2.4 Let C, C' be two categories. We shall say that C' is a *subcategory* of C if the following conditions are satisfied:

(i) Ob $C' \subseteq$ Ob C, i.e., any object of C' is an object of C;
(ii) $\mathrm{Hom}_{C'}(A, B) \subseteq \mathrm{Hom}_C(A, B)$ for every $A, B \in$ Ob C';
(iii) the composition of morphisms in C' is induced by the composition of morphisms in C;
(iv) the identity morphisms in C' are identity morphisms in C.

Moreover, C' is said to be a *full subcategory* of C if for every pair (A, B) of objects of C' we have

$$\mathrm{Hom}_{C'}(A, B) = \mathrm{Hom}_C(A, B).$$

Examples 1.2.5 (1) The category **FinSet** is a full subcategory of **Set**.
(2) The category **Ab** is a full subcategory of **Grp**.
(3) The category $_R\mathcal{M}^{S-\mathrm{aut}}$ is a full subcategory of $_R\mathcal{M}$.
(4) The category **Haus** is a full subcategory of **Top**.
(5) **Ring** is a subcategory of **Rng** but not a full subcategory as not all morphisms in **Rng** between unitary rings are unit preserving.
(6) **Set** is a subcategory of **RelSet** but not a full subcategory since not every subset of a cartesian product defines a function. □

1.3 Special Objects and Morphisms in a Category

The notions of monomorphism and epimorphism which we will introduce next are generalizations to arbitrary categories of the familiar injective and surjective functions from **Set**.

Definition 1.3.1 Let C be a category and $f \in \mathrm{Hom}_C(A, B)$.

(1) f is called a *monomorphism* if for any $g_1, g_2 \in \mathrm{Hom}_C(C, A)$ such that $f \circ g_1 = f \circ g_2$ we have $g_1 = g_2$;
(2) f is called an *epimorphism* if for any $h_1, h_2 \in \mathrm{Hom}_C(B, C)$ such that $h_1 \circ f = h_2 \circ f$ we have $h_1 = h_2$;
(3) f is called an *isomorphism* if there exists an $f' \in \mathrm{Hom}_C(B, A)$ such that $f \circ f' = 1_B$ and $f' \circ f = 1_A$. In this case we say that A and B are *isomorphic objects*.

Although in **Set** monomorphisms (resp. epimorphisms) coincide with injective (resp. surjective) functions, this is no longer true in an arbitrary category whose objects and morphisms are sets and functions, respectively, as we will see in Example 1.3.2, (4) and (5).

Examples 1.3.2

(1) In each of the categories **Set**, **Grp**, **Ab**, $_R\mathcal{M}$ monomorphisms coincide with the injective homomorphisms, while in **Top** and **KHaus** monomorphisms coincide with the injective continuous maps. We will only prove here that monomorphisms in **Set** coincide with injective functions. Indeed, suppose $f : A \to B$ is an injective map and $g, h : C \to A$ are such that $f \circ h = f \circ g$. Then, we have $f(h(c)) = f(g(c))$ for any $c \in C$ and since f is injective we get $h(c) = g(c)$ for any $c \in C$, i.e., $g = h$ as desired.

Assume now that $f : A \to B$ is a monomorphism and let $a, a' \in A$ be such that $f(a) = f(a')$. We denote by $i_a : \{*\} \to A$, respectively $i_{a'} : \{*\} \to A$, the maps given by $i_a(*) = a$, $i_{a'}(*) = a'$. This implies that $f \circ i_a = f \circ i_{a'}$ and since f is a monomorphism we obtain $i_a = i_{a'}$. Therefore $a = a'$ and f is indeed injective.

Note that the proof above can be carried over verbatim to the categories **Top** and **KHaus** by simply considering on $\{*\}$ the indiscrete topology.[9]

(2) Similarly, in each of the categories **Set**, **Grp**, **Ab**, $_R\mathcal{M}$ epimorphisms coincide with the surjective homomorphisms, while in **Top** and **KHaus** epimorphisms coincide with the surjective continuous functions. We prove here that epimorphisms in **Set** and **Grp** coincide with surjective functions and surjective homomorphisms, respectively.

Consider first an epimorphism $f : A \to B$ in **Set**. Define $g, h : B \to \{0, 1\}$ as follows:

$$g(b) = \begin{cases} 1, & \text{if } b \in f(A) \\ 0, & \text{if } b \notin f(A) \end{cases}, \qquad h(b) = 1, \text{ for all } b \in B.$$

Then, for any $b \in B$, we have $(h \circ f)(b) = 1 = (g \circ f)(b)$ and since f is an epimorphism we obtain $g = h$. This shows that the image of f is the entire B, as desired.

Next we look at epimorphisms in **Grp**. Let $f : G \to H$ be an epimorphism in **Grp** and denote by $K = \text{Im}(f)$ the image of f. Assume that $K \neq H$. If K is a normal subgroup of H we can form the quotient group H/K. Consider now two group homomorphisms $\pi, u : H \to H/K$, where π is the canonical projection and u is the trivial homomorphism defined by $\pi(h) = hK, u(h) = K$ for all $h \in H$. Obviously $u \circ f = \pi \circ f$ and since f is an epimorphism we obtain $u = \pi$. Therefore, $K = H$ which contradicts our assumption. Hence, K cannot be a normal subgroup of H. In particular, note that the index of K in H is at least 3, otherwise K would be a normal subgroup of H ([49, Proposition 2.62]). This allows us to choose three distinct right cosets K, Kx and Ky, for some $x, y \in H$. Let $S(H)$ be the set of permutations on H and define $\sigma \in S(H)$

---

[9] The topology consisting only of the set itself and the empty set is called the *indiscrete topology*.

as follows:

$$\sigma(h) = \begin{cases} kx, & \text{if } h = ky \\ ky, & \text{if } h = kx \\ h, & \text{if } h \notin Kx \cup Ky \end{cases}, \quad k \in K.$$

Consider now the group homomorphisms $\psi, \xi \colon H \to S(H)$ defined as follows for all $t, h \in H$:

$$\psi(t)(h) = th, \quad \xi(t)(h) = \sigma^{-1} \circ \psi(t) \circ \sigma(h).$$

First, we show that $\psi \circ f = \xi \circ f$. Indeed, for any $g \in G$ and $h \in H$, we have $\psi(f(g))(h) = f(g)h$ and respectively $\xi(f(g))(h) = \sigma^{-1} \circ \psi(f(g)) \circ \sigma(h)$.

Keeping in mind that $f(g) \in K$ for all $g \in G$, if $h = ky$ for some $k \in K$, we have

$$\xi(f(g))(ky) = \sigma^{-1} \circ \psi(f(g))(kx) = \sigma^{-1}(f(g)kx) = f(g)ky = \psi(f(g))(ky).$$

Similarly, if $h = kx$ for some $k \in K$, we obtain

$$\xi(f(g))(kx) = \sigma^{-1} \circ \psi(f(g))(ky) = \sigma^{-1}(f(g)ky) = f(g)kx = \psi(f(g))(kx).$$

Finally, $h \notin Kx \cup Ky$ yields

$$\xi(f(g))(h) = \sigma^{-1} \circ \psi(f(g))(h) = \sigma^{-1}(f(g)h) = f(g)h = \psi(f(g))(h).$$

To conclude, we have proved that $\psi \circ f = \xi \circ f$ and since f is an epimorphism in **Grp** we obtain $\psi = \xi$. However, this is not true as we have

$$\psi(y^{-1})(x) = y^{-1}x,$$

$$\xi(y^{-1})(x) = \sigma^{-1} \circ \psi(y^{-1}) \circ \sigma(x) = \sigma^{-1} \circ \psi(y^{-1})(y) = \sigma^{-1}(1_H) = 1_H.$$

Clearly $y^{-1}x \neq 1_H$ as Kx and Ky are distinct cosets. Therefore $\psi \neq \xi$ and we have reached a contradiction. We can now conclude that $K = H$ and f is surjective.

Conversely, assume now that $f \colon A \to B$ is a surjective map in **Set** and let $t_1, t_2 \colon B \to C$ be two maps such that $t_1 \circ f = t_2 \circ f$. As f is surjective, for any $b \in B$ there exists some $a \in A$ such that $f(a) = b$ and we obtain

$$t_1(b) = t_1(f(a)) = t_2(f(a)) = t_2(b),$$

i.e., $t_1 = t_2$, which shows that f is an epimorphism in **Set**. Note that the argument above can be used verbatim for the categories **Grp**, **Ab**, $_RM$ in order to show that surjective morphisms are epimorphisms.

(3) In each of the categories **Set**, **Grp**, **Ab**, $_R\mathcal{M}$, isomorphisms coincide with the bijective homomorphisms. In **Top**, isomorphisms are exactly the *homeomorphisms*, i.e., continuous bijections whose inverses are also continuous.

(4) In the category **Div** of divisible groups, the quotient map $q\colon \mathbb{Q} \to \mathbb{Q}/\mathbb{Z}$ is obviously not injective but it is a monomorphism. Indeed, let G be another divisible group and $f, g\colon G \to \mathbb{Q}$ be two morphisms of groups such that $q \circ f = q \circ g$. Denoting $f - g$ by h we obtain $q \circ h = 0$. Now for any $x \in G$ we have $q(h(x)) = 0$ and thus $h(x) \in \mathbb{Z}$. Suppose there exists some $x_0 \in G$ such that $h(x_0) \neq 0$. We can assume without loss of generality that $h(x_0) \in \mathbb{N}\backslash\{0\}$. Since we are working with divisible groups, we can find some $y_0 \in G$ such that $x_0 = 2h(x_0)y_0$. Applying h to the above equality we obtain

$$h(x_0) = 2h(x_0)h(y_0),$$

which is an obvious contradiction since $h(x) \in \mathbb{Z}$ for all $x \in G$ and $h(x_0) \neq 0$. Hence we get $h = 0$, which implies that $f = g$. This proves that q is indeed a monomorphism in **Div**.

(5) In the category **Ring**^{c} of unitary commutative rings, the inclusion $i\colon \mathbb{Z} \to \mathbb{Q}$ is obviously not surjective but it is an epimorphism. Indeed, let R be another commutative ring together with two ring morphisms $f, g\colon \mathbb{Q} \to R$ such that $f \circ i = g \circ i$. Consider now $z \in \mathbb{Z}\backslash\{0\}$; then we have $1 = f(1) = f(z)f(1/z)$ and therefore $f(1/z) = 1/f(z)$. Similarly we can prove that $g(1/z) = 1/g(z)$ and since f and g coincide on $\mathbb{Z}$ we get $f(1/z) = g(1/z)$. Now for any $z' \in \mathbb{Z}$ we have

$$f(z'/z) = f(z')f(1/z) = g(z')g(1/z) = g(z'/z).$$

Therefore $f = g$, which implies that i is an epimorphism in **Ring**^{c}.

In a similar manner one can show that if R is a commutative ring with unity and $(S^{-1}R, j)$ is its localization with respect to the multiplicative set $S \subset R$ then $j\colon R \to S^{-1}R$ defined by $j(r) = \frac{r}{1}$, for all $r \in R$, is also an epimorphism in **Ring**^{c}. We refer to [3, Chapter 11] for more details on localization of rings.

(6) It can be easily seen that the inclusion $i\colon \mathbb{Z} \to \mathbb{Q}$ is a monomorphism in the category **Ring**^{c} of unitary commutative rings and also an epimorphism by Example 1.3.2, (5). Therefore, it provides an example of a morphism which is both a monomorphism and an epimorphism but not an isomorphism.

(7) Let (T, τ) be a topological space such that τ is different from the discrete topology.[10] Consider now the set T endowed with the discrete topology $\mathcal{P}(T)$. Then the identity $\mathrm{Id}_T\colon (T, \mathcal{P}(T)) \to (T, \tau)$ is obviously bijective and a continuous map between the two topological spaces. Therefore, Id_T is a morphism in **Top** but not an isomorphism as the inverse map $\mathrm{Id}_T\colon (T, \tau) \to (T, \mathcal{P}(T))$ is obviously not continuous.

---

[10] The topology consisting of all subsets of T is called the *discrete topology on T*.

(8) A rather special situation occurs in **KHaus**, the subcategory of **Top** consisting of all compact Hausdorff topological spaces. As opposed to the category of topological spaces, in **KHaus** any bijective continuous map $f \in$ $\text{Hom}_{\textbf{KHaus}}(K, H)$ is automatically an isomorphism. Indeed, it will suffice to show that the inverse map $f^{-1}: H \rightarrow K$ is continuous too. To this end, we need to show that images of closed sets of K under f are closed in H ([39, Theorem 18.1]). Consider U to be a closed subset of K; as K is compact it follows that U is compact as well ([39, Theorem 26.2]). Moreover, as the image of a compact space under a continuous map is compact ([39, Theorem 26.5]) we obtain $f(U)$ compact. Now recall that compact subspaces of Hausdorff spaces are closed ([39, Theorem 26.3]). Therefore $f(U)$ is closed, as desired.

(9) In the category $\text{PO}(X, \leqslant)$ associated to a partially ordered set $(X, \leqslant)$, any isomorphism is an identity morphism. Indeed suppose $f: x \rightarrow y$ is an isomorphism; this implies that $x \leqslant y$. If $g: y \rightarrow x$ is the inverse of f then we also have $y \leqslant x$. Due to the antisymmetry of $\leqslant$ we obtain $x = y$. Therefore $f: x \rightarrow x$ must be the identity on x. □

It can be easily seen that an isomorphism is in particular a monomorphism and an epimorphism. However, the converse is not necessarily true: a morphism that is both a monomorphism and an epimorphism need not be an isomorphism, as we have seen, for instance, in Example 1.3.2, (6). This motivates the following definition:

Definition 1.3.3 A morphism that is both a monomorphism and an epimorphism is called a *bimorphism*. A category with the property that every bimorphism is an isomorphism is called *balanced*.

Examples 1.3.4

(1) The inclusion $i: \mathbb{Z} \rightarrow \mathbb{Q}$ is a bimorphism in the category **Ring**^{c} of unitary commutative rings. Consequently, **Ring**^{c} is not a balanced category.
(2) The identity map $\text{Id}_T: (T, \mathcal{P}(T)) \rightarrow (T, \tau)$ defined in Example 1.3.2, (7) is obviously a bimorphism in **Top** but not an isomorphism. Therefore, **Top** is not a balanced category.
(3) The categories **Set**, **Grp**, **Ab**, $_R\mathcal{M}$ are balanced. □

Definition 1.3.5 A category in which every morphism is an isomorphism is called a *groupoid*.

Examples 1.3.6

(1) The category associated to a group as in Example 1.2.2, (3) is perhaps the first obvious example of a groupoid.
(2) More generally, we can associate a groupoid to any category. Indeed, given a category C we consider its subcategory C^{grp} consisting of all objects and all isomorphisms; in other words, when constructing C^{grp} we leave out all morphisms of C which are not isomorphisms. C^{grp} is usually called the *core groupoid* of C. □

For the remaining of this section we focus on various properties of objects.

Definition 1.3.7 Let C be a category and $A \in \mathrm{Ob}\,C$.

(1) A is called an *initial object* if the set $\mathrm{Hom}_C(A, B)$ has exactly one element for each $B \in \mathrm{Ob}\,C$.
(2) A is called a *final object* if the set $\mathrm{Hom}_C(C, A)$ has exactly one element for each $C \in \mathrm{Ob}\,C$.
(3) If A is both an initial and a final object we say that A is a *zero-object*. A category which admits a zero-object is called *pointed*.

Remark 1.3.8 Note that in a pointed category there exists a morphism between any two objects. Indeed, if C is a pointed category with zero-object C_0 and A, B are two arbitrary objects of C, then we have a unique morphism $0_A \in \mathrm{Hom}_C(A, C_0)$ and also a unique morphism $0^B \in \mathrm{Hom}_C(C_0, B)$. By composing the two aforementioned morphisms we obtain a morphism $0_{A, B} = 0^B \circ 0_A$ called the *zero-morphism* from A to B.

Proposition 1.3.9 *If A and B are initial (resp. final) objects in a category C then A is isomorphic to B.*

Proof Since A is initial there exists a unique morphism $f : A \to B$ and a unique morphism from A to A which must be the identity 1_A. The same applies for B: there exists a unique morphism $g : B \to A$ and a unique morphism from B to B, namely the identity 1_B. Now $g \circ f \in \mathrm{Hom}_C(A, A)$ and thus $g \circ f = 1_A$. Similarly we get $f \circ g = 1_B$ and we have proved that A and B are isomorphic. The statement about final objects can be proved in a similar manner. □

Examples 1.3.10

(1) In the category **Set** of sets the initial object is the empty set while the final objects are the singletons, i.e., the one-element sets $\{x\}$. Thus **Set** has infinitely many final objects and they are all isomorphic.
(2) The category **Set** of sets has no zero-objects. In the categories **Grp**, **Ab** and $_R\mathcal{M}$ the trivial group, respectively trivial module, is the zero-object.
(3) The category **Ring** of unitary rings has the ring of integers $\mathbb{Z}$ as its initial object while the zero ring $\{0\}$ is its final object (note that $\{0\}$ becomes an object in **Ring** by assuming that $1 = 0$).
(4) The category **Field** of fields has neither an initial nor a final object since there are no morphisms between fields of different characteristics.
(5) Let $(X, \leqslant)$ be a pre-ordered set and $\mathrm{PO}(X, \leqslant)$ the associated category (see Example 1.2.2, (2)). Then $\mathrm{PO}(X, \leqslant)$ has an initial object if and only if $(X, \leqslant)$ has a least element (i.e., some element $0 \in X$ such that $0 \leq x$ for any $x \in X$). Similarly, $\mathrm{PO}(X, \leqslant)$ has a final object if and only if $(X, \leqslant)$ has a greatest element (i.e., some element $1 \in X$ such that $x \leq 1$ for any $x \in X$).
(6) The category **Set**$(\subseteq)$ defined in Example 1.2.2, (2) has an initial object and no final object. Indeed, note first that the empty set is the initial object of **Set**$(\subseteq)$ as for any set X we have $\emptyset \subset X$ and therefore there exists a unique morphism between $\emptyset$ and X, namely $u_{\emptyset, X}$.

The category **Set**($\subseteq$) has no final object. Indeed, U being a final object in **Set**($\subseteq$) would imply that for all sets X there exists a unique morphism $u_{X,U}$. Consequently, $X \subseteq U$ for all sets X, which is a contradiction to the power-set axiom ([37, Axiom 3.12]). $\square$

It is well known that for any object in **Set** one can define the notion of a subset. The corresponding concept in an arbitrary category is called a subobject:

Definition 1.3.11 Let C be a category and $C \in \mathrm{Ob}\,C$. An equivalence class of monomorphisms with codomain C is called a *subobject* of C, where two monomorphisms $f \in \mathrm{Hom}_C(A, C)$ and $g \in \mathrm{Hom}_C(B, C)$ are equivalent if there exists an isomorphism $u \in \mathrm{Hom}_C(A, B)$ such that $g \circ u = f$.

Note that if the category C is not small then the subobjects of a given object C might form a class rather than a set.

Examples 1.3.12

(1) In the category **Set**, the class of subobjects of a set X is in bijection with the power set of X. Indeed, if we denote by $SO(X)$ the class of all subobjects of X, the map $\psi_X : \mathcal{P}(X) \to SO(X)$ defined as follows is bijective:

$$\psi_X(Y) = \widehat{i_Y}, \quad Y \subseteq X,$$

where $\widehat{i_Y}$ denotes the equivalence class of the inclusion monomorphism $i_Y : Y \to X$. First we show that ψ_X is surjective. To this end, we prove that any monomorphism $f : V \to X$ whose image is the subset Y of X belongs to the equivalence class of the inclusion monomorphism i_Y. Recall that monomorphisms in **Set** coincide with injective functions and denote by $u : V \to Y$ the map obtained by restricting the range of f to Y, i.e., $u(v) = f(v)$ for all $v \in V$. Then u is obviously a bijection and moreover we have $i_Y \circ u = f$, which shows that $\widehat{i_Y} = \widehat{f}$, as desired.

We are left to prove that ψ_X is also injective. Consider now $Y, Z \subseteq X$ such that $\widehat{i_Y} = \widehat{i_Z}$. Hence, there exists a bijective map $u : Y \to Z$ such that $i_Z \circ u = i_Y$. This comes down to $u(y) = y$ for all $y \in Y$ and therefore $Y \subseteq Z$. In the same manner, $i_Z = i_Y \circ u^{-1}$ leads to $Z \subseteq Y$. To conclude, we proved that $Y = Z$, as desired.

Similarly one can see that in **Grp** subobjects of a group G are in bijection with the subgroups of G.

(2) In the category **Top** the situation is somewhat different. Recall that the subspaces of a topological space (X, τ) are pairs (Y, τ_Y) where $Y \subseteq X$ and $\tau_Y = \{Y \cap U \mid U \in \tau\}$ is the subspace topology on Y. On the other hand, the subobjects of (X, τ) are in bijection with the topological spaces $(Z, \mathcal{U})$ where $Z \subseteq X$ and $\mathcal{U}$ is a topology finer than τ_Z.[11] If

---

[11] Let τ_1, τ_2 be two topologies on a set X. Then τ_2 is called *finer* than τ_1 (or τ_1 is called *coarser* than τ_2) if $\tau_1 \subseteq \tau_2$.

we denote by $SO(X)$ the class of all subobjects of X and by $T(X) = \{(Z, \mathcal{U}) \mid (Z, \mathcal{U})$ is a topological space, where $Z \subseteq X$ and $\tau_Z \subseteq \mathcal{U}\}$, then the map $\psi_X \colon T(X) \to SO(X)$ defined as follows is a bijection:

$$\psi_X(Z, \mathcal{U}) = \widehat{i_Z}, \quad (Z, \mathcal{U}) \in T(X),$$

where $\widehat{i_Z}$ denotes the equivalence class of the inclusion monomorphism $i_Z \colon (Z, \mathcal{U}) \to (X, \tau)$. Note that i_Z is obviously continuous as $\mathcal{U}$ contains τ_Z, the coarsest topology on Z for which the inclusion map is continuous.

We show first that ψ_X is injective. To this end, consider $(Y, \mathcal{U}), (Z, \mathcal{V}) \in T(X)$ such that $\widehat{i_Y} = \widehat{i_Z}$. Hence, there exists an isomorphism $u \colon (Y, \mathcal{U}) \to (Z, \mathcal{V})$ in **Top** such that $i_Z \circ u = i_Y$. This implies that $u(y) = y$ for all $y \in Y$ and therefore $Y \subseteq Z$ and $\mathcal{V} \subseteq \mathcal{U}$. Furthermore, we also have $i_Z = i_Y \circ u^{-1}$ and since u is a homeomorphism (see Example 1.3.2, (3)) we obtain $Z \subseteq Y$ and $\mathcal{U} \subseteq \mathcal{V}$. Therefore, $Y = Z$ and $\mathcal{U} = \mathcal{V}$, as desired.

We are left to show that ψ_X is surjective. Consider $f \colon (Z, \mathcal{V}) \to (X, \tau)$ to be a monomorphism in **Top**. By Example 1.3.2, (1) f is an injective continuous map. Let $Y = \text{Im}(f)$ and denote by $g \colon Z \to Y$ the bijection induced by restricting the range of f, i.e., $f = i_Y \circ g$, where $i_Y \colon Y \to X$ denotes the inclusion. Define the topology $\mathcal{U}_f$ on Y by letting its open subsets be the images of open subsets of Z under f, i.e., $\mathcal{U}_f = \{f(V) \mid V \in \mathcal{V}\}$. Note that this is indeed a topology on Y as the injectivity of f implies that $f(V_1 \cap V_2) = f(V_1) \cap f(V_2)$ for all $V_1, V_2 \in \mathcal{V}$ ([37, Exercise 1.4]). Therefore, $g \colon (Z, \mathcal{V}) \to (Y, \mathcal{U}_f)$ is a homeomorphism. The inclusion $i_Y \colon (Y, \mathcal{U}_f) \to (X, \tau)$ is obviously continuous. Indeed, let $U \in \tau$; as $f \colon (Z, \mathcal{V}) \to (X, \tau)$ is continuous, we have $f^{-1}(U) = V \in \mathcal{V}$. We obtain:

$$i_Y^{-1}(U) = \{y \in Y \mid i_Y(y) \in U\} = Y \cap U = f(\{z \in Z \mid f(z) \in U\})$$
$$= f(f^{-1}(U)) = f(V).$$

As the subspace topology τ_Y is the coarsest topology on Y for which the inclusion map i_Y is continuous, we have $\tau_Y \subseteq \mathcal{U}_f$. This shows that $(Y, \mathcal{U}_f) \in T(X)$. Furthermore, as $g \colon (Z, \mathcal{V}) \to (Y, \mathcal{U}_f)$ is a homeomorphism and $f = i_Y \circ g$, we can conclude that $\widehat{i_Y} = \widehat{f}$ and therefore $\psi_X(Y, \mathcal{U}_f) = \widehat{i_Y} = \widehat{f}$. This shows that ψ_X is also surjective, as desired.

(3) As opposed to **Top**, in **KHaus** subobjects of a given object K are in bijection with the closed subsets of K. Indeed, if we denote by $SS^c(K)$ and $SO(K)$ the set of all closed subsets and respectively the class of all subobjects of K, the map $\psi_K \colon SS^c(K) \to SO(K)$ defined as follows is a bijection:

$$\psi_K(Y) = \widehat{i_Y}, \quad Y \in SS^c(K),$$

where $\widehat{i_Y}$ denotes the equivalence class of the inclusion monomorphism $i_Y \colon (Y, \tau_Y) \to (K, \tau)$ and τ_Y is the subspace topology on Y.

Consider first Y, $Z \subseteq K$ such that $\widehat{i_Y} = \widehat{i_Z}$. Therefore, we have an isomorphism $u\colon (Y, \tau_Y) \to (Z, \tau_Z)$ in **KHaus** such that $i_Z \circ u = i_Y$, which implies, as in the case of the category **Top**, that $Y = Z$. This shows that ψ_K is injective.

Consider now a monomorphism $f\colon (Z, \mathcal{V}) \to (X, \tau)$ in **KHaus**. In particular f is injective and if we denote $\mathrm{Im}(f)$ by Y then the induced map $u\colon (Z, \mathcal{V}) \to (Y, \tau_Y)$ obtained by restricting the range of f is obviously bijective. Moreover, u is also continuous; indeed, if $W \in \tau_Y$ then there exists $U \in \tau$ such that $W = U \cap Y$ and we obtain

$$u^{-1}(W) = u^{-1}(U \cap Y) = \{z \in Z \mid u(z) \in U \cap Y\} = \{z \in Z \mid f(z) \in U \cap Y\}$$

$$= \{z \in Z \mid f(z) \in U\} = f^{-1}(U) \in \mathcal{V}.$$

Now recall that any subspace of a Hausdorff space is also Hausdorff; this shows that (Y, τ_Y) is a Hausdorff space. It can be easily seen, as in Example 1.3.2, (8) that u is in fact an isomorphism in **KHaus**. We are left to show that (Y, τ_Y) is a closed subspace of (X, τ). First note that since $(Z, \mathcal{V})$ is in particular a compact space, its image under the continuous map f is compact ([39, Theorem 26.5]) as well. Therefore (Y, τ_Y) is a compact subspace in the Hausdorff space (K, τ) and is closed by [39, Theorem 26.3]. To summarize, we have an isomorphism $u\colon (Z, \mathcal{V}) \to (Y, \tau_Y)$ in **KHaus** such that $i_Y \circ u = f$, which shows that $\widehat{i_Y} = \widehat{f}$ and therefore ψ_K is also surjective.

$\square$

Similarly, we can define quotient objects:

Definition 1.3.13 Let C be a category and $C \in \mathrm{Ob}\, C$. An equivalence class of epimorphisms with domain C is called a *quotient* of C, where two epimorphisms $f \in \mathrm{Hom}_C(C, A)$ and $g \in \mathrm{Hom}_C(C, B)$ are equivalent if there exists an isomorphism $u \in \mathrm{Hom}_C(A, B)$ such that $u \circ f = g$.

As in the case of subobjects, if the category C is not small then the quotients of a given object C might form a class, not a set.

Examples 1.3.14

(1) In the category **Set**, the quotients of a set X are in bijection with the set of all equivalence relations on X. Indeed, if we denote by $\mathcal{E}(X)$ and $Q(X)$ the set of equivalence relations and respectively the quotient objects on X, then the map $\psi_X\colon Q(X) \to \mathcal{E}(X)$ defined as follows is a bijection:

$$\psi_X(\overline{f}) = E_f, \quad \text{for all epimorphisms } f \in \mathrm{Hom}_{\mathbf{Set}}(X, Y),$$

where E_f is the equivalence relation on X induced by f. Recall that $x\, E_f\, x'$ if and only if $f(x) = f(x')$. We check first that ψ_X is well-defined. Consider two epimorphisms $f \in \mathrm{Hom}_{\mathbf{Set}}(X, Y)$ and $g \in \mathrm{Hom}_{\mathbf{Set}}(X, Z)$ such that $\overline{f} = \overline{g}$.

Hence, there exists an isomorphism $u \in \mathrm{Hom}_{\mathbf{Set}}(Y, Z)$ such that $u \circ f = g$. As u is a set bijection, we clearly have $f(x) = f(x')$ if and only if $g(x) = g(x')$, which shows that $E_f = E_g$, as desired.

Next, we prove that ψ_X is injective. To this end, let $f \in \mathrm{Hom}_{\mathbf{Set}}(X, Y)$ and $g \in \mathrm{Hom}_{\mathbf{Set}}(X, Z)$ be two epimorphisms such that $E_f = E_g$. Since f is surjective (see Example 1.3.2, (2)) we can define a map $u \in \mathrm{Hom}_{\mathbf{Set}}(Y, Z)$ by

$$u\big(f(x)\big) = g(x) \text{ for all } x \in X. \tag{1.1}$$

Then $E_f = E_g$ implies, in particular, that if $f(x) = f(x')$ then $g(x) = g(x')$. This shows that u is well-defined. Furthermore, it follows from (1.1) and the surjectivity of g (Example 1.3.2, (2)) that u is also surjective. We show that u is injective as well. Indeed, let $y, y' \in Y$ be such that $u(y) = u(y')$. As f is surjective, we can find $x_y, x_{y'} \in X$ such that $f(x_y) = y$ and $f(x_{y'}) = y'$. Therefore we have $u\big(f(x_y)\big) = u\big(f(x_{y'})\big)$ and by (1.1) we obtain $g(x_y) = g(x_{y'})$. Now $E_f = E_g$ implies that $f(x_y) = f(x_{y'})$ i.e., $y = y'$. Thus u is a bijection, which proves that $\overline{f} = \overline{g}$ and therefore ψ_X is injective.

We are left to show that ψ_X is surjective. Indeed, consider an equivalence relation E on X. We will show that $E = E_\pi$, where $\pi : X \to X/E$, $\pi(x) = \widehat{x}$ for all $x \in X$, and X/E denotes the set of equivalence classes of X by E. Assume first that $x \, E \, x'$ for some $x, x' \in X$. Then x and x' belong to the same equivalence class in X/E. This shows that $\pi(x) = \pi(x')$ and therefore $x \, E_\pi \, x'$. Conversely, if $x \, E_\pi \, x'$ then $\pi(x) = \pi(x')$, which implies that $\widehat{x} = \widehat{x'}$. Thus x and x' are in the same equivalence class in X/E, i.e., $x \, E \, x'$. Putting all of this together, we have proved that $E = E_\pi$ and $\psi_X(\overline{\pi}) = E$.

(2) In the category **Grp**, the quotient objects of a group G are in bijection with the set of normal subgroups of G. Indeed, if we denote by $\mathcal{N}(G)$ and $\mathcal{Q}(G)$ the set of normal subgroups and respectively quotient objects of G, the map $\psi_G : \mathcal{N}(G) \to \mathcal{Q}(G)$ defined as follows is a bijection:

$$\psi_G(K) = \overline{\pi_K}, \quad \text{for all normal subgroups } K \text{ of } G,$$

where $\overline{\pi_K}$ denotes the equivalence class of the canonical projection $\pi_K : G \to G/K$.

We show first that ψ_G is injective. Consider two normal subgroups K, K' of G such that $\overline{\pi_K} = \overline{\pi_{K'}}$. Hence, there exists an isomorphism $u : G/K' \to G/K$ in **Grp** such that $u \circ \pi_{K'} = \pi_K$. Now $x \in K$ implies that $u\big(\pi_{K'}(x)\big) = 1$ and since u is in particular injective we obtain that $\pi_{K'}(x) = 1$. Thus $x \in K'$, which leads to $K \subset K'$. Furthermore, if $x \in K'$ then $1 = u\big(\pi_{K'}(x)\big) = \pi_K(x)$. This implies that $x \in K$ and therefore $K' \subset K$. Hence $K = K'$, as desired.

Consider now an epimorphism $f : G \to H$ in **Grp** and consider $K = \ker(f)$, which is a normal subgroup of H. We will show that f is equivalent to π_K. This shows that $\psi_G(K) = \overline{f}$ and therefore ψ_G is surjective as well.

Indeed, as in particular we have $K \subseteq \ker(f)$, the universal property of the quotient group yields a unique group homomorphism $u \colon G/K \to H$ such that

$$u \circ \pi_K = f. \tag{1.2}$$

The proof will be finished once we show that u is an isomorphism. To start with, let $\widehat{x}, \widehat{y} \in G/K$ be such that $u(\widehat{x}) = u(\widehat{y})$. Now (1.2) implies that $f(x) = f(y)$ and therefore $xy^{-1} \in \mathrm{Ker}(f) = K$. Thus $\widehat{x} = \widehat{y}$ and u is injective. We are left to show that u is surjective as well. To this end recall from Example 1.3.2, (2) that f is surjective as a consequence of being an epimorphism in **Grp**. The surjectivity of f together with (1.2) implies that u is surjective too and the proof is finished.
□

Categories for which the subobjects (resp. quotients) of any given object form a set are particularly important.

Definition 1.3.15 A category C is called *well-powered* if the subobjects of any object form a set. Similarly, C is called *co-well-powered* if the quotients of any object form a set.

Example 1.3.16 The discussion in Examples 1.3.12 and 1.3.14 immediately implies that **Set** and **Grp** are both well-powered and co-well-powered. Furthermore, in light of Example 1.3.12, **Top** and **KHaus** are well-powered.   □

1.4   Some Constructions of Categories

In this section we provide several methods of constructing new categories. The first one relies on formally reversing the direction of the morphisms in the given category leading to what is called the dual category.

Definition 1.4.1 Given a category C, the *dual (or opposite) category* of C, denoted by C^{op}, is defined as follows:

(i) $\mathrm{Ob}\, C^{op} = \mathrm{Ob}\, C$;
(ii) $\mathrm{Hom}_{C^{op}}(A, B) = \mathrm{Hom}_C(B, A)$; in order to avoid any confusion we write $f^{op} \colon A \to B$ for the morphism of C^{op} corresponding to the morphism $f \colon B \to A$ of C;
(iii) the composition map

$$\circ^{op} \colon \mathrm{Hom}_{C^{op}}(A, B) \times \mathrm{Hom}_{C^{op}}(B, C) \to \mathrm{Hom}_{C^{op}}(A, C)$$

is defined as follows:

$$g^{op} \circ^{op} f^{op} = (f \circ g)^{op}, \text{ for all } f^{op} \in \mathrm{Hom}_{C^{op}}(A, B), \ g^{op} \in \mathrm{Hom}_{C^{op}}(B, C);$$

(iv) the identities are the same as in C, i.e., $1_C^{op} = 1_C$ for all $C \in \mathrm{Ob}\,C$.

Examples 1.4.2

(1) Obviously $(C^{op})^{op} = C$ for any category C.
(2) Let $\mathrm{PO}(X, \leqslant)$ be the category associated to the pre-ordered set $(X, \leqslant)$. Then $\mathrm{PO}(X, \leqslant)^{op} = \mathrm{PO}(X, \geqslant)$, where $\geqslant$ is the pre-order on X defined as follows: $x \geqslant y$ if and only if $y \leqslant x$. Indeed, this follows immediately by observing that there exists a morphism in $\mathrm{PO}(X, \leqslant)^{op}$ from x to y if and only if there exists a morphism in $\mathrm{PO}(X, \leqslant)$ from y to x.   □

The dual category introduced above suggests that we can assign a *dual* to any categorical concept. More precisely, the dual of a certain concept will be obtained by considering this concept in the dual category. To illustrate this, we highlight several dual concepts we have encountered so far.

Proposition 1.4.3 *Let C be a category and $f \in \mathrm{Hom}_C(A, B)$.*

(1) f is a monomorphism in C if and only if f^{op} is an epimorphism in C^{op}.
(2) I is an initial object in C if and only if I is a final object in C^{op}.
(3) f is an isomorphism in C if and only if f^{op} is a isomorphism in C^{op}.

Proof

(1) Let $h_1^{op}, h_2^{op} \in \mathrm{Hom}_{C^{op}}(A, C)$ be such that $h_1^{op} \circ^{op} f^{op} = h_2^{op} \circ^{op} f^{op}$. In other words, we have $f \circ h_1 = f \circ h_2$ and the desired conclusion follows easily.
(2) If I is an initial object in C then $\mathrm{Hom}_C(I, C)$ has exactly one element for all $C \in \mathrm{Ob}\,C$. Since $\mathrm{Hom}_{C^{op}}(C, I) = \mathrm{Hom}_C(I, C)$, the conclusion follows.
(3) f is an isomorphism if and only if there exists a morphism $g \in \mathrm{Hom}_C(B, A)$ such that $f \circ g = 1_B$ and $g \circ f = 1_A$. This is equivalent to the existence of a morphism $g^{op} \in \mathrm{Hom}_{C^{op}}(A, B)$ such that $g^{op} \circ^{op} f^{op} = 1_B$ and $f^{op} \circ^{op} g^{op} = 1_A$. Hence f^{op} is an isomorphism in C^{op}.
                                                                                      □

To conclude, the notions of epimorphisms and monomorphisms are dual to each other; similarly, initial objects are dual to final objects while the notion of an isomorphism is *self-dual*. We will discuss the duality principle in more depth in Sect. 1.8.

Next we introduce the product of two given categories. The construction can be easily extended to any family of categories indexed by a set.

Definition 1.4.4 Let C and $\mathcal{D}$ be two categories. We define the *product category* $C \times \mathcal{D}$ as follows:

(i) $\mathrm{Ob}\,(C \times \mathcal{D}) = \mathrm{Ob}\,C \times \mathrm{Ob}\,\mathcal{D}$, i.e., the objects of $C \times \mathcal{D}$ are pairs of the form (C, D) with $C \in \mathrm{Ob}\,C$ and $D \in \mathrm{Ob}\,\mathcal{D}$;
(ii) $\mathrm{Hom}_{C \times \mathcal{D}}((C, D), (C', D')) = \mathrm{Hom}_C(C, C') \times \mathrm{Hom}_\mathcal{D}(D, D')$;

(iii) the composition map is defined as follows:

$$\mathrm{Hom}_{C\times D}\big((C,D),\,(C',D')\big) \times \mathrm{Hom}_{C\times D}\big((C',D'),\,(C'',D'')\big)$$

$$\to \mathrm{Hom}_{C\times D}\big((C,D),\,(C'',D'')\big),$$

$$(f',g')\circ(f,g) = (f'\circ f,\, g'\circ g),$$

$$\text{for all } (f,g)\in \mathrm{Hom}_{C\times D}\big((C,D),\,(C',D')\big),$$

$$(f',g')\in \mathrm{Hom}_{C\times D}\big((C',D'),\,(C'',D'')\big);$$

(iv) $1_{(C,D)} = (1_C,\, 1_D)$.

Example 1.4.5 Let G_1 and G_2 be two groups and consider the associated categories (as described in Example 1.2.2, (3)) denoted by $\mathcal{G}_1$ and $\mathcal{G}_2$, respectively.
Then the product category $\mathcal{G}_1 \times \mathcal{G}_2$ is given by the associated category of the direct
product of groups $G_1 \times G_2$.   □

Another way of constructing categories involves directed graphs. We start by
reviewing these first.

Definition 1.4.6 A *graph* consists of a class $\mathcal{V}$ whose elements are called *vertices*
and for each pair $(A,B)\in\mathcal{V}\times\mathcal{V}$ a set $\mathcal{E}(A,B)$ whose elements are called *edges*.
A graph is called *small* if $\mathcal{V}$ is a set.

A *path* in a graph is a non-empty finite sequence $(A_1, f_1, A_2, \ldots, A_n)$ of
vertices and edges succeeding one another such that the first and the last terms are
vertices and each edge $f_i \in \mathcal{E}(A_i, A_{i+1})$. A path of the form (A) is called the *trivial
path on A*.

Notice that every category is in particular a graph; this can be easily seen by
leaving aside the composition law of the given category and by forgetting which
morphisms are identities. Conversely, we will be able to construct a category out of
a graph as follows:

Definition 1.4.7 Let $G = (\mathcal{V}, \mathcal{E})$ be a small graph.[12] The *free category on the
graph G*, denoted by $\mathcal{G}$, is constructed by considering:

(i) $\mathrm{Ob}\,\mathcal{G} = \mathcal{V}$ as class of objects;
(ii) $\mathrm{Hom}_{\mathcal{G}}(A,B) = \mathcal{P}(A,B)$ the set of paths between the vertices A and B, for
      any $A, B \in \mathrm{Ob}\,\mathcal{G}$;
(iii) the composition of morphisms is given by concatenation of paths:

$$(A_n, f_n, \ldots, A_m)\circ(A_1, f_1, \ldots, A_n) = (A_1, f_1, \ldots, A_n, f_n, \ldots, A_m);$$

(iv) the identity maps are given by the trivial paths on each object.

---

[12] The smallness assumption on the graph G is needed in order for the paths between any two
vertices to form a set.

Example 1.4.8 Let G be the oriented graph depicted below:

$$v_1 \xrightarrow{\ \ e_1\ \ } v_2 \xrightarrow{\ \ e_2\ \ } v_3$$

Then the free category $\mathcal{G}$ on the graph G is given as follows:

$\mathrm{Ob}\,\mathcal{G} = \{v_1,\ v_2,\ v_3\}$,

$\mathrm{Hom}_{\mathcal{G}}(v_1,\ v_1) = \{(v_1)\}$, $\mathrm{Hom}_{\mathcal{G}}(v_1,\ v_2) = \{(v_1\ e_1\ v_2)\}$,

$\mathrm{Hom}_{\mathcal{G}}(v_1,\ v_3) = \{(v_1\ e_1\ v_2\ e_2\ v_3)\}$,

$\mathrm{Hom}_{\mathcal{G}}(v_2,\ v_1) = \emptyset$, $\mathrm{Hom}_{\mathcal{G}}(v_2,\ v_2) = \{(v_2)\}$, $\mathrm{Hom}_{\mathcal{G}}(v_2,\ v_3) = \{(v_2\ e_2\ v_3)\}$,

$\mathrm{Hom}_{\mathcal{G}}(v_3,\ v_1) = \emptyset$, $\mathrm{Hom}_{\mathcal{G}}(v_3,\ v_2) = \emptyset$, $\mathrm{Hom}_{\mathcal{G}}(v_3,\ v_3) = \{(v_3)\}$.

□

Definition 1.4.9 Let C be a category. A *diagram* in C is a graph whose vertices and edges are objects and respectively morphisms of C. A diagram in C will be called *commutative* if for each pair of vertices, every two paths between them are equal as morphisms.

Examples 1.4.10 Let C be a category and $A, B, C, D \in \mathrm{Ob}\,C$.

(1) The following diagram is commutative if and only if $g \circ f = h$:

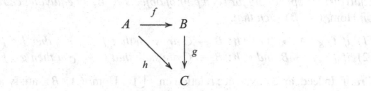

(2) The following diagram is commutative if and only if $g \circ f = h \circ k$:

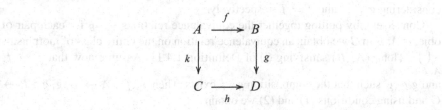

(3) The following diagram is commutative if and only if $h \circ g \circ f = k$:

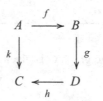

$\square$

    The last construction we introduce is that of a quotient category; it involves a certain kind of equivalence relation on the class of all morphisms of a given category.

Definition 1.4.11 Let C be a category. An equivalence relation $\sim$ on the class of all morphisms $\bigcup_{A,\,B\in\mathrm{Ob}\,C} \mathrm{Hom}_C(A,\,B)$ of C is called a *congruence* if the following conditions are fulfilled:

(i) if $f \in \mathrm{Hom}_C(A,\,B)$ and $f \sim f'$ then $f' \in \mathrm{Hom}_C(A,\,B)$;
(ii) if $f \sim f'$, $g \sim g'$ and the composition $g \circ f$ exists then $g \circ f \sim g' \circ f'$.

    In fact, a congruence on a given category can be built up from certain equivalence relations on all hom sets of that category. We make this precise in the following:

Proposition 1.4.12 *Defining a congruence relation on a given category C is equivalent to specifying for each pair of objects A, B, an equivalence relation $\sim_{A,B}$ on $\mathrm{Hom}_C(A,\,B)$ such that:*

(1) if $f, g : A \to B$ and $h : B \to C$ are such that $f \sim_{A,B} g$ then $h \circ f \sim_{A,C} h \circ g$;
(2) if $f : A \to B$ and $g, h : B \to C$ are such that $g \sim_{B,C} h$ then $g \circ f \sim_{A,C} h \circ f$.

Proof Indeed, an equivalence relation on $\bigcup_{A,\,B\in C} \mathrm{Hom}_C(A,\,B)$ satisfying (i) of Definition 1.4.11 restricts to an equivalence relation on each $\mathrm{Hom}_C(A,\,B)$. Furthermore, (ii) of Definition 1.4.11 shows that the two conditions listed above are fulfilled by considering $g = g'$ and $f = f'$ respectively.

    Conversely, by putting together the equivalence relations $\sim_{A,B}$ for each pair of objects A, B in C we obtain an equivalence relation on the entire class of morphisms $\bigcup_{A,\,B\in C} \mathrm{Hom}_C(A,\,B)$ satisfying (i) of Definition 1.4.11. Assume now that $f \sim f'$ and $g \sim g'$ such that the composition $g \circ f$ exists. Then $f, f' : A \to B, g, g' : B \to C$ and using conditions (1) and (2) we obtain

$$g \circ f \sim_{A,C} g \circ f' \sim_{A,C} g' \circ f',$$

which proves that (2) holds, as desired. $\square$

    We can now introduce the quotient category:

Proposition 1.4.13 *Let $\sim$ be a congruence on a category C and denote by $\overline{f}$ the equivalence class of a morphism f of C. Then $C/\sim$ defined below is a category called a* quotient category *of C:*

(i) $\mathrm{Ob}\, C/\sim\, =\mathrm{Ob}\, C;$

(ii) $\mathrm{Hom}_{C/\sim}(A,\, B) = \{\overline{f} \mid f \in \mathrm{Hom}_C(A,\, B)\}$ *for any* $A,\, B \in \mathrm{Ob}\, C/\sim;$

(iii) the composition map $\mathrm{Hom}_{C/\sim}(A,\, B) \times \mathrm{Hom}_{C/\sim}(B,\, C) \to \mathrm{Hom}_{C/\sim}(A,\, C)$ *is defined as follows:*

$$\overline{g} \circ \overline{f} = \overline{g \circ f}, \text{ for all } \overline{f} \in \mathrm{Hom}_{C/\sim}(A,\, B),\ \overline{g} \in \mathrm{Hom}_{C/\sim}(B,\, C); \qquad (1.3)$$

(iv) the identity on $A \in \mathrm{Ob}\, C/\sim$ *is* $\overline{1_A}$.

Proof Proposition 1.4.12 shows that $\sim$ induces a partition on each hom set $\mathrm{Hom}_C(A,\, B)$ and therefore $\mathrm{Hom}_{C/\sim}(A,\, B)$ is also a set. Furthermore, Definition 1.4.11, (ii) shows that the composition law in $C/\sim$ is well-defined. □

Examples 1.4.14

(1) Consider a group G and let $\mathcal{G}$ be the associated category (as in Example 1.2.2, (3)) whose unique object we denote by $*$. There is a bijection between normal subgroups of G and congruence relations on $\mathcal{G}$. Furthermore, for a normal subgroup N of G, the quotient category by the congruence relation induced by N is the quotient group G/N.

Indeed, suppose first that N is a normal subgroup of G and define the following equivalence relation on $\mathrm{Hom}_{\mathcal{G}}(*,\, *) = G$:

$$g \sim_N h \text{ if and only if } gh^{-1} \in N.$$

The equivalence relation defined above is in fact a congruence on G in the sense of Definition 1.4.11. To this end, we show that the two conditions in Proposition 1.4.12 are fulfilled. Indeed, let $g,\, h,\, t \in G$ be such that $g \sim_N h$. Then $gh^{-1} \in N$ and since any normal subgroup is invariant under conjugation we obtain

$$tg(th)^{-1} = tgh^{-1}t^{-1} \in N, \qquad gt(ht)^{-1} = gtt^{-1}h^{-1} = gh^{-1} \in N.$$

Therefore $tg \sim_N th$ and $gt \sim_N ht$ and we can conclude that $\sim_N$ is a congruence on G. Obviously, the quotient category $C/\sim_N$ coincides with the quotient group G/N.

Consider now $\sim$ to be a congruence relation on the morphisms of $\mathcal{G}$ and let $N_\sim = \{xy^{-1} \mid x,\, y \in G$ such that $x \sim y\}$. Note that the reflexivity and symmetry of $\sim$ imply that $1_G \in N_\sim$ and respectively $g^{-1} \in N_\sim$ whenever $g \in N_\sim$. Consider now $g,\, g' \in N_\sim$, i.e., $g = xy^{-1}$ and $g' = zt^{-1}$ for some $x,\, y,\, z,\, t \in G$ such that $x \sim y$ and $z \sim t$. Using conditions (1) and (2) of Proposition 1.4.12 we obtain $xy^{-1} \sim 1_G$ and $1_G \sim tz^{-1}$. By transitivity of $\sim$

it follows that $xy^{-1} \sim tz^{-1}$ and therefore $gg' = xy^{-1}zt^{-1} = xy^{-1}(tz^{-1})^{-1} \in N_\sim$. Hence $N_\sim$ is a subgroup of G. Consider now $h \in G$ and $g \in N_\sim$ with $g = xy^{-1}$, $x, y \in G$ such that $x \sim y$. Then we have $hgh^{-1} = hxy^{-1}h^{-1} = hx(hy)^{-1} \in N_\sim$ where the last term belongs to $N_\sim$ due to condition (1) of Proposition 1.4.12. We have proved that $N_\sim$ is a normal subgroup of G, as desired.

In order to conclude that there is a bijection between the normal subgroups of G and the congruence relations on $\mathcal{G}$ we are left to show that for all normal subgroups N of G and all congruence relations $\sim$ on $\mathcal{G}$ we have

$$N_{\sim_N} = N \quad \text{and} \quad \sim_{N_\sim} = \sim .$$

To start with, we have

$$N_{\sim_N} = \{xy^{-1} \mid x, y \in G \text{ such that } x \sim_N y\}$$
$$= \{xy^{-1} \mid x, y \in G \text{ such that } xy^{-1} \in N\} = N$$

and our first claim is proved.

Assume now that $g \sim h$; then $gh^{-1} \in N_\sim$ which implies that $g \sim_{N_\sim} h$. Conversely, $g \sim_{N_\sim} h$ gives $gh^{-1} \in N_\sim = \{uv^{-1} \mid u, v \in G \text{ such that } u \sim v\}$. Hence $gh^{-1} = xy^{-1}$ for some $x, y \in G$ such that $x \sim y$. Now since $x \sim y$ and $y^{-1} \sim y^{-1}$, Proposition 1.4.12, (2) implies $xy^{-1} \sim yy^{-1} = 1$. Putting all the above together yields

$$g = (xy^{-1})h \sim 1h = h, \quad \text{i.e., } g \sim h.$$

To summarize, we have proved that for all $g, h \in G$ we have $g \sim h$ if and only if $g \sim_{N_\sim} h$.

(2) On the category **Top** we consider the relation $\sim$, called *homotopy*, defined as follows for all $f, g \in \mathrm{Hom}_{\mathbf{Top}}(X, Y)$: $f \sim_{X,Y} g$ if and only if there exists a continuous map $F: X \times [0, 1] \to Y$ satisfying $F(x, 0) = f(x)$ and $F(x, 1) = g(x)$ for all $x \in X$. We will prove that homotopy is a congruence on **Top**.

To start with, we show that for all topological spaces X and Y, $\sim_{X,Y}$ is an equivalence relation on $\mathrm{Hom}_{\mathbf{Top}}(X, Y)$. Indeed, first note that for any $f \in \mathrm{Hom}_{\mathbf{Top}}(X, Y)$ we have $f \sim_{X,Y} f$ by considering the continuous map $F: X \times [0, 1] \to Y$ defined by $F(x, t) = f(x)$. Furthermore, if $f \sim_{X,Y} g$ then there exists a continuous map $F: X \times [0, 1] \to Y$ such that $F(x, 0) = f(x)$ and $F(x, 1) = g(x)$ for all $x \in X$. Then, the continuous map $G: X \times [0, 1] \to Y$ defined by $G(x, t) = F(x, 1 - t)$ shows that $g \sim_{X,Y} f$.

Finally, assume that $f \sim_{X,Y} g$ and $g \sim_{X,Y} h$. Thus, there exist two continuous maps $F: X \times [0, 1] \to Y$ and $G: X \times [0, 1] \to Y$ such that for all $x \in X$ we have

$$F(x, 0) = f(x), \quad F(x, 1) = g(x), \quad G(x, 0) = g(x), \quad G(x, 1) = h(x).$$

Since $F(x, 1) = G(x, 0) = g(x)$ we can consider the map $H \colon X \times [0, 1] \to Y$ defined as follows:

$$H(x, t) = \begin{cases} F(x, 2t), & \text{if } 0 \leqslant t \leqslant 1/2 \\ G(x, 2t - 1), & \text{if } 1/2 \leqslant t \leqslant 1. \end{cases}$$

The pasting lemma[13] implies that H is continuous. Moreover, we have

$$H(x, 0) = F(x, 0) = f(x), \quad H(x, 1) = G(x, 1) = h(x), \quad \text{for all } x \in X.$$

Therefore, we obtain $f \sim_{X,Y} h$ and we can conclude that $\sim_{X,Y}$ is an equivalence relation on $\mathrm{Hom}_{\mathbf{Top}}(X, Y)$. We are left to prove that the conditions in Proposition 1.4.12 also hold. To this end, let $f, f' \in \mathrm{Hom}_{\mathbf{Top}}(X, Y)$ and $g, g' \in \mathrm{Hom}_{\mathbf{Top}}(Y, Z)$.

If $f \sim_{X,Y} f'$, then there exists a continuous map $F \colon X \times [0, 1] \to Y$ such that

$$F(x, 0) = f(x), \quad F(x, 1) = f'(x).$$

We can now define $F' \colon X \times [0, 1] \to Z$ by $F'(x, t) = g\big(F(x, t)\big)$. This yields

$$F'(x, 0) = g\big(F(x, 0)\big) = g\big(f(x)\big) = (g \circ f)(x),$$
$$F'(x, 1) = g\big(F(x, 1)\big) = g\big(f'(x)\big) = (g \circ f')(x),$$

which shows that $g \circ f \sim_{X,Z} g \circ f'$. Assume now that $g \sim_{Y,Z} g'$ and consequently there exists a continuous map $G \colon Y \times [0, 1] \to Z$ such that

$$G(x, 0) = g(x), \quad G(x, 1) = g'(x).$$

Define $G' \colon X \times [0, 1] \to Z$ by $G'(x, t) = G\big(f(x), t\big)$. This yields

$$G'(x, 0) = G\big(f(x), 0\big) = g\big(f(x)\big) = (g \circ f)(x),$$
$$G'(x, 1) = G\big(f(x), 1\big) = g'\big(f(x)\big) = (g' \circ f)(x),$$

which shows that $g \circ f \sim_{X,Z} g' \circ f$. We can now conclude that the homotopy relation $\sim$ is a congruence on **Top**. The resulting quotient category **Top**/ $\sim$ is called the *homotopy category* and will be denoted by **HTop**. For a thorough introduction to homotopy theory, one of the cornerstones of algebraic topology, we refer the reader to [4]. □

---

[13] The pasting lemma: Let $X = A \bigcup B$, where A and B are closed in X, and consider two continuous maps $f \colon A \to Y$, $g \colon B \to Y$. If $f(x) = g(x)$ for all $x \in A \bigcap B$, then f and g combine to give a continuous function $h \colon X \to Y$, defined by $h(x) = \begin{cases} f(x), & \text{if } x \in A \\ g(x), & \text{if } x \in B \end{cases}$ (see [39, Theorem 18.3]).

1.5   Functors

Functors are structure preserving maps which will be used to relate different categories in the way morphisms do with objects.

Definition 1.5.1 Let C and $\mathcal{D}$ be two categories. A *covariant functor* (respectively *contravariant functor*) $F: C \to \mathcal{D}$ consists of the following data:

(1) a mapping $A \mapsto F(A): \mathrm{Ob}\,C \to \mathrm{Ob}\,\mathcal{D}$;
(2) for each pair of objects $A, B \in \mathrm{Ob}\,C$, a mapping

$$f \mapsto F(f): \mathrm{Hom}_C(A, B) \to \mathrm{Hom}_{\mathcal{D}}(F(A), F(B))$$

$$(\text{respectively } f \mapsto F(f): \mathrm{Hom}_C(A, B) \to \mathrm{Hom}_{\mathcal{D}}(F(B), F(A)))$$

subject to the following conditions:

(1) for every $A \in \mathrm{Ob}\,C$ we have $F(1_A) = 1_{F(A)}$;
(2) for every $f \in \mathrm{Hom}_C(A, B)$, $g \in \mathrm{Hom}_C(B, C)$ we have

$$F(g \circ f) = F(g) \circ F(f) \text{ (respectively } F(g \circ f) = F(f) \circ F(g)).$$

A functor $F: \mathcal{A} \times \mathcal{B} \to C$ defined on the product of two categories is called a *bifunctor* (functor of two variables).

Throughout, the term *functor* will denote a *covariant functor*. Any reference to contravariant functors will be explicitly stated.

Remark 1.5.2 Note that the image[14] of a functor need not be a category. Indeed, consider the following two categories C and $\mathcal{D}$:

koh

---

[14] The *image of a functor* (or *values of a functor* as defined in [23]) $F: C \to \mathcal{D}$ consists of a class $\{F(C) \mid C \in \mathrm{Ob}\,C\}$ together with all sets $\{F(f) \mid f \in \mathrm{Hom}_C(A, B)\}$ for any $A, B \in \mathrm{Ob}\,C$.

The image of the functor $F: C \to \mathcal{D}$ defined below is not a category:

$$F(C_1) = D_1, \quad F(C_2) = F(C_3) = D_2, \quad F(C_4) = D_3,$$

$$F(f) = h, \quad F(g) = k, F(1_{C_1}) = 1_{D_1}, \quad F(1_{C_2}) = F(1_{C_3}) = 1_{D_2}, \quad F(1_{C_4}) = 1_{D_3}.$$

Indeed, the morphisms h and k are contained in the image of F while their composition $k \circ h$ is not. However, if F is injective on objects then it can be easily seen that its image is indeed a category.

Examples 1.5.3

(1) If C' is a subcategory of C we can define the *inclusion functor* $I: C' \to C$ which sends every object as well as every morphism to itself. If $C' = C$ then I is just the identity functor 1_C on C.

(2) Let $\sim$ be a congruence on a category C and $C/\sim$ the corresponding quotient category. Then, we can define a *quotient functor* $\Pi: C \to C/\sim$ as follows for all $C \in \mathrm{Ob}\, C$ and $f \in \mathrm{Hom}_C(A, B)$:

$$\Pi(C) = C, \qquad \Pi(f) = \overline{f},$$

where $\overline{f}$ denotes the equivalence class of the morphism f of C. For all $f \in \mathrm{Hom}_C(A, B)$ and $g \in \mathrm{Hom}_C(B, C)$ we have

$$\Pi(g \circ f) = \overline{g \circ f} \overset{(1.3)}{=} \overline{g} \circ \overline{f} = \Pi(g) \circ \Pi(f),$$

which shows that Π is indeed a functor.

(3) If $F: C \to \mathcal{D}$ and $G: \mathcal{D} \to \mathcal{E}$ are two functors, we can define a new functor $G \circ F: C \to \mathcal{E}$ by *pointwise composition*, i.e., $(G \circ F)(C) = G\big(F(C)\big)$ and $(G \circ F)(f) = G\big(F(f)\big)$ for any $C \in \mathrm{Ob}\, C$ and f morphism in C. It is straightforward to check that $G \circ F$ respects compositions and identity maps. For brevity, the pointwise composition of functors will sometimes be denoted by juxtaposition, i.e., we write GF instead of $G \circ F$.

   If F, G are both covariant or contravariant functors, then the pointwise composition defined above yields a covariant functor. On the other hand, if one of the functors is covariant and the other one contravariant then their pointwise composition is a contravariant functor.

(4) For any category C we can define a functor $O_C: C \to C^{op}$ which sends each object to itself and a morphism $f \in \mathrm{Hom}_C(C, C')$ to the opposite morphism $f^{op} \in \mathrm{Hom}_{C^{op}}(C', C)$. The functor O_C is obviously contravariant.

   It can be easily seen that a functor $F: C \to \mathcal{D}$ is contravariant if and only if $F \circ O_{C^{op}}: C^{op} \to \mathcal{D}$ (or $O_{\mathcal{D}} \circ F: C \to \mathcal{D}^{op}$) is a covariant functor.

(5) Let C and $\mathcal{D}$ be two categories and $D_0 \in \mathrm{Ob}\,\mathcal{D}$ a fixed object. The *constant functor* at D_0, denoted by Δ_{D_0}, assigns to every object $C \in \mathrm{Ob}\, C$ the object D_0 and to every morphism f in C the identity morphism 1_{D_0}.

(6) Let $C, \mathcal{D}$ be two categories and consider $C \times \mathcal{D}$ to be the product category as defined in Definition 1.4.4. We can define two *projection functors* as follows:

$$p_C: C \times \mathcal{D} \to C, \quad p_C(C, D) = C, \quad p_C(f, g) = f,$$
$$p_\mathcal{D}: C \times \mathcal{D} \to \mathcal{D}, \quad p_\mathcal{D}(C, D) = D, \quad p_\mathcal{D}(f, g) = g$$

for all $(C, D) \in \mathrm{Ob}\, C \times \mathcal{D}$ and $(f, g) \in \mathrm{Hom}_{C \times \mathcal{D}}\big((C, D), (C', D')\big)$.

(7) If I is a small discrete category, then a functor $F: I \to C$ is uniquely defined by a family of objects $(C_i)_{i \in I}$ indexed by I. More precisely, such a functor is completely determined by the images of each object $i \in \mathrm{Ob}\, I$, say C_i. Indeed, note that for all $i \in \mathrm{Ob}\, I$ the image of the identity morphism 1_i is forced to be 1_{C_i}. If I is the empty set, then there exists a unique functor $F: I \to C$, called the *empty functor*.

(8) Let G, H be two groups and $\mathcal{G}$, respectively $\mathcal{H}$ the corresponding associated categories (see Example 1.2.2, (3)). Then, defining a functor $\mathcal{G} \to \mathcal{H}$ is the same as providing a group homomorphism $G \to H$.

(9) For any category C we can define the bifunctor $\mathrm{Hom}_C(-, -): C^{op} \times C \to \mathbf{Set}$ as follows:

$\mathrm{Hom}_C(A, B) = \mathrm{Hom}_C(A, B) \in \mathrm{Ob}\,\mathbf{Set}$ for all $(A, B) \in \mathrm{Ob}\big(C^{op} \times C\big)$;

if $(f^{op}, g) \in \mathrm{Hom}_{C^{op} \times C}\big((A, B), (C, D)\big) = \mathrm{Hom}_{C^{op}}(A, C) \times \mathrm{Hom}_C(B, D)$

then $\mathrm{Hom}_C(f^{op}, g): \mathrm{Hom}_C(A, B) \to \mathrm{Hom}_C(C, D)$ is defined by

$\mathrm{Hom}_C(f^{op}, g)(h) = g \circ h \circ f$, for all $h \in \mathrm{Hom}_C(A, B)$.

Indeed, for any $h \in \mathrm{Hom}_C(A, B)$ we have $\mathrm{Hom}_C(1_A^{op}, 1_B)(h) = 1_B \circ h \circ 1_A = h$, which shows that $\mathrm{Hom}_C(-, -)$ respects identities.

Furthermore, we have

$$\mathrm{Hom}_C\big((r^{op}, t) \circ (f^{op}, g)\big)(h) = \mathrm{Hom}_C\big((f \circ r)^{op}, t \circ g\big)(h) = t \circ g \circ h \circ f \circ r$$
$$= \mathrm{Hom}_C(r^{op}, t)(g \circ h \circ f) = \mathrm{Hom}_C(r^{op}, t) \circ \mathrm{Hom}_C(f^{op}, g)(h)$$

for all $(f^{op}, g) \in \mathrm{Hom}_{C^{op} \times C}\big((A, B), (C, D)\big) = \mathrm{Hom}_{C^{op}}(A, C) \times \mathrm{Hom}_C(B, D)$ and $(r^{op}, t) \in \mathrm{Hom}_{C^{op} \times C}\big((C, D), (E, F)\big) = \mathrm{Hom}_{C^{op}}(C, E) \times \mathrm{Hom}_C(D, F)$.

Hence $\mathrm{Hom}_C\big((r^{op}, t) \circ (f^{op}, g)\big) = \mathrm{Hom}_C(r^{op}, t) \circ \mathrm{Hom}_C(f^{op}, g)$ and $\mathrm{Hom}_C(-, -)$ is indeed a functor, called the *hom bifunctor*.

(10) Given a category C and a fixed object $C \in \mathrm{Ob}\, C$, we can define two functors, one of them being covariant and the other one contravariant, called the *hom*

functors. Indeed, define $\mathrm{Hom}_C(C, -)$, $\mathrm{Hom}_C(-, C): C \to$ **Set** as follows:

(i) $\mathrm{Hom}_C(C, A) = \mathrm{Hom}_C(C, A) \in \mathrm{Ob\,Set}$ for all $A \in \mathrm{Ob}\,C$;

if $f \in \mathrm{Hom}_C(A, B)$ then $\mathrm{Hom}_C(C, f): \mathrm{Hom}_C(C, A) \to \mathrm{Hom}_C(C, B)$

is defined by $\mathrm{Hom}_C(C, f)(g) = f \circ g$, for all $g \in \mathrm{Hom}_C(C, A)$.

(ii) $\mathrm{Hom}_C(A, C) = \mathrm{Hom}_C(A, C) \in \mathrm{Ob\,Set}$ for all $A \in \mathrm{Ob}\,C$;

if $f \in \mathrm{Hom}_C(A, B)$ then $\mathrm{Hom}_C(f, C): \mathrm{Hom}_C(B, C) \to \mathrm{Hom}_C(A, C)$

is defined by $\mathrm{Hom}_C(f, C)(g) = g \circ f$, for all $g \in \mathrm{Hom}_C(B, C)$.

In certain cases, the hom sets $\mathrm{Hom}_C(A, B)$ can inherit some extra structure from the objects of C, as can be seen in the following examples. This simple observation is the main idea behind the concept of an *enriched category*. For the precise definition and further details on enriched category theory we refer the reader to [33].

(11) If $X, Y \in \mathrm{Ob}\,\mathbf{Top}$ then the set of continuous maps $\mathrm{Hom}_{\mathbf{Top}}(X, Y)$ can be endowed with the so-called *compact-open topology*,[15] which will turn out to be particularly important when dealing with adjoint functors in Chap. 3. We check first that if $f \in \mathrm{Hom}_{\mathbf{Top}}(Y, Z)$ then $\mathrm{Hom}_{\mathbf{Top}}(X, f): \mathrm{Hom}_{\mathbf{Top}}(X, Y) \to \mathrm{Hom}_{\mathbf{Top}}(X, Z)$ is a continuous map with respect to the compact-open topology. Indeed, let $K \subseteq X$ be a compact subset, $U \subseteq Z$ an open subset and $W(K, U)$ a sub-basis open set of the compact-open topology on $\mathrm{Hom}_{\mathbf{Top}}(X, Z)$. As $f^{-1}(U)$ is an open subset of Y, we have

$$
\begin{aligned}
\mathrm{Hom}_{\mathbf{Top}}&(X, f)^{-1}\big(W(K, U)\big) \\
&= \{t \in \mathrm{Hom}_{\mathbf{Top}}(X, Y) \mid \mathrm{Hom}_{\mathbf{Top}}(X, f)(t) \subseteq W(K, U)\} \\
&= \{t \in \mathrm{Hom}_{\mathbf{Top}}(X, Y) \mid \mathrm{Hom}_{\mathbf{Top}}(X, f)(t)(K) \subseteq U\} \\
&= \{t \in \mathrm{Hom}_{\mathbf{Top}}(X, Y) \mid (f \circ t)(K) \subseteq U\} \\
&= \{t \in \mathrm{Hom}_{\mathbf{Top}}(X, Y) \mid t(K) \subseteq f^{-1}(U)\} \\
&= W\big(K, f^{-1}(U)\big).
\end{aligned}
$$

Since continuity of a map need only be checked on a sub-basis of the codomain (see the discussion in [39, page 103]), we can conclude that $\mathrm{Hom}_{\mathbf{Top}}(X, f)$ is continuous, as desired. Hence, we have a functor $\mathrm{Hom}_{\mathbf{Top}}(X, -): \mathbf{Top} \to \mathbf{Top}$.

---

[15] The *compact-open topology* on $\mathrm{Hom}_{\mathbf{Top}}(X, Y)$ is the topology generated by the sub-basis $W(K, U) = \{f \in \mathrm{Hom}_{\mathbf{Top}}(X, Y) \mid f(K) \subseteq U\}$, where $K \subseteq X$ is compact and $U \subseteq Y$ is open.

(12) If $A, B \in \mathrm{Ob}\,\mathbf{Ab}$ then $\mathrm{Hom}_{\mathbf{Ab}}(A,\ B)$ has an abelian group structure given by

$$(f+g)(a) = f(a) + g(a) \tag{1.4}$$

for all $f, g \in \mathrm{Hom}_{\mathbf{Ab}}(A,\ B)$ and $a \in A$. Moreover, if $f \in \mathrm{Hom}_{\mathbf{Ab}}(B,\ C)$, it can be easily seen that $\mathrm{Hom}_{\mathbf{Ab}}(A,\ f)$ is a group homomorphism. Therefore, for any $A \in \mathrm{Ob}\,\mathbf{Ab}$ we have a functor $\mathrm{Hom}_{\mathbf{Ab}}(A,\ -) \colon \mathbf{Ab} \to \mathbf{Ab}$.

(13) Let K be a field. If $M, N \in \mathrm{Ob}\,{}_K\mathcal{M}$ then $\mathrm{Hom}_{{}_K\mathcal{M}}(M,\ N)$ has the abelian group structure given as in (1.4) and a K-vector space structure defined by scalar multiplication, i.e., given $k \in K$, $f \in \mathrm{Hom}_{{}_K\mathcal{M}}(M,\ N)$ define the linear map kf as follows:

$$(kf)(m) = kf(m), \quad \text{for all}\ \ m \in M.$$

Therefore, any $M \in \mathrm{Ob}\,{}_K\mathcal{M}$ yields a functor $\mathrm{Hom}_{{}_K\mathcal{M}}(A,\ -)\colon {}_K\mathcal{M} \to {}_K\mathcal{M}$.

(14) Consider the field K as an object in ${}_K\mathcal{M}$ and denote the corresponding contravariant hom functor $\mathrm{Hom}_{{}_K\mathcal{M}}(-,\ K)$ by $(-)^* \colon {}_K\mathcal{M} \to \mathbf{Set}$. That is, we denote $\mathrm{Hom}_{{}_K\mathcal{M}}(V,\ K)$ by V^* and $\mathrm{Hom}_{{}_K\mathcal{M}}(u,\ K)$ by u^*. Given a vector space V, the set V^* of linear maps from V to K can be endowed with a vector space structure as follows: for any $f, g \in V^*$ and $a, b \in K$ the linear map $af + bg$ is defined by $(af + bg)(v) = af(v) + bg(v)$. Furthermore, given a linear map $u \colon V \to W$ it can be easily seen that $u^* \colon W^* \to V^*$ defined by $u^*(w) = w \circ u$ for all $w \in W^*$ is also a linear map. Therefore, the functor $(-)^*$ maps into the category ${}_K\mathcal{M}$ and is called the *dual space (contravariant) functor*.

(15) By composing the dual space functor with itself we obtain a covariant functor denoted by $(-)^{**} \colon {}_K\mathcal{M} \to {}_K\mathcal{M}$ and called the *double dual space functor*. We only point out for further use that if $u \colon U \to V$ is a linear map then $u^{**} \colon U^{**} \to V^{**}$ is defined by

$$u^{**}(\phi) = \phi \circ u^* \text{ for all } \phi \colon U^* \to K. \tag{1.5}$$

(16) The *cartesian product bifunctor* $- \times - \colon \mathbf{Set} \times \mathbf{Set} \to \mathbf{Set}$ is defined as follows for all $X, Y \in \mathrm{Ob}\,\mathbf{Set}$ and $f \in \mathrm{Hom}_{\mathbf{Set}}(A,\ C)$, $g \in \mathrm{Hom}_{\mathbf{Set}}(B,\ D)$:

$$(- \times -)(A,\ B) = A \times B;$$

$$(- \times -)(f,\ g) = f \times g \colon A \times B \to C \times D,$$

where $(f \times g)(a,\ b) = \big(f(a),\ g(b)\big)$ for all $a \in A, b \in B$.

(17) For any set X we can define the corresponding *cartesian product functor* $- \times X \colon \mathbf{Set} \to \mathbf{Set}$ as follows for all $Y, Z \in \mathrm{Ob}\,\mathbf{Set}$ and $f \in \mathrm{Hom}_{\mathbf{Set}}(Y,\ Z)$:

$$(- \times X)(Y) = Y \times X;$$

$$(- \times X)(f) = f \times 1_X \colon Y \times X \to Z \times X,$$

where $(f \times 1_X)(y, x) = \big(f(y), x\big)$ for all $y \in Y$, $x \in X$. Similarly we can define the cartesian product functor $X \times -: \textbf{Set} \to \textbf{Set}$.

Exactly as in the case of the hom functor, in certain cases, when the given sets are endowed with some extra structures, the corresponding cartesian product inherits this structure.

(18) Any $X \in \text{Ob}\,\textbf{Top}$ defines a functor $- \times X: \textbf{Top} \to \textbf{Top}$, where for all $Y \in \text{Ob}\,\textbf{Top}$ we consider on $Y \times X$ the product topology.[16]

(19) Similarly, any $G \in \text{Ob}\,\textbf{Grp}$ defines a functor $- \times G: \textbf{Grp} \to \textbf{Grp}$, where for all $H \in \text{Ob}\,\textbf{Grp}$ the group structure on $H \times G$ is defined component-wise.

(20) Let K be a field and for simplicity denote the tensor product over K by $\otimes$ (i.e., $\otimes = \otimes_K$). For any $X \in \text{Ob}\,_K\mathcal{M}$ we can define a functor $- \otimes X: {}_K\mathcal{M} \to {}_K\mathcal{M}$, called the *tensor product functor*, as follows:

$$(- \otimes X)(M) = M \otimes X, \text{ for all } M \in \text{Ob}\,_K\mathcal{M};$$

$$(- \otimes X)(f) = f \otimes 1_X, \text{ for all } f \in \text{Hom}_{\text{Ob}\,_K\mathcal{M}}(M, N),$$

where $f \otimes 1_X: M \otimes X \to N \otimes X$ is the K-linear map defined by $(f \otimes 1_X)(m \otimes x) = f(m) \otimes x$ and the K vector space structure on $M \otimes X$ is given by $k(m \otimes x) = km \otimes x$ for all $k \in K$, $m \in M$ and $x \in X$. For more details on the tensor product of vector spaces (or modules) we refer the reader to [45].

(21) For any set X we denote by $\mathcal{P}(X) = \{Y \mid Y \subseteq X\}$ the power set of X. We can define two *power set functors* $P: \textbf{Set} \to \textbf{Set}$ and respectively $P^c: \textbf{Set} \to \textbf{Set}$, the first one being covariant and the second one contravariant, as follows:

i) $P: \textbf{Set} \to \textbf{Set}$, $P(A) = \mathcal{P}(A) \in \text{Ob}\,\textbf{Set}$, for all $A \in \text{Ob}\,\textbf{Set}$;

if $f \in \text{Hom}_{\textbf{Set}}(A, B)$ then $P(f): \mathcal{P}(A) \to \mathcal{P}(B)$ is defined by

$P(f)(U) = f(U)$, for all $U \subseteq A$.

ii) $P^c: \textbf{Set} \to \textbf{Set}$, $P^c(A) = \mathcal{P}(A) \in \text{Ob}\,\textbf{Set}$, for all $A \in \text{Ob}\,\textbf{Set}$;

if $f \in \text{Hom}_{\textbf{Set}}(A, B)$ then $P^c(f): \mathcal{P}(B) \to \mathcal{P}(A)$ is defined by

$P^c(f)(V) = f^{-1}(V)$, for all $V \subseteq B$.

(22) The so-called *forgetful functors* are functors which *forget* (some of) the structure on objects of the domain category. For instance the categories in Example 1.2.2, (7)–(13) allow for a forgetful functor to the category **Set** of sets, which sends the objects of that category to the underlying set, and the

---

[16] Let $(X_i)_{i \in I}$ be a family of topological spaces indexed by the set I and consider the product of the underlying sets $\big(\prod_{i \in I} X_i, (\pi_i)_{i \in I}\big)$. The topology generated by the sub-basis $\mathcal{S} = \bigcup_{\beta \in I} \mathcal{S}_\beta$ is called the *product topology*, where $\mathcal{S}_\beta = \{\pi_\beta^{-1}(U_\beta) \mid U_\beta \text{ open in } X_\beta\}$ ([39, Definition, page 114]).

homomorphisms to the underlying mappings between the underlying sets. Similarly, we have many other examples of forgetful functors which forget only some of the structure in objects of the domain category:

Grp → **Set**, forgets about the group structure;

Top → **Set**, forgets about the topological structure;

Rng → **Ab**, forgets about the product;

Ab → **Grp**, forgets about the commutativity;

Ring → **Rng**, forgets about the unit;

Top$_*$ → **Top**, forgets about the base point;

Haus → **Top**, forgets about the Hausdorff property.

(23) For a group G we denote by $[G, G]$ its commutator subgroup; in other words $[G, G]$ is the subgroup of G generated by all elements of the form $xyx^{-1}y^{-1}$, where $x, y \in G$. The corresponding quotient group $G_{ab} = G/[G, G]$ is obviously an abelian group called the *abelianization* of G. Furthermore, if $f \in \text{Hom}_{\textbf{Grp}}(G, H)$ is a group homomorphism and $\pi_G \colon G \to G_{ab}$, $\pi_H \colon H \to H_{ab}$ denote the corresponding projections, we have $[G, G] \subseteq \ker(\pi_H \circ f)$. Hence, the universal property of the quotient group G_{ab} yields a unique group homomorphism $f_{ab} \colon G_{ab} \to H_{ab}$ which makes the following diagram commutative:

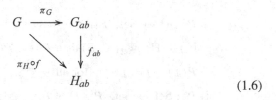

$$\begin{array}{ccc} G & \xrightarrow{\pi_G} & G_{ab} \\ & \searrow^{\pi_H \circ f} & \downarrow^{f_{ab}} \\ & & H_{ab} \end{array}$$

(1.6)

This construction allows us to define a functor $F \colon \textbf{Grp} \to \textbf{Ab}$, called the *abelianization functor*, as follows:

$$F(G) = G_{ab}, \text{ for all } G \in \text{Ob}\,\textbf{Grp};$$

$$F(f) = f_{ab}, \text{ for all } f \in \text{Hom}_{\textbf{Grp}}(G, H).$$

(24) For a topological space X we denote by $(\widetilde{X}, i_X)$ the corresponding *Stone–Čech compactification*, where $\widetilde{X}$ is a compact Hausdorff topological space while $i_X \colon X \to \widetilde{X}$ is a continuous function (see [39, Section 38] for further details). Given $f \in \text{Hom}_{\textbf{Top}}(X, Y)$, the universal property of the Stone–Čech compactification yields a unique $\widetilde{f} \in \text{Hom}_{\textbf{KHaus}}(\widetilde{X}, \widetilde{Y})$ such that the

following diagram is commutative:

$$X \xrightarrow{\; i_X \;} \widetilde{X} \qquad \text{i.e., } \widetilde{f} \circ i_X = i_Y \circ f.$$

with $i_Y \circ f$ going diagonally to $\widetilde{Y}$ and $\widetilde{f}: \widetilde{X} \to \widetilde{Y}$.

(1.7)

Now we can define a functor $\mathcal{S}$: **Top** $\rightarrow$ **KHaus**, called the *Stone–Čech compactification functor*, as follows:

$$\mathcal{S}(X) = \widetilde{X}, \quad \text{for all } X \in \text{Ob } \mathbf{Top};$$

$$\mathcal{S}(f) = \widetilde{f}, \quad \text{for all } f \in \text{Hom}_{\mathbf{Top}}(X, Y).$$

(25) Given a monoid $(M, \cdot)$, we denote by $\big(G(M),\ i_M\big)$ the *universal enveloping group* of M (also called the *group completion* or the *Grothendieck group* of M), where $G(M)$ is a group while $i_M\colon M \to G(M)$ is a homomorphism of monoids (see [6, Section 4.11]).

For any $f \in \text{Hom}_{\mathbf{Mon}}(M,\, N)$, the universal property of the universal enveloping group yields a unique $\widetilde{f} \in \text{Hom}_{\mathbf{Grp}}(G(M),\, G(N))$ such that the following diagram is commutative:

$$M \xrightarrow{\; i_M \;} G(M) \qquad \text{i.e., } \widetilde{f} \circ i_M = i_N \circ f.$$

with $i_N \circ f$ going diagonally to $G(N)$ and $\widetilde{f}: G(M) \to G(N)$.

(1.8)

The functor $\mathcal{G}$: **Mon** $\rightarrow$ **Grp** defined below is called the *universal enveloping group functor*:

$$\mathcal{G}(M) = G(M), \quad \text{for all } M \in \text{Ob } \mathbf{Mon};$$

$$\mathcal{G}(f) = \widetilde{f}, \quad \text{for all } f \in \text{Hom}_{\mathbf{Mon}}(M,\, N).$$

(26) Given a monoid $(M, \cdot)$, we denote by $U(M)$ the set of all invertible elements of M. Furthermore, if $f \in \text{Hom}_{\mathbf{Mon}}(M,\, N)$, it can be easily seen that $f_{|U(M)} \subseteq U(N)$. This gives rise to a functor $\mathcal{U}$: **Mon** $\rightarrow$ **Grp** defined as follows:

$$\mathcal{U}(M) = U(M), \quad \text{for all } M \in \text{Ob } \mathbf{Mon};$$

$$\mathcal{U}(f) = f_{|U(M)}, \quad \text{for all } f \in \text{Hom}_{\mathbf{Mon}}(M,\, N).$$

(27) Similarly, given a ring $(R, +, \cdot)$ we denote by $U(R)$ the set of invertible elements of $(R, \cdot)$. This yields a functor $\mathcal{U}\colon \mathbf{Ring} \to \mathbf{Grp}$ defined as follows:

$$\mathcal{U}(R) = U(R), \text{ for all } R \in \mathrm{Ob}\,\mathbf{Ring};$$

$$\mathcal{U}(f) = f_{|U(R)}, \text{ for all } f \in \mathrm{Hom}_{\mathbf{Ring}}(R, R'),$$

where $f_{|U(R)}$ denotes the restriction of f to the subset $U(R)$ of R.

(28) Let R be a ring, $M \in \mathrm{Ob}\,\mathcal{M}_R$ and $N \in \mathrm{Ob}\,_R\mathcal{M}$. We can define the functor $\mathrm{Bil}_{M, N}\colon \mathbf{Ab} \to \mathbf{Ab}$ of R-bilinear maps as follows for all $A \in \mathbf{Ab}$ and $f \in \mathrm{Hom}_{\mathbf{Ab}}(A, B)$:[17]

$$\mathrm{Bil}_{M, N}(A) = \{\alpha\colon M \times N \to A \mid \alpha \text{ is } R-\text{bilinear}\};$$

$$\mathrm{Bil}_{M, N}(f)(u) = f \circ u, \quad u \in \mathrm{Bil}_{M, N}(A).$$

Note that each $\mathrm{Bil}_{M, N}(A)$ can be made into an abelian group as in (1.4), i.e., for all $\alpha, \beta \in \mathrm{Bil}_{M, N}(A)$ and $m \in M, n \in N$ define

$$(\alpha + \beta)(m, n) = \alpha(m, n) + \beta(m, n).$$

(29) Let R be a commutative ring with unity, S a multiplicative subset of R, and $(S^{-1}R, j)$ the corresponding localization ring, where $j\colon R \to S^{-1}R$ is the ring homomorphism defined by $j(r) = \frac{r}{1}$, for all $r \in R$. Furthermore, if $M \in \mathrm{Ob}\,_R\mathcal{M}$, we denote by $(S^{-1}M, \varphi_M)$ the corresponding localization module, where $S^{-1}M \in \mathrm{Ob}\,_{S^{-1}R}\mathcal{M}$ and $\varphi_M\colon M \to S^{-1}M$ is the R-module homomorphism defined by $\varphi_M(m) = \frac{m}{1}$, for all $m \in M$. We refer the reader to [3, Chapter 12] for more details on localization of modules.

For any $f \in \mathrm{Hom}_{_R\mathcal{M}}(M, N)$, the universal property of the localization module ([3, Theorem 12.3]) yields a unique $\widetilde{f} \in \mathrm{Hom}_{_{S^{-1}R}\mathcal{M}}(S^{-1}M, S^{-1}N)$ such that the following diagram is commutative:

$$M \xrightarrow{\varphi_M} S^{-1}M \qquad \text{i.e., } \widetilde{f} \circ \varphi_M = \varphi_N \circ f.$$

$$\varphi_N \circ f \searrow \quad \downarrow \widetilde{f}$$

$$S^{-1}N$$

$$(1.9)$$

---

[17] $\alpha\colon M \times N \to A$ is called R-*bilinear* if for all $r \in R, m, m' \in M$ and $n, n' \in N$ we have

$$\alpha(m + m', n) = \alpha(m, n) + \alpha(m', n),$$

$$\alpha(m, n + n') = \alpha(m, n) + \alpha(m, n'),$$

$$\alpha(mr, n) = \alpha(m, rn).$$

The functor $\mathcal{L} \colon {}_R\mathcal{M} \to {}_{S^{-1}R}\mathcal{M}$ defined below is called the *localization with respect to* S:

$$\mathcal{L}(M) = S^{-1}M, \quad \text{for all } M \in \text{Ob}\,{}_R\mathcal{M};$$

$$\mathcal{L}(f) = \widetilde{f}, \quad \text{for all } f \in \text{Hom}_{{}_R\mathcal{M}}(M, N).$$

(30) For a topological space X we denote by $\big(H(X), q_X\big)$ the corresponding *Hausdorff quotient*, where $H(X)$ is a Hausdorff topological space while $q_X \colon X \to H(X)$ is a continuous function (see [44] for more details).

Given $f \in \text{Hom}_{\textbf{Top}}(X, Y)$, the universal property of the Hausdorff quotient (as stated, for instance, in [40, page 81]) yields a unique $\widetilde{f} \in \text{Hom}_{\textbf{Haus}}(H(X), H(Y))$ such that the following diagram is commutative:

$$
\begin{array}{ccc}
X & \xrightarrow{\;q_X\;} & H(X) \\
& \searrow{\scriptstyle q_Y \circ f} & \big\downarrow{\scriptstyle \widetilde{f}} \\
& & H(Y)
\end{array}
\qquad \text{i.e.,} \quad \widetilde{f} \circ i_X = i_Y \circ f.
$$

$$(1.10)$$

Now we can define a functor $\mathcal{H} \colon \textbf{Top} \to \textbf{Haus}$, called the *Hausdorff quotient functor*, as follows:

$$\mathcal{H}(X) = H(X), \quad \text{for all } X \in \text{Ob}\,\textbf{Top};$$

$$\mathcal{H}(f) = \widetilde{f}, \quad \text{for all } f \in \text{Hom}_{\textbf{Top}}(X, Y).$$

(31) For an arbitrary ring R we denote by $\big(D(R), j_R\big)$ the corresponding *Dorroh extension*, where $D(R)$ is a unitary ring while $j_R \colon R \to D(R)$ is a morphism in **Rng**.[18]

Given $f \in \text{Hom}_{\textbf{Rng}}(R, S)$, the universal property of the Dorroh extension (see [18, Theorem 3.1.1])[19] yields a unique $h \in \text{Hom}_{\textbf{Ring}}(D(R), D(S))$ such

---

[18] The *Dorroh extension* of an arbitrary ring R is defined as $D(R) = \mathbb{Z} \times R$ with componentwise addition and multiplication given for all (n_1, r_1), $(n_2, r_2) \in \mathbb{Z} \times R$ by $(n_1, r_1)(n_2, r_2) = (n_1 n_2, n_1 r_2 + n_2 r_1 + r_1 r_2)$ and $j_R \colon R \to D(R)$ is given by $j_R(r) = (0, r)$ for all $r \in R$ (we refer to [15, Theorem 2.12] for more details).

[19] For any $f \in \text{Hom}_{\textbf{Rng}}(R, S)$, where $S \in \text{Ob}\,\textbf{Ring}$, there exists a unique $\widetilde{f} \in \text{Hom}_{\textbf{Ring}}(D(R), S)$ such that $h \circ j_R = f$.

that the following diagram is commutative:

$$R \xrightarrow{\ j_R\ } D(R) \qquad \text{i.e., } \tilde{f} \circ j_R = j_S \circ f.$$

$$\begin{array}{c} \text{\small} j_S \circ f \searrow \quad \downarrow \tilde{f} \\ D(S) \end{array}$$

$$(1.11)$$

Now we can define a functor $\mathcal{D} \colon \mathbf{Rng} \to \mathbf{Ring}$, called the *Dorroh extension functor*, as follows:

$$\mathcal{D}(R) = D(R), \ \text{ for all } R \in \mathrm{Ob}\, \mathbf{Rng};$$

$$\mathcal{D}(f) = \tilde{f}, \ \text{ for all } f \in \mathrm{Hom}_{\mathbf{Rng}}(R, S).$$

(32) For any $u \in \mathrm{Hom}_{\mathbf{Ring}^c}(R, S)$ we can define a functor $F_u \colon {}_S\mathcal{M} \to {}_R\mathcal{M}$, called *restriction of scalars*, as follows for all $M \in \mathrm{Ob}\, {}_S\mathcal{M}$ and $f \in \mathrm{Hom}_{{}_S\mathcal{M}}(M, N)$:

$$F_u(M) = M, \text{ where } M \in \mathrm{Ob}\, {}_R\mathcal{M} \text{ via } rm = f(r)m, \text{ for all } r \in R, \ m \in M;$$

$$F_u(f) = f.$$

(33) Given a pre-ordered set $(X, \leqslant_X)$, we denote by $(X, \mathcal{A}_{\leqslant_X})$ the *Alexandroff (or Alexandrov) topology*[20] on X with respect to the pre-order $\leqslant_X$. Furthermore, if $f \colon (X, \leqslant_X) \to (Y, \leqslant_Y)$ is order preserving then $f \colon (X, \mathcal{A}_{\leqslant_X}) \to (Y, \mathcal{A}_{\leqslant_Y})$ is continuous. Indeed, let $U \subseteq Y$ be an open subset, $x \in f^{-1}(U)$ and $x' \in X$ such that $x \leqslant_X x'$. As f is order preserving, we have $f(x) \leqslant_X f(x')$ and since $f(x) \in U$ we obtain $f(x') \in U$. Therefore, $x' \in f^{-1}(U)$, which shows that $f^{-1}(U) \subseteq X$ is an open set. This yields a functor $F \colon \mathbf{PreOrd} \to \mathbf{Top}$ defined as follows:

$$F(X, \leqslant_X) = (X, \mathcal{A}_{\leqslant_X});$$

$$F(f) = f.$$

(34) A contravariant functor $F \colon \mathbf{Top} \to \mathbf{Poset}$ can be defined as follows for all topological spaces (X, τ), (Y, γ) and continuous maps $f \colon (X, \tau) \to (Y, \gamma)$:

$$F(X, \tau) = \{U \subseteq X \mid U \text{ is open in } \tau\};$$

$$F(f) = f^{-1},$$

---

[20] Given a pre-ordered set $(X, \leqslant_X)$, the *Alexandroff topology* on X with respect to $\leqslant_X$ is defined by considering a subset U of X to be open if $x \leqslant_X x'$ and $x \in U$ imply $x' \in U$.

where $\{U \subseteq X \mid U$ is open in $\tau\}$ is partially ordered by set inclusion and $f^{-1} \colon \{V \subseteq Y \mid V$ is open in $\gamma\} \to \{U \subseteq X \mid U$ is open in $\tau\}$ denotes the preimage of f. Indeed, note that since $f \colon (X, \tau) \to (Y, \gamma)$ was assumed to be a continuous map we have $f^{-1}(V) \in \tau$ for any $V \in \gamma$.

(35) Given two pre-ordered sets $(X, \leqslant)$ and $(Y, \ll)$, a functor between the corresponding induced categories $\mathrm{PO}(X, \leqslant)$ and $\mathrm{PO}(Y, \ll)$, respectively, as defined in Example 1.2.2, (2) is nothing but an order preserving function between $(X, \leqslant)$ and $(Y, \ll)$. Consequently, a functor between $\mathrm{PO}(X, \leqslant)^{op}$ and $\mathrm{PO}(Y, \ll)$ is an order-reversing function between $(X, \leqslant)$ and $(Y, \ll)$. $\square$

Having defined a functor as a type of *morphism* between categories, it is only natural to consider the following category:

Proposition 1.5.4 *The small categories and functors between them constitute a category which we will denote by* **Cat**, *called the* category of small categories.

Proof Given two functors $F \colon C \to \mathcal{D}$ and $G \colon \mathcal{D} \to \mathcal{E}$ we obtain a new functor $G \circ F \colon C \to \mathcal{E}$ by pointwise composition. The composition law is obviously associative and the identity functor on a category is an identity for this composition.

Finally, if C and $\mathcal{D}$ are small categories, i.e., $\mathrm{Ob}\,C$ and $\mathrm{Ob}\,\mathcal{D}$ are sets, then $\mathrm{Hom}_{\mathbf{Cat}}(C, \mathcal{D})$ is also a set. $\square$

Proposition 1.5.5 *Let* $F \colon \mathcal{A} \times \mathcal{B} \to C$ *be a bifunctor. Then for any* $A \in \mathrm{Ob}\,\mathcal{A}$ *there exists a functor* $F_A \colon \mathcal{B} \to C$, *called the* right associated functor *with respect to* A, *defined as follows for all* $B, B' \in \mathrm{Ob}\,\mathcal{B}$ *and* $f \in \mathrm{Hom}_{\mathcal{B}}(B, B')$:

$$F_A(B) = F(A, B), \qquad F_A(f) = F(1_A, f).$$

Similarly, for any $B \in \mathrm{Ob}\,\mathcal{B}$ *there exists a functor* $F^B \colon \mathcal{A} \to C$, *called the* left associated functor *with respect to* B, *defined as follows for all* $A, A' \in \mathrm{Ob}\,\mathcal{A}$ *and* $f \in \mathrm{Hom}_{\mathcal{A}}(A, A')$:

$$F^B(A) = F(A, B), \qquad F^B(f) = F(f, 1_B).$$

Proof We only show the first assertion. To this end, for any $B \in \mathrm{Ob}\,\mathcal{B}$ we have $F_A(1_B) = F(1_A, 1_B) = F(1_{(A, B)}) = 1_{F(A, B)}$, where the last equality holds because F is a functor. Thus F_A respects identities. Furthermore, for any $f \in \mathrm{Hom}_{\mathcal{B}}(B, B')$ and $g \in \mathrm{Hom}_{\mathcal{B}}(B', B'')$ we have

$$F_A(g \circ f) = F(1_A, g \circ f) = F\big((1_A, g) \circ (1_A, f)\big) = F(1_A, g) \circ F(1_A, f)$$
$$= F_A(g) \circ F_A(f).$$

We have proved that F_A respects compositions as well and the proof is now finished. $\square$

Examples 1.5.6

(1) For any object C in a category C, the hom functor $\mathrm{Hom}_C(C, -)$ is the right associated functor of the Hom bifunctor with respect to C. Similarly, the contravariant hom functor $\mathrm{Hom}_C(-, C)$ is the left associated functor of the Hom bifunctor with respect to C.

(2) For any set X, the cartesian product functors $X \times -$ and $- \times X$ are the right and respectively the left associated functors with respect to X of the cartesian bifunctor.   □

1.6 Isomorphisms of Categories

We start by introducing the following notions, which are weaker than isomorphism but very useful.

Definition 1.6.1 Let $F: C \to \mathcal{D}$ be a functor and for all $A, B \in \mathrm{Ob}\,C$ consider the following induced mapping:

$$\mathcal{F}_{A, B}: \mathrm{Hom}_C(A, B) \to \mathrm{Hom}_{\mathcal{D}}\big(F(A), F(B)\big), \quad f \mapsto F(f). \tag{1.12}$$

(1) The functor F is called *faithful* if the mappings $\mathcal{F}_{A, B}$ are injective for all $A, B \in \mathrm{Ob}\,C$.

(2) The functor F is called *full* if the mappings $\mathcal{F}_{A, B}$ are surjective for all $A, B \in \mathrm{Ob}\,C$.

(3) The functor F is called *fully faithful* if the mappings $\mathcal{F}_{A, B}$ are bijective for all $A, B \in \mathrm{Ob}\,C$.

(4) The functor F is called *essentially surjective* if each object D of $\mathcal{D}$ is isomorphic to an object of the form $F(C)$ for some $C \in \mathrm{Ob}\,C$.

Examples 1.6.2

(1) The inclusion functor is automatically faithful. If the subcategory is full then the inclusion functor is also full.

(2) The inclusion functor $I: \mathbf{Ab} \to \mathbf{Grp}$ is fully faithful.

(3) The quotient functor $\Pi: C \to C/\sim$ is always full, where $\sim$ is a congruence on the category C and $C/\sim$ denotes the corresponding quotient category.

(4) The quotient functor $\Pi: \mathbf{Top} \to \mathbf{HTop}$ is full but not faithful.

(5) The inclusion functor $I: C^{\mathrm{grp}} \to C$ is faithful but not full unless C itself is a groupoid, where C^{grp} is the core groupoid of the category C constructed in Example 1.3.6, (2).

(6) Given $f \in \mathrm{Hom}_{\mathbf{Ring}^c}(A, B)$, the restriction of scalars functor $F_f: {}_B\mathcal{M} \to {}_A\mathcal{M}$ defined in Example 1.5.3, (32) is obviously faithful. Furthermore, if f is an epimorphism in $\mathbf{Ring}^c$ then the corresponding restriction of scalars functor is also full ([53, Proposition XI.1.2]). Indeed, let $M, N \in \mathrm{Ob}\,{}_B\mathcal{M}$ and $u \in$

$\text{Hom}_A{}_\mathcal{M}(F_f(M), F_f(N))$, i.e., for all $a \in A$ and $m \in M$ we have

$$u\big(f(a)m\big) = f(a)u(m). \tag{1.13}$$

Note that throughout this example all module actions will be denoted by juxtaposition and we will see B as an A-module via f, i.e., $ab = f(a)b$ for all $a \in A$ and $b \in B$.

Now let $m \in M$ and consider the map $v \colon B \otimes_A B \to N$ defined for all b, $b' \in B$ as follows:

$$v(b \otimes_A b') = b'u(bm). \tag{1.14}$$

It can be easily seen that v is well-defined; indeed, for instance we have

$$v(ab \otimes_A b') = v\big(f(a)b \otimes_A b'\big) \overset{(1.14)}{=} b'u\big(f(a)bm\big) \overset{(1.13)}{=} f(a)b'u(bm)$$

$$\overset{(1.14)}{=} v\big(b \otimes_A f(a)b'\big) = v(b \otimes_A ab').$$

Moreover, consider $\alpha, \beta \in \text{Hom}_{\mathbf{Ring}^c}(B, B \otimes_A B)$ defined for all $b \in B$ by

$$\alpha(b) = b \otimes_A 1_B, \qquad \beta(b) = 1_B \otimes_A b.$$

This shows that for all $a \in A$ we have

$$\alpha\big(f(a)\big) = f(a) \otimes_A 1_B = a1_B \otimes_A 1_B = 1_B \otimes_A a1_B = 1 \otimes_A f(a) = \beta\big(f(a)\big).$$

As f was assumed to be an epimorphism, it follows that $b \otimes_A 1_B = 1_B \otimes_A b$ in $B \otimes_A B$ for all $b \in B$. This gives $u(bm) = v(b \otimes_A 1_B) = v(1_B \otimes_A b) = bu(m)$. We can now conclude that u is in fact a morphism in ${}_B\mathcal{M}$ and therefore F_f is full.

(7) The forgetful functor $U \colon \mathbf{Grp} \to \mathbf{Set}$ is faithful but not full as not any function between two given groups is a group homomorphism.

(8) The functor $\mathcal{U} \colon \mathbf{Ring} \to \mathbf{Grp}$ defined in Example 1.5.3, (27) is neither full nor faithful. To start with, recall that $U(\mathbb{Z}) = \mathbb{Z}_2$ and $U(\mathbb{F}_p) = \mathbb{Z}_{p-1}$ for any prime number p, where $\mathbb{F}_p$ denotes the field with p elements and $\mathbb{Z}_n$ is the group of integers modulo n. Then, given an odd prime number p, the following induced map is not surjective:

$$\mathcal{U}_{\mathbb{Z}, \mathbb{F}_p} \colon \text{Hom}_{\mathbf{Ring}}(\mathbb{Z}, \mathbb{F}_p) \to \text{Hom}_{\mathbf{Grp}}(\mathbb{Z}_2, \mathbb{Z}_{p-1}).$$

This follows easily by noticing that since $\mathbb{Z}$ is the initial object in $\mathbf{Ring}$, the set $\text{Hom}_{\mathbf{Ring}}(\mathbb{Z}, \mathbb{F}_p)$ has only one element while the cardinality of the set $\text{Hom}_{\mathbf{Grp}}(\mathbb{Z}_2, \mathbb{Z}_{p-1})$ is $\gcd(2, p - 1) = 2$ (see [27]).

In order to show that $\mathcal{U}$ is not faithful either, consider the polynomial ring $k[X]$ over a field k and recall that $U(k[X]) = k \backslash \{0\}$. Then, the following induced map is not injective:

$$\mathcal{U}_{\mathbb{F}_2[X], \, \mathbb{F}_2[X]} : \operatorname{Hom}_{\mathbf{Ring}}(\mathbb{F}_2[X], \, \mathbb{F}_2[X]) \to \operatorname{Hom}_{\mathbf{Grp}}(\{1\}, \, \{1\}),$$

where $\{1\}$ denotes the trivial group. Indeed, the set $\operatorname{Hom}_{\mathbf{Grp}}(\{1\}, \, \{1\})$ obviously has only one element while the cardinality of $\operatorname{Hom}_{\mathbf{Ring}}(\mathbb{F}_2[X], \, \mathbb{F}_2[X])$ is at least two. $\qquad\square$

Definition 1.6.3 A category C is said to be *concrete* if there exists a faithful functor $F : C \to \mathbf{Set}$.

Example 1.6.4 The categories **FinSet**, **Grp**, **Ab**, **Rng**, **Ring**, **Top**, $_R\mathcal{M}$ are all concrete categories due to the existence of forgetful functors from any of the above categories to **Set**, which are obviously faithful. $\qquad\square$

In fact, we have a lot more examples of concrete categories, as can be seen from the next result which, as noted in [6, Theorem 7.5.6], resembles Cayley's theorem from group theory.

Theorem 1.6.5 *Any small category is concrete.*

Proof For any small category C we construct a faithful functor $F : C \to \mathbf{Set}$ as follows. Given $C \in \operatorname{Ob} C$ and $u \in \operatorname{Hom}_C(C, C')$ we define

$$F(C) = \{(Y, \alpha) \mid Y \in \operatorname{Ob} C, \, \alpha \in \operatorname{Hom}_C(Y, C)\} \in \operatorname{Ob}\mathbf{Set};$$

$$F(u) : F(C) \to F(C'), \quad F(u)(Y, \alpha) = (Y, u \circ \alpha).$$

It is straightforward to see that F defined above is a functor; we only point out that $F(C)$ is a set due to the fact that C is a small category. We will prove that it is faithful. Indeed, if $u_1, u_2 \in \operatorname{Hom}_C(C, C')$ such that $F(u_1) = F(u_2)$ we obtain $(Y, u_1 \circ \alpha) = (Y, u_2 \circ \alpha)$ for any $(Y, \alpha) \in F(C)$. Now for $(C, 1_C) \in F(C)$ we get $u_1 = u_2$, as desired. $\qquad\square$

Not every category admits a faithful functor to **Set**. The interested reader may find such an example in [24], where it is shown that the homotopy category of pointed spaces is not concrete.

Definition 1.6.6 A functor $F : C \to \mathcal{D}$ is called an *isomorphism of categories* if there exists another functor $G : \mathcal{D} \to C$ such that $F \circ G = 1_{\mathcal{D}}$ and $G \circ F = 1_C$. In this case we say that the categories C and $\mathcal{D}$ are *isomorphic* and G is called the *inverse* of F. A contravariant isomorphism of categories is called an *anti-isomorphism of categories*.

Let $F : C \to \mathcal{D}$ be a functor and for all $A, B \in \operatorname{Ob} C$ consider the induced mapping $\mathcal{F}_{A, B}$ defined in (1.12). If F is an isomorphism of categories with inverse $G : \mathcal{D} \to C$ then each $\mathcal{F}_{A, B}$ is a bijective map with inverse given by $\mathcal{G}_{F(A), F(B)}$, where

$$\mathcal{G}_{F(A), F(B)} \colon \mathrm{Hom}_{\mathcal{D}}\big(F(A),\ F(B)\big) \to \mathrm{Hom}_{C}(A,\ B),\quad \mathcal{G}_{F(A), F(B)}(g) = G(g)$$

for all $g \in \mathrm{Hom}_{\mathcal{D}}\big(F(A),\ F(B)\big)$.

In particular, an isomorphism of categories takes initial (resp. final) objects to initial (resp. final) objects. Indeed, it follows from the above discussion that there is a bijection between the sets $\mathrm{Hom}_{C}(A,\ G(D))$ and respectively $\mathrm{Hom}_{\mathcal{D}}(F(A),\ D)$; hence, if A is an initial object in C then $F(A)$ is an initial object in $\mathcal{D}$.

Examples 1.6.7

(1) The forgetful functor $F \colon {}_{\mathbb{Z}}\mathcal{M} \to \mathbf{Ab}$ is an isomorphism of categories. Indeed, the inverse of F is the functor $G \colon \mathbf{Ab} \to {}_{\mathbb{Z}}\mathcal{M}$ defined by $G(M) = M, G(u) = u$, where $M \in \mathbf{Ab}$ has a left $\mathbb{Z}$-module structure as follows:

$$t \cdot m = \begin{cases} \underbrace{m + m + \cdots + m}_{t \text{ times}} & \text{if } t > 0 \\ 0_M & \text{if } t = 0 \\ \underbrace{-m - m - \cdots - m}_{-t \text{ times}} & \text{if } t < 0 \end{cases}, \quad \text{for every } t \in \mathbb{Z},\ m \in M.$$

(2) Let R be a ring and denote by R^{op} the opposite ring.[21] Then we have an isomorphism of categories $F \colon {}_{R}\mathcal{M} \to \mathcal{M}_{R^{op}}$ given by

$$F(M) = M \in \mathcal{M}_{R^{op}} \text{ via } m * r = rm, \text{ for all } m \in M,\ r \in R;$$

$$F(u) = u,$$

with the inverse constructed in the same manner.

(3) The categories $\mathbf{Set}$ and $\mathbf{Set}^{op}$ are not isomorphic. Indeed, recall that $\mathbf{Set}$ has one initial object, namely the empty set, and infinitely many final objects, the singletons. Therefore, in $\mathbf{Set}^{op}$ we have infinitely many initial objects and one final object. The conclusion now follows since a potential isomorphism between the two categories should take initial (final) objects to initial (final) objects.

(4) Let C be an arbitrary category. The opposite functor $O_C \colon C \to C^{op}$ is an anti-isomorphism of categories. $\square$

Definition 1.6.8 Let $F \colon C \to \mathcal{D}$ be a functor.

(1) We say that F *preserves a property* P of morphisms if whenever f has the property P in C so does $F(f)$ in $\mathcal{D}$.

---

[21] In R^{op} we have $(R^{op}, +) = (R, +)$ and the multiplication is given by $r \cdot_{op} r' = r'r$, for all $r, r' \in (R^{op}, +)$.

(2) Similarly, *F reflects a property P* of morphisms if whenever $F(f)$ has the property P in $\mathcal{D}$ so does f in C.

Proposition 1.6.9 *The following hold:*

(1) Any functor preserves isomorphisms.
(2) Any fully faithful functor reflects isomorphisms.
(3) Any fully faithful functor reflects initial and final objects.
(4) Any faithful functor reflects monomorphisms and epimorphisms.

Proof

(1) Let $f \in \mathrm{Hom}_C(A, B)$ be an isomorphism and $g \in \mathrm{Hom}_C(B, A)$ its inverse. Hence, we have $f \circ g = 1_B$ and $g \circ f = 1_A$. Applying the functor $F: C \to \mathcal{D}$ to these identities we obtain $F(f) \circ F(g) = 1_{F(B)}$ and $F(g) \circ F(f) = 1_{F(A)}$, i.e., $F(f)$ is an isomorphism.

(2) Let $F: C \to \mathcal{D}$ be a fully faithful functor and $f \in \mathrm{Hom}_C(A, B)$ such that $F(f)$ is an isomorphism in $\mathcal{D}$. Then, there exists an $h \in \mathrm{Hom}_{\mathcal{D}}(F(B), F(A))$ such that

$$F(f) \circ h = 1_{F(B)} \text{ and } h \circ F(f) = 1_{F(A)}.$$

Since F is full we can find $g \in \mathrm{Hom}_C(B, A)$ such that $F(g) = h$. Therefore, the above identities come down to

$$1_{F(B)} = F(f) \circ F(g) = F(f \circ g) \text{ and } 1_{F(A)} = F(g) \circ F(f) = F(g \circ f).$$

Now F is faithful and $F(1_A) = 1_{F(A)}$, respectively $F(1_B) = 1_{F(B)}$, yield $f \circ g = 1_B$ and $g \circ f = 1_A$, as desired.

(3) Let $F: C \to \mathcal{D}$ be a fully faithful functor and $I, C \in \mathrm{Ob}\,C$. We have the following bijection of sets:

$$\mathrm{Hom}_C(I, C) \cong \mathrm{Hom}_{\mathcal{D}}(F(I), F(C)).$$

If $F(I)$ is an initial object in $\mathcal{D}$ then $1 = |\mathrm{Hom}_{\mathcal{D}}(F(I), F(C))|$ for any $C \in \mathrm{Ob}\,C$ and the above isomorphism gives $1 = |\mathrm{Hom}_C(I, C)|$, i.e., I is an initial object in C. The last statement follows in a similar manner.

(4) Let $F: C \to \mathcal{D}$ be a faithful functor and $f \in \mathrm{Hom}_C(A, B)$ such that $F(f)$ is a monomorphism in $\mathcal{D}$. Consider now $C \in \mathrm{Ob}\,C$ and $g, h: C \to A$ such that $f \circ g = f \circ h$. Applying F to this equality gives $F(f) \circ F(g) = F(f) \circ F(h)$. Since $F(f)$ is a monomorphism this implies that $F(g) = F(h)$ and by the faithfulness of F we get $g = h$ as desired. A similar argument proves that F reflects epimorphisms as well.

$\square$

Examples 1.6.10

(1) The forgetful functor U: **Grp** $\to$ **Set** reflects isomorphisms. Indeed, recall that a group homomorphism is an isomorphism if and only if it is a bijection.
(2) The forgetful functor U: **Top** $\to$ **Set** does not reflect isomorphisms, as can be easily seen from Example 1.3.2, (7). $\qquad\Box$

Corollary 1.6.11 *Let C be a concrete category, $F: C \to$ **Set** the corresponding faithful functor and $f \in \mathrm{Hom}_C(A, B)$.*

(1) If $F(f)$ is injective then f is a monomorphism in C.
(2) If $F(f)$ is surjective then f is an epimorphism in C.

Proof 1) We have already proved that monomorphisms in **Set** coincide with injective maps (see Example 1.3.2, (1)). Since $F(f)$ is a monomorphism in **Set** and F is faithful it follows by Proposition 1.6.9, (4) that f is also a monomorphism. The second statement follows by similar arguments. $\qquad\Box$

Remark 1.6.12 Each of the following categories **Grp**, **Ab**, **Rng**, **Ring**, **Ring**^{c}, **Div**, $_R\mathcal{M}$, **Top** allow for a forgetful functor into **Set**. Hence, we can conclude that in the above mentioned categories all injective maps are monomorphisms and respectively all surjective maps are epimorphisms. However, the converse is not necessarily true, as can be seen from Example 1.3.2, (4) and (5) respectively.

1.7 Natural Transformations: Representable Functors

Natural transformations are in some sense *morphisms between functors*, as we will see in Proposition 1.9.1. The naturality of a certain transformation is meant to be understood in the sense that its definition does not depend on any arbitrary choices such as choosing a basis, a set of generators, etc.

Definition 1.7.1 Let $F, G: C \to \mathcal{D}$ be two functors. A *natural transformation* $\alpha: F \to G$ consists of a family of morphisms $\alpha_C: F(C) \to G(C)$ in $\mathcal{D}$, indexed by $C \in \mathrm{Ob}\,C$, such that for every morphism $f \in \mathrm{Hom}_C(C, C')$ the following diagram is commutative:

$$
\begin{array}{ccc}
F(C) & \xrightarrow{\;\alpha_C\;} & G(C) \\
{\scriptstyle F(f)}\Big\downarrow & & \Big\downarrow{\scriptstyle G(f)} \\
F(C') & \xrightarrow[\;\alpha_{C'}\;]{} & G(C')
\end{array}
\qquad \text{i.e.,}\quad \alpha_{C'} \circ F(f) = G(f) \circ \alpha_C.
$$

$$(1.15)$$

If all components α_C are isomorphisms then $\alpha: F \to G$ is called a *natural isomorphism*. In this case we say that the functors F and G are naturally isomorphic and we use the notation $F \cong G$. We denote by $\mathrm{Nat}(F, G)$ the class of all natural transformations between the functors F and G.

Examples 1.7.2

(1) Let G_1 and G_2 be two groups, $\mathcal{G}_1$, respectively $\mathcal{G}_2$, the corresponding categories (see Example 1.2.2, (3)) and $F, H \colon \mathcal{G}_1 \to \mathcal{G}_2$ two functors. We denote by f_F, respectively f_H, the morphisms of groups from G_1 to G_2 corresponding to the two functors F and H (see Example 1.5.3, (8)). Then $\mathrm{Nat}(F, H)$ is in bijective correspondence with the set $\{g \in G_2 \mid f_H(g') = g f_F(g') g^{-1}$, for all $g' \in G_1\}$.

   To this end denote $\mathrm{Ob}\,\mathcal{G}_1$ by $\{*\}$ and $\mathrm{Ob}\,\mathcal{G}_2$ by $\{\star\}$. Since $F(*) = \star$ and $H(*) = \star$, any natural transformation $\varphi \colon F \to H$ is completely determined by a morphism $\varphi_* \colon \star \to \star$ in $\mathcal{G}_2$ (i.e., an element of the group G_2) which makes the following diagram commute for all morphisms $t \colon * \to *$ in $\mathcal{G}_1$ (i.e., an element of the group G_1):

$$
\begin{array}{ccc}
F(*) = \star & \xrightarrow{\;\varphi_*\;} & H(*) = \star \\
{\scriptstyle F(t)} \downarrow & & \downarrow {\scriptstyle H(t)} \\
F(*) = \star & \xrightarrow[\;\varphi_*\;]{} & H(*) = \star
\end{array}
\qquad \text{i.e.,}\quad H(t) \circ \varphi_* = \varphi_* \circ F(t).
$$

   Having in mind that the composition of morphisms in $\mathcal{G}_2$ is given by the multiplication of the group G_2, we can conclude that the natural transformations from F to H are in bijection with elements $g \in G_2$ such that for any $g' \in G_1$ we have $f_H(g')g = g f_F(g')$. Hence, we have an isomorphism of sets between $\mathrm{Nat}(F, H)$ and $\{g \in G_2 \mid f_H(g')g = g f_F(g')$, for all $g' \in G_1\}$, as desired.

(2) The concept of a natural transformation, respectively a natural isomorphism, is very well illustrated by looking at double duals of vector spaces. Recall that a classical algebraic result states that any finite-dimensional vector space is *naturally* isomorphic to its double dual. Loosely speaking, this means precisely that by putting together the isomorphisms $V \to V^{**}$ for all finite-dimensional vector spaces V, we obtain a natural isomorphism in the sense of the above definition. In more rigorous categorical terms, this comes down to defining a natural transformation between the identity functor on the category of vector spaces ${}_K\mathcal{M}$ and the double dual functor introduced in Example 1.5.3, (15). To this end, let $\eta \colon 1_{{}_K\mathcal{M}} \to (-)^{**}$ be the natural transformation whose components $\eta_V \colon V \to V^{**}$ for a given vector space V are the K-linear maps defined as follows:

$$
\eta_V(v)(f) = f(v) \quad \text{for all } v \in V \text{ and } f \in V^*. \tag{1.16}
$$

We will show that for any $u \in \mathrm{Hom}_{K\mathcal{M}}(V, W)$ the following diagram is commutative:

$$
\begin{array}{ccc}
V & \xrightarrow{\eta_V} & V^{**} \\
\downarrow{\scriptstyle u} & & \downarrow{\scriptstyle u^{**}} \\
W & \xrightarrow[\eta_W]{} & W^{**}
\end{array}
$$

Indeed, for all $v \in V$ and $f \in W^*$ we have

$$
\left[u^{**} \circ \eta_V(v)\right](f) \overset{(1.5)}{=} \left[\eta_V(v) \circ u^*\right](f) = \eta_V(v)\left(u^*(f)\right)
$$

$$
\overset{(1.16)}{=} u^*(f)(v) = (f \circ u)(v) = f\left(u(v)\right)
$$

$$
\overset{(1.16)}{=} \eta_W\left(u(v)\right)(f) = \left[(\eta_W \circ u)(v)\right](f).
$$

This shows that $\eta: 1_{K\mathcal{M}} \to (-)^{**}$ is indeed a natural transformation. Furthermore, if we restrict the two functors, the identity functor and the double dual functor, to the category of finite-dimensional vector spaces $_K\mathcal{M}^{fd}$ we obtain the natural isomorphism $\eta: 1_{_K\mathcal{M}^{fd}} \to (-)^{**}$ mentioned in the beginning.

(3) In contrast, the identity functor $1_{_K\mathcal{M}^{fd}}$ and the dual functor $(-)^*: {_K\mathcal{M}^{fd}} \to {_K\mathcal{M}^{fd}}$ introduced in Example 1.5.3, (14) are not naturally isomorphic. In fact, strictly speaking, the question itself of whether there are natural transformations between the aforementioned functors does not make sense, as the dual functor is contravariant while the identity functor is covariant.

(4) For any functor $F: C \to \mathcal{D}$ we have a natural transformation $1_F: F \to F$ called the identity natural transformation defined by $1_F = \left(1_{F(C)}\right)_{C \in \mathrm{Ob}\, C}$.

(5) If $F, G, H: C \to \mathcal{D}$ are functors and $\alpha: F \to G$, $\beta: G \to H$ are natural transformations then we can define a new natural transformation $\beta \circ \alpha: F \to H$ by the formula

$$
(\beta \circ \alpha)_C : F(C) \to H(C), \quad (\beta \circ \alpha)_C = \beta_C \circ \alpha_C \text{ for all } C \in \mathrm{Ob}\, C. \quad (1.17)
$$

The composition defined above is called the *vertical composition of natural transformations*.

(6) If $F, G: C \to \mathcal{D}$ are functors and $\alpha: F \to G$ is a natural isomorphism then $\alpha^{-1}: G \to F$ defined by

$$
\alpha_C^{-1} = (\alpha_C)^{-1} \text{ for all } C \in \mathrm{Ob}\, C
$$

is also a natural isomorphism, called the *inverse natural transformation*.

(7) Let $F, G: C \to \mathcal{D}$, $H: \mathcal{D} \to \mathcal{E}$ be two functors and $\alpha: F \to G$ a natural transformation as in the picture below:

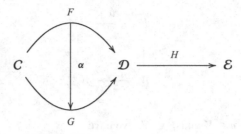

Then we can define a new natural transformation $H\alpha: HF \to HG$ as follows:

$$(H\alpha)_C = H(\alpha_C) \text{ for all } C \in \text{Ob}\,C,$$

called the *whiskering of the natural transformation α on the right by H*.

Similarly, let $K: \mathcal{B} \to C$, $F, G: C \to \mathcal{D}$ be functors and $\alpha: F \to G$ a natural transformation as in the picture below:

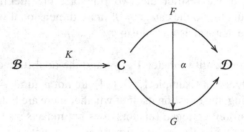

We can define a new natural transformation $\alpha K: FK \to GK$ as follows:

$$(\alpha K)_B = \alpha_{K(B)} \text{ for all } B \in \text{Ob}\,\mathcal{B},$$

called the *whiskering of the natural transformation α on the left by the functor K*.

Furthermore, if α is a natural isomorphism then both $H\alpha$ and αK are natural isomorphisms. Indeed, the first assertion follows by Proposition 1.6.9, (1) while the second one is an easy consequence of α itself being a natural isomorphism.
□

The examples above describing the whiskering of a natural transformation are both special cases of a more general construction called the *Godement product* or *horizontal composition of natural transformations*.

Proposition 1.7.3 *Let* $F, G: C \rightarrow \mathcal{D}, H, K: \mathcal{D} \rightarrow \mathcal{E}$ *functors and* $\alpha: F \rightarrow G$, $\beta: H \rightarrow K$ *two natural transformations as depicted below:*

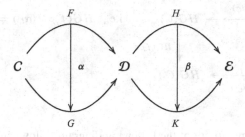

Then $\beta * \alpha: HF \rightarrow KG$ *defined as follows for all* $C \in \mathrm{Ob}C$:

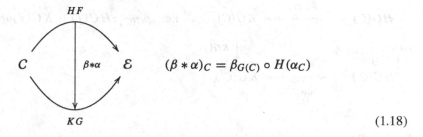

$$(\beta * \alpha)_C = \beta_{G(C)} \circ H(\alpha_C) \tag{1.18}$$

is a natural transformation called the Godement product of α *and* β.

Proof To start with, the naturally of β applied to the morphism $\alpha_C: F(C) \rightarrow G(C)$ yields the following commutative diagram:

$$
\begin{array}{ccc}
HF(C) & \xrightarrow{\beta_{F(C)}} & KF(C) \\
\downarrow{\scriptstyle H(\alpha_C)} & & \downarrow{\scriptstyle K(\alpha_C)} \\
HG(C) & \xrightarrow{\beta_{G(C)}} & KG(C)
\end{array}
\qquad \text{i.e.,} \quad \beta_{G(C)} \circ H(\alpha_C) = K(\alpha_C) \circ \beta_{F(C)}.
$$

Therefore, we have $(\beta * \alpha)_C = \beta_{G(C)} \circ H(\alpha_C) = K(\alpha_C) \circ \beta_{F(C)}$ for all $C \in \mathrm{Ob}C$. Let $f \in \mathrm{Hom}_C(C, C')$; showing that $\beta * \alpha$ is a natural transformation comes down to proving the commutativity of the following diagram:

$$
\begin{array}{ccc}
HF(C) & \xrightarrow{\beta_{G(C)} \circ H(\alpha_C)} & KG(C) \\
\downarrow{\scriptstyle HF(f)} & & \downarrow{\scriptstyle KG(f)} \\
HF(C') & \xrightarrow{\beta_{G(C')} \circ H(\alpha_{C'})} & KG(C')
\end{array}
\tag{1.19}
$$

To this end, the naturality of α and functoriality of H render the following diagram commutative:

$$
\begin{array}{ccc}
HF(C) & \xrightarrow{\;\;H(\alpha_C)\;\;} & HG(C) \\
{\scriptstyle HF(f)}\Big\downarrow & & \Big\downarrow{\scriptstyle HG(f)} \\
HF(C') & \xrightarrow[\;\;H(\alpha_{C'})\;\;]{} & HG(C')
\end{array}
\qquad \text{i.e.,}\quad HG(f)\circ H(\alpha_C) = H(\alpha_{C'})\circ HF(f).
$$

$$(1.20)$$

Moreover, the commutativity of the following diagram follows by the naturality of β:

$$
\begin{array}{ccc}
HG(C) & \xrightarrow{\;\;\beta_{G(C)}\;\;} & KG(C) \\
{\scriptstyle HG(f)}\Big\downarrow & & \Big\downarrow{\scriptstyle KG(f)} \\
HG(C') & \xrightarrow[\;\;\beta_{G(C')}\;\;]{} & KG(C')
\end{array}
\qquad \text{i.e.,}\quad \beta_{G(C')}\circ HG(f) = KG(f)\circ\beta_{G(C)}.
$$

$$(1.21)$$

Putting all the above together yields

$$
\beta_{G(C')} \circ \underline{H(\alpha_{C'}) \circ HF(f)} \overset{(1.20)}{=} \beta_{G(C')} \circ \underline{HG(f) \circ H(\alpha_C)}
$$

$$
\overset{(1.21)}{=} KG(f) \circ \beta_{G(C)} \circ H(\alpha_C),
$$

which proves that (1.19) holds and the proof is now finished.   □

As we will see in Chap. 3, naturality in the case of bifunctors turns out to be important when dealing with adjoint functors. For this reason, we conclude our discussion on natural transformations with the following result, which shows that naturality for bifunctors is equivalent to the naturality of both the left and the right associated functors as defined in Proposition 1.5.5.

Proposition 1.7.4 *Let F, $G : \mathcal{A} \times \mathcal{B} \to C$ be two bifunctors. A family of morphisms $\alpha_{(A,B)} : F(A,B) \to G(A,B)$ in C, indexed by $(A,B) \in \mathrm{Ob}(\mathcal{A} \times \mathcal{B})$, form a natural transformation $\alpha : F \to G$ if and only if the following conditions are fulfilled:*

(1) for all $A_0 \in \mathrm{Ob}\,\mathcal{A}$, the family of morphisms $(\alpha_{A_0})_B : F_{A_0}(B) \to G_{A_0}(B)$ in C, indexed by $B \in \mathrm{Ob}\,\mathcal{B}$, form a natural transformation $\alpha_{A_0} : F_{A_0} \to G_{A_0}$, where $(\alpha_{A_0})_B = \alpha_{(A_0,B)}$;

(2) for all $B_0 \in \text{Ob}\,\mathcal{B}$, the family of morphisms $(\alpha^{B_0})_A \colon F^{B_0}(A) \to G^{B_0}(A)$ in C, indexed by $A \in \text{Ob}\,\mathcal{A}$, form a natural transformation $\alpha^{B_0} \colon F^{B_0} \to G^{B_0}$, where $(\alpha^{B_0})_A = \alpha_{(A, B_0)}$.

Proof Assume first that $\alpha \colon F \to G$ is a natural transformation; therefore, for all $(f, g) \in \text{Hom}_{\mathcal{A} \times \mathcal{B}}\big((A, B), (A', B')\big)$ the following diagram is commutative:

$$
\begin{array}{ccc}
F(A, B) & \xrightarrow{\ \alpha_{(A,B)}\ } & G(A, B) \\[4pt]
{\scriptstyle F(f,g)}\Big\downarrow & & \Big\downarrow{\scriptstyle G(f,g)} \\[4pt]
F(A', B') & \xrightarrow[\ \alpha_{(A',B')}\ ]{} & G(A', B')
\end{array}
\qquad \text{i.e., } \ \alpha_{(A',B')} \circ F(f,g) = G(f,g) \circ \alpha_{(A,B)}.
$$

$$(1.22)$$

Now considering $A = A'$ and $f = 1_A$ in (1.22) yields $\alpha_{(A,B')} \circ F(1_A, g) = G(1_A, g) \circ \alpha_{(A,B)}$. The last compatibility is equivalent to $(\alpha_A)_{B'} \circ F_A(g) = G_A(g) \circ (\alpha_A)_B$, which in turn implies that the following diagram is commutative for all $g \in \text{Hom}_{\mathcal{B}}(B, B')$:

$$
\begin{array}{ccc}
F(A, B) & \xrightarrow{\ (\alpha_A)_B\ } & G(A, B) \\[4pt]
{\scriptstyle F_A(g)}\Big\downarrow & & \Big\downarrow{\scriptstyle G_A(g)} \\[4pt]
F(A, B') & \xrightarrow[\ (\alpha_A)_{B'}\ ]{} & G(A, B')
\end{array}
$$

$$(1.23)$$

This shows that $\alpha_A \colon F_A \to G_A$ is a natural transformation for all $A \in \text{Ob}\,\mathcal{A}$. Similarly, by setting $B = B'$ and $g = 1_B$ in (1.22) we obtain $\alpha_{(A',B)} \circ F(f, 1_B) = G(f, 1_B) \circ \alpha_{(A,B)}$. Consequently, we have $(\alpha^B)_{A'} \circ F^B(f) = G^B(f) \circ (\alpha^B)_A$, which implies the commutativity of the following diagram for all $f \in \text{Hom}_{\mathcal{B}}(A, A')$:

$$
\begin{array}{ccc}
F(A, B) & \xrightarrow{\ (\alpha^B)_A\ } & G(A, B) \\[4pt]
{\scriptstyle F^B(f)}\Big\downarrow & & \Big\downarrow{\scriptstyle G^B(f)} \\[4pt]
F(A', B) & \xrightarrow[\ (\alpha^B)_{A'}\ ]{} & G(A', B)
\end{array}
$$

$$(1.24)$$

Hence $\alpha^B \colon F^B \to G^B$ is a natural transformation for all $B \in \text{Ob}\,\mathcal{B}$.

Conversely, assume that the conditions *1)* and *2)* in the statement hold; thus for all $f \in \text{Hom}_{\mathcal{B}}(A, A')$ and $g \in \text{Hom}_{\mathcal{B}}(B, B')$, the compatibilities (1.23) and (1.24) are fulfilled. Consider $(u, v) \in \text{Hom}_{\mathcal{A} \times \mathcal{B}}\big((X, Y), (X', Y')\big)$; we have

$$G(u, v) \circ \alpha_{(X,Y)} \doteq G\big((u, 1_{Y'}) \circ (1_X, v)\big) \circ \alpha_{(X,Y)}$$

$$= G(u, 1_{Y'}) \circ G(1_X, v) \circ \alpha_{(X,Y)}$$

$$= G^{Y'}(u) \circ G_X(v) \circ \alpha_{(X,Y)}$$

$$= G^{Y'}(u) \circ G_X(v) \circ \big(\alpha_X\big)_Y$$

$$\overset{(1.23)}{=} G^{Y'}(u) \circ (\alpha_X)_{Y'} \circ F_X(v)$$

$$= G^{Y'}(u) \circ (\alpha^{Y'})_X \circ F_X(v)$$

$$\overset{(1.24)}{=} \big(\alpha^{Y'}\big)_{X'} \circ F^{Y'}(u) \circ F_X(v)$$

$$= \alpha_{(X',Y')} \circ F(u, 1_{Y'}) \circ F(1_X, v)$$

$$= \alpha_{(X',Y')} \circ F(u, v).$$

This shows that for all $(u, v) \in \mathrm{Hom}_{\mathcal{A} \times \mathcal{B}}\big((X, Y), (X', Y')\big)$ the following diagram is commutative:

$$
\begin{array}{ccc}
F(X, Y) & \xrightarrow{\;\;\alpha_{(X,Y)}\;\;} & G(X, Y) \\
{\scriptstyle F(u,v)}\big\downarrow & & \big\downarrow{\scriptstyle G(u,v)} \\
F(X', Y') & \xrightarrow[\;\;\alpha_{(X',Y')}\;\;]{} & G(X', Y')
\end{array}
$$

and we can conclude that $\alpha \colon F \to G$ is a natural transformation, as desired.   □

We are now ready to introduce an important concept: *representable functors*.

Definition 1.7.5 Let C be a category. We say that a functor $F \colon C \to \mathbf{Set}$ is *representable* if there exist $C \in \mathrm{Ob}\, C$ and a natural isomorphism $F \cong \mathrm{Hom}_C(C, -)$. Similarly, a contravariant functor $G \colon C \to \mathbf{Set}$ is representable if there exist $C \in \mathrm{Ob}\, C$ and a natural isomorphism $G \cong \mathrm{Hom}_C(-, C)$. In this case, C is called the *representing object* of F.

More precisely, Definition 1.7.5 reads as follows: $F \colon C \to \mathbf{Set}$ is representable if and only if there exist an object $C \in \mathrm{Ob}\, C$ and a family of isomorphisms $\big(\alpha_A \colon \mathrm{Hom}_C(C, A) \to F(A)\big)_{A \in \mathrm{Ob}\, C}$ in $\mathbf{Set}$ (set bijections) such that for any $f \in \mathrm{Hom}_C(X, Y)$ the following diagram is commutative:

$$\text{Hom}_C(C, X) \xrightarrow{\ \alpha_X\ } F(X) \qquad \text{i.e.,} \quad F(f)\circ\alpha_X = \alpha_Y\circ\text{Hom}_C(C, f).$$

$$
\begin{array}{ccc}
\text{Hom}_C(C, X) & \xrightarrow{\ \alpha_X\ } & F(X) \\
{\scriptstyle\text{Hom}_C(C, f)}\Big\downarrow & & \Big\downarrow{\scriptstyle F(f)} \\
\text{Hom}_C(C, Y) & \xrightarrow[\ \alpha_Y\ ]{} & F(Y)
\end{array}
$$

Note that the representing object of a functor will prove to be unique up to isomorphism as a consequence of Yoneda's lemma (see Proposition 1.10.5). This explains why the object C in Definition 1.7.5 is referred to as *the* representing object of F.

Examples 1.7.6

(1) The forgetful functor $U\colon \mathbf{Grp} \to \mathbf{Set}$ is representable and the representing object is the abelian group $(\mathbb{Z}, +)$. Indeed, for any $X \in \text{Ob}\,\mathbf{Grp}$ and any $g \in \text{Hom}_{\mathbf{Grp}}(\mathbb{Z}, X)$ we define $\alpha_X\colon \text{Hom}_{\mathbf{Grp}}(\mathbb{Z}, X) \to U(X) = X$ by $\alpha_X(g) = g(1)$. Note that each α_X is a bijection as any group homomorphism $g\colon \mathbb{Z} \to X$ is uniquely defined by its value in 1. The above diagram is now obviously commutative for any $f \in \text{Hom}_{\mathbf{Grp}}(X, Y)$:

$$
\begin{array}{ccc}
\text{Hom}_{\mathbf{Grp}}(\mathbb{Z}, X) & \xrightarrow{\ \alpha_X\ } & U(X) = X \\
{\scriptstyle\text{Hom}_{\mathbf{Grp}}(\mathbb{Z}, f)}\Big\downarrow & & \Big\downarrow{\scriptstyle U(f)=f} \\
\text{Hom}_{\mathbf{Grp}}(\mathbb{Z}, Y) & \xrightarrow[\ \alpha_Y\ ]{} & U(Y) = Y
\end{array}
$$

Indeed, for any $g \in \text{Hom}_{\mathbf{Grp}}(\mathbb{Z}, X)$ we have $\alpha_Y \circ \text{Hom}_{\mathbf{Grp}}(\mathbb{Z}, f)(g) = \alpha_Y\big(f \circ g\big) = (f \circ g)(1)$ and $f \circ \alpha_X(g) = f(g(1))$.

(2) The forgetful functor $U\colon \mathbf{Top} \to \mathbf{Set}$ is representable and the representing object is any singleton topological space $\{x_0\}$ (with the discrete topology). To this end, for any $X \in \text{Ob}\,\mathbf{Top}$ and any $h \in \text{Hom}_{\mathbf{Top}}(\{x_0\}, X)$ we define $\alpha_X\colon \text{Hom}_{\mathbf{Top}}(\{x_0\}, X) \to U(X) = X$ by $\alpha_X(h) = h(x_0) \in X$. Each α_X is obviously bijective. Furthermore, it can be easily seen that the above diagram is commutative for any $f \in \text{Hom}_{\mathbf{Top}}(X, Y)$:

$$
\begin{array}{ccc}
\text{Hom}_{\mathbf{Top}}(\{x_0\}, X) & \xrightarrow{\ \alpha_X\ } & U(X) = X \\
{\scriptstyle\text{Hom}_{\mathbf{Top}}(\{x_0\}, f)}\Big\downarrow & & \Big\downarrow{\scriptstyle U(f)=f} \\
\text{Hom}_{\mathbf{Top}}(\{x_0\}, Y) & \xrightarrow[\ \alpha_Y\ ]{} & U(Y) = Y
\end{array}
$$

Indeed, for any $h \in \mathrm{Hom}_{\mathbf{Top}}(\{x_0\}, X)$ we have

$$f \circ \alpha_X(h) = f(h(x_0))$$

$$\alpha_Y \circ \mathrm{Hom}_{\mathbf{Top}}(\{x_0\}, f)(h) = \alpha_Y(f \circ h) = (f \circ h)(x_0).$$

Therefore, $f \circ \alpha_X = \alpha_Y \circ \mathrm{Hom}_{\mathbf{Top}}(\{x_0\}, f)$, as desired.

(3) The constant functor $F_{x_0} : C \to \mathbf{Set}$ which sends every object of C to the singleton $\{x_0\}$ and every morphism in C to the identity map on $\{x_0\}$ is representable if and only if C has an initial object. Moreover, in this case the representing object is the initial object. Indeed, suppose the functor F_{x_0} is represented by $I \in \mathrm{Ob}\,C$. Then, for any $C \in \mathrm{Ob}\,C$ we have an isomorphism in $\mathbf{Set}$ denoted by $\alpha_C : \mathrm{Hom}_C(I, C) \to \{x_0\}$. This implies that $\mathrm{Hom}_C(I, C)$ has exactly one element for any $C \in \mathrm{Ob}\,C$, i.e., I is initial in C.

(4) The covariant power-set functor $\mathcal{P} : \mathbf{Set} \to \mathbf{Set}$ is not representable. To this end, assume that $\mathcal{P}$ is representable; consider A to be the representing object and $\tau : \mathrm{Hom}_{\mathbf{Set}}(A, -) \to \mathcal{P}$ the corresponding natural isomorphism. Let $\{*\}$ be an arbitrary singleton. Since τ is a natural isomorphism we obtain a bijective map $\tau : \mathrm{Hom}_{\mathbf{Set}}(A, \{*\}) \to \mathcal{P}(\{*\})$. This leads to a contradiction since $|\mathrm{Hom}_{\mathbf{Set}}(A, \{*\})| = 1$ and $|\mathcal{P}(\{*\})| = 2$. Therefore, $\mathcal{P}$ is not representable. $\square$

The following result provides an important criterion for deciding whether a functor is representable or not.

Proposition 1.7.7 *Let $F : C \to \mathbf{Set}$ be a functor. Then F is representable if and only if there exists a pair (A, a) with $A \in \mathrm{Ob}\,C$ and $a \in F(A)$ satisfying the following property: for any other pair (B, b) with $B \in \mathrm{Ob}\,C$ and $b \in F(B)$ there exists a unique morphism $f \in \mathrm{Hom}_C(A, B)$ such that $F(f)(a) = b$. In this case (A, a) is called a* representing pair.

Proof Suppose first that F is representable, i.e., there exists a natural isomorphism $\varphi : \mathrm{Hom}_C(A, -) \to F$ for some $A \in \mathrm{Ob}\,C$. In particular we have a bijection of sets $\varphi_A : \mathrm{Hom}_C(A, A) \to F(A)$ and we denote $\varphi_A(1_A) \in F(A)$ by a. We will show that (A, a) is a representing pair. Indeed, consider $B \in \mathrm{Ob}\,C$ and $b \in F(B)$. As before, we have a bijective map $\varphi_B : \mathrm{Hom}_C(A, B) \to F(B)$ so there exists a unique $f \in \mathrm{Hom}_C(A, B)$ such that $\varphi_B(f) = b$. We are left to prove that $F(f)(a) = b$. Since φ is a natural transformation, the following diagram is commutative:

$$
\begin{array}{ccc}
\mathrm{Hom}_C(A, A) & \xrightarrow{\;\;\varphi_A\;\;} & F(A) \\
{\scriptstyle \mathrm{Hom}_C(A, f)} \downarrow & & \downarrow {\scriptstyle F(f)} \\
\mathrm{Hom}_C(A, B) & \xrightarrow[\;\;\varphi_B\;\;]{} & F(B)
\end{array}
$$

i.e., $\varphi_B \circ \text{Hom}_C(A, f) = F(f) \circ \varphi_A$. This yields $F(f) = \varphi_B \circ \text{Hom}_C(A, f) \circ \varphi_A^{-1}$
and we obtain

$$
\begin{aligned}
F(f)(a) &= \varphi_B \circ \text{Hom}_C(A, f) \circ \varphi_A^{-1}(a) \\
&= \varphi_B \circ \text{Hom}_C(A, f)(1_A) \\
&= \varphi_B(f \circ 1_A) \\
&= \varphi_B(f) = b.
\end{aligned}
$$

Assume now that (A, a) is a representing pair. Let $\psi : \text{Hom}_C(A, -) \to F$ be the natural isomorphism defined as follows for any $B \in \text{Ob}\,C$:

$$
\psi_B : \text{Hom}_C(A, B) \to F(B), \quad \psi_B(f) = F(f)(a), \quad f \in \text{Hom}_C(A, B).
$$

The property assumed to be satisfied by (A, a) implies that each such map ψ_B is bijective. The proof will be finished once we show that ψ is a natural transformation, i.e., for any $g \in \text{Hom}_C(B, C)$ the following diagram is commutative:

$$
\begin{CD}
\text{Hom}_C(A, B) @>{\psi_B}>> F(B) \\
@V{\text{Hom}_C(A, g)}VV @VV{F(g)}V \\
\text{Hom}_C(A, C) @>>{\psi_C}> F(C)
\end{CD}
$$

Indeed, for any $h \in \text{Hom}_C(A, B)$ we have

$$
F(g)\big(\psi_B(h)\big) = F(g)\big(F(h)(a)\big) = F(g \circ h)(a) = \psi_C(g \circ h) = \psi_C \circ \text{Hom}_C(A, g)(h),
$$

as desired. This finishes the proof. $\qquad\square$

Example 1.7.8 Let A be a group and $H \trianglelefteq A$ a normal subgroup. Consider the functor $F : \mathbf{Grp} \to \mathbf{Set}$ defined as follows:

$$
F(G) = \{f \in \text{Hom}_{\mathbf{Grp}}(A, G) \mid H \subseteq \ker f\},
$$

$$
F(u)(g) = u \circ g,
$$

for any $G \in \text{Ob}\,\mathbf{Grp}$ and any $u \in \text{Hom}_{\mathbf{Grp}}(G, G')$, $g \in F(G)$. Then F is representable and $\big(A/H, \pi : A \to A/H\big)$ is the representing pair, where π is the canonical projection. Indeed, consider another pair $(G, f : A \to G)$ with

$G \in \mathrm{Ob}\,\mathbf{Grp}$ and $f \in F(G)$:

$$
\begin{array}{ccc}
A & \xrightarrow{\ \pi\ } & A/H \\
{\scriptstyle f}\downarrow & \swarrow {\scriptstyle \overline{f}} & \\
G & &
\end{array}
$$

Since $H \subseteq \ker f$, the universal property of the quotient group A/H yields a unique $\overline{f} \in \mathrm{Hom}_{\mathbf{Grp}}(A/H,\, G)$ such that the above diagram is commutative, i.e., $\overline{f}\circ\pi = f$. The last equality is equivalent to $F(\overline{f})(\pi) = f$ and the desired conclusion now follows from Proposition 1.7.7. □

1.8 The Duality Principle

The dual category allows us not only to define a dual notion for every concept but also to state a dual result for any theorem. This new result requires no proof and is obtained by reversing all morphisms and consequently the order of composition in the given theorem. Indeed, if a given statement T is valid in any category then the dual statement T^{op} is also valid in any category. This can be easily seen by noticing that proving the statement T^{op} in a category C is equivalent to proving the statement T in the category C^{op}, which is assumed to be valid. To illustrate this duality principle we look at the following statement proved in an equivalent form in Proposition 1.3.9:

When it exists, the initial object of a category is unique up to isomorphism.

By simply applying the duality principle we obtain the following dual statement:

When it exists, the final object of a category is unique up to isomorphism.

A certain care is required, however, when dealing with statements which involve functors. More precisely, in the process of dualizing these statements, all categories are replaced by their duals, and all morphisms are reversed, while functors $C \to \mathcal{D}$ are not reversed but replaced by *dual functors* $C^{op} \to \mathcal{D}^{op}$ as defined below.

Proposition 1.8.1 *Let $F: C \to \mathcal{D}$ be a functor. There exists a functor $F^{op}: C^{op} \to \mathcal{D}^{op}$, called the* dual functor, *defined as follows for any $C,\, D \in \mathrm{Ob}\,C^{op}$ and any $f^{op} \in \mathrm{Hom}_{C^{op}}(C,\, D)$:*

$$
F^{op}(C) = F(C), \qquad F^{op}(f^{op}) = F(f)^{op}.
$$

Furthermore, $(F^{op})^{op} = F$ and if $G: \mathcal{D} \to \mathcal{E}$ is another functor we have

$$
(G \circ F)^{op} = G^{op} \circ F^{op}, \tag{1.25}
$$

where the functor composition in (1.25) is defined as in Example 1.5.3, (3).

Proof For any $C \in \mathrm{Ob}\,C^{op}$ we have $F^{op}(1_C^{op}) = F(1_C)^{op} = 1_{F(C)}^{op}$ and therefore F^{op} preserves unit morphisms. Furthermore, for all $f^{op} \in \mathrm{Hom}_{C^{op}}(C, D)$, $g^{op} \in \mathrm{Hom}_{C^{op}}(D, E)$ we have

$$F^{op}(g^{op} \circ^{op} f^{op}) = F^{op}((f \circ g)^{op}) = F(f \circ g)^{op}$$
$$= (F(f) \circ F(g))^{op} = F(g)^{op} \circ^{op} F(f)^{op},$$

which shows that F^{op} is indeed a functor.

Furthermore, we have

$$(G \circ F)^{op}(C) = (G \circ F)(C) = G(F(C)) = G^{op}(F^{op}(C))$$
$$= (G^{op} \circ F^{op})(C) \text{ and}$$
$$(G \circ F)^{op}(f^{op}) = ((G \circ F)(f))^{op} = (G(F(f)))^{op} = G^{op}(F(f)^{op})$$
$$= G^{op}(F^{op}(f^{op})) = (G^{op} \circ F^{op})(f^{op}).$$

Therefore, (1.25) holds as well.   □

Consider now the following statement proved in Proposition 1.6.9, (4):

Any faithful functor reflects monomorphisms.

We have already established that a morphism f of a given category C is a monomorphism if and only if f^{op} is an epimorphism in C^{op}. Moreover, it can be easily seen that a functor F is faithful (resp. full) if and only if the dual functor F^{op} is faithful (full). Therefore, the duality principle shows that the following statement also holds:

Any faithful functor reflects epimorphisms.

In light of the duality principle, any statement about covariant functors has a correspondent for contravariant functors obtained by replacing the given functor $F: C \to D$ by $F \circ O_{C^{op}}: C^{op} \to D$ (or $O_D \circ F: C \to D^{op}$). For example, in order to obtain the contravariant version of Proposition 1.7.7 we consider $G: C \to$ **Set** to be a contravariant functor. Then $G \circ O_{C^{op}}: C^{op} \to$ **Set** is a covariant functor and Proposition 1.7.7 implies that $G \circ O_{C^{op}}: C^{op} \to$ **Set** is representable if and only if there exists a pair (A, a) with $A \in \mathrm{Ob}\,C^{op}$ and $a \in G \circ O_{C^{op}}(A)$ satisfying the following property: for any other pair (B, b) with $B \in \mathrm{Ob}\,C^{op}$ and $b \in G \circ O_{C^{op}}(B)$ there exists a unique $f^{op} \in \mathrm{Hom}_{C^{op}}(A, B)$ such that $G \circ O_{C^{op}}(f^{op})(a) = b$. Since $\mathrm{Ob}\,C^{op} = \mathrm{Ob}\,C$ and $\mathrm{Hom}_{C^{op}}(A, B) = \mathrm{Hom}_C(B, A)$ the statement concerning contravariant functors is the following:

Corollary 1.8.2 *A contravariant functor $G: C \to$ **Set** is representable if and only if there exists a pair (A, a) with $A \in \mathrm{Ob}\,C$ and $a \in G(A)$ satisfying the following* *property: for any other pair* (B, b) *with* $B \in \mathrm{Ob}\,C$ *and* $b \in G(B)$ *there exists a unique* $f \in \mathrm{Hom}_C(B, A)$ *such that* $G(f)(a) = b$.

Finally, the *dual of a natural transformation* is defined as follows:

Proposition 1.8.3 *Let* $F, G: C \to \mathcal{D}$ *be two functors and* $\psi: F \to G$ *a natural transformation. There exists a natural transformation* $\psi^{op}: G^{op} \to F^{op}$, *called the* opposite or dual natural transformation, *defined as follows for all* $C \in \mathrm{Ob}\,C$:

$$\left(\psi^{op}\right)_C = \left(\psi_C\right)^{op} \tag{1.26}$$

and any natural transformation between G^{op} *and* F^{op} *appears as the dual of some natural transformation between* F *and* G. *Furthermore,* ψ *is a natural isomorphism if and only if* ψ^{op} *is a natural isomorphism.*

Proof For any $f^{op} \in \mathrm{Hom}_{C^{op}}(C, C')$, the naturality of ψ renders the following diagram commutative:

$$
\begin{array}{ccc}
F(C') & \xrightarrow{\psi_{C'}} & G(C') \\
{\scriptstyle F(f)}\downarrow & & \downarrow{\scriptstyle G(f)} \\
F(C) & \xrightarrow{\psi_C} & G(C)
\end{array}
\qquad \text{i.e.,} \quad \psi_C \circ F(f) = G(f) \circ \psi_{C'}.
$$

$$\tag{1.27}$$

We will show that this implies the naturality of ψ^{op}. Indeed, (1.27) takes the following equivalent forms:

$$\psi_C \circ F(f) = G(f) \circ \psi_{C'}$$
$$\Leftrightarrow \left(\psi_C \circ F(f)\right)^{op} = \left(G(f) \circ \psi_{C'}\right)^{op}$$
$$\Leftrightarrow F(f)^{op} \circ^{op} (\psi_C)^{op} = (\psi_{C'})^{op} \circ^{op} G(f)^{op}$$
$$\Leftrightarrow F^{op}\left(f^{op}\right) \circ^{op} (\psi_C)^{op} = (\psi_{C'})^{op} \circ^{op} G^{op}\left(f^{op}\right),$$

which shows that the following diagram is commutative:

$$
\begin{array}{ccc}
G^{op}(C) & \xrightarrow{(\psi_C)^{op}} & F^{op}(C) \\
{\scriptstyle G^{op}(f^{op})}\downarrow & & \downarrow{\scriptstyle F^{op}(f^{op})} \\
G^{op}(C') & \xrightarrow{(\psi_{C'})^{op}} & F^{op}(C')
\end{array}
$$

and therefore $\psi^{op}: G^{op} \to F^{op}$ is a natural transformation.

The second claim follows by noticing that for any natural transformation we have $(\psi^{op})^{op} = \psi$.   □

Numerous examples of duality arguments will be used in proofs throughout the book.

Comma Categories

Comma categories are constructed using two functors with the same codomain. We will see many instances of comma categories at work later on when dealing with (co)limits or adjoint functors.

Theorem 1.8.4 *Any two functors* $F \colon \mathcal{A} \to C$ *and* $G \colon \mathcal{B} \to C$ *define a* comma category *denoted by* $(F \downarrow G)$ *as follows:*

(1) the objects are triples (A, f, B) *consisting of two objects* $A \in \mathrm{Ob}\,\mathcal{A}$, $B \in \mathrm{Ob}\,\mathcal{B}$ *and a morphism* $f \in \mathrm{Hom}_C(F(A), G(B))$;

(2) a morphism in $(F \downarrow G)$ *from* (A, f, B) *to* (A', f', B') *is a pair* (a, b), *where* $a \in \mathrm{Hom}_{\mathcal{A}}(A, A')$, $b \in \mathrm{Hom}_{\mathcal{B}}(B, B')$ *such that the following diagram is commutative:*

$$
\begin{array}{ccc}
F(A) & \xrightarrow{\;\;F(a)\;\;} & F(A') \\
{\scriptstyle f}\downarrow & & \downarrow{\scriptstyle f'} \\
G(B) & \xrightarrow[\;\;G(b)\;\;]{} & G(B')
\end{array}
\qquad \text{i.e., } \; G(b) \circ f = f' \circ F(a);
$$

$$(1.28)$$

(3) the composition law in $(F \downarrow G)$ *is that induced by the composition laws of* $\mathcal{A}$ *and* $\mathcal{B}$, *i.e.:*

$$(a, b) \circ (a', b') = (a \circ a', b \circ b');$$

(4) the identities are $1_{(A, f, B)} = (1_A, 1_B)$.

Proof First note that for any $(A, f, B) \in \mathrm{Ob}\,(F \downarrow G)$, the following diagram is trivially fulfilled:

$$
\begin{array}{ccc}
F(A) & \xrightarrow{\;\;1_{F(A)}\;\;} & F(A) \\
{\scriptstyle f}\downarrow & & \downarrow{\scriptstyle f} \\
G(B) & \xrightarrow[\;\;1_{G(B)}\;\;]{} & G(B)
\end{array}
$$

Furthermore, if (A, f, B), (A', f', B'), $(A'', f'', B'') \in \mathrm{Ob}\,(F \downarrow G)$ and (a, b), (a', b') are morphisms in $(F \downarrow G)$ between (A, f, B) and (A', f', B') and respectively (A', f', B') and (A'', f'', B''), the following diagrams are commutative:

$$
\begin{array}{ccc}
F(A) & \xrightarrow{\ F(a)\ } & F(A') \\
{\scriptstyle f}\big\downarrow & & \big\downarrow{\scriptstyle f'} \\
G(B) & \xrightarrow[\ G(b)\ ]{} & G(B')
\end{array}
\qquad
\begin{array}{ccc}
F(A') & \xrightarrow{\ F(a')\ } & F(A'') \\
{\scriptstyle f'}\big\downarrow & & \big\downarrow{\scriptstyle f''} \\
G(B') & \xrightarrow[\ G(b')\ ]{} & G(B'')
\end{array}
$$

Since F and G are functors and as a consequence they respect composition, we obtain the commutativity of the following diagram:

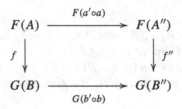

$$
\begin{array}{ccc}
F(A) & \xrightarrow{\ F(a'\circ a)\ } & F(A'') \\
{\scriptstyle f}\big\downarrow & & \big\downarrow{\scriptstyle f''} \\
G(B) & \xrightarrow[\ G(b'\circ b)\ ]{} & G(B'')
\end{array}
$$

$\square$

Examples 1.8.5

(1) Let $\mathcal{A}$ and $\mathcal{B}$ be discrete categories with only one object, say $\mathrm{Ob}\,\mathcal{A} = \{A\}$ and $\mathrm{Ob}\,\mathcal{B} = \{B\}$. As we have seen in Example 1.5.3, (7), defining two functors $F : \mathcal{A} \to C$ and $G : \mathcal{B} \to C$ comes down to choosing two objects C and C' in C. Then, the comma category $(F \downarrow G)$ is isomorphic to the discrete category on the set $\mathrm{Hom}_C(C, C')$. Indeed, the objects of the corresponding comma category $(F \downarrow G)$ are triples (A, f, B), where $f \in \mathrm{Hom}_C(C, C')$. The morphisms in $(F \downarrow G)$ between two objects (A, f, B) and (A, f', B) are pairs (a, b) where $a \in \mathrm{Hom}_{\mathcal{A}}(A, A)$ and $b \in \mathrm{Hom}_{\mathcal{B}}(B, B)$. Now since the only morphisms in $\mathcal{A}$ and $\mathcal{B}$ are identities on A and B respectively, we get $(a, b) = (1_A, 1_B)$. So the only morphisms in $(F \downarrow G)$ are the identities on each object. Therefore the functor from $(F \downarrow G)$ to the discrete category on $\mathrm{Hom}_C(C, C')$ sending any object (A, f, B) to f and any morphism $(1_A, 1_B)$ to the identity on f is obviously an isomorphism of categories.

(2) Let $\mathcal{A} = \mathbf{1}$ be a discrete category with only one object, say A, and $\mathcal{B} = C = \mathbf{Top}$. Furthermore, consider the functors $F : \mathbf{1} \to \mathbf{Top}$, $G : \mathbf{Top} \to \mathbf{Top}$ defined as follows:

$$
F(A) = \{\star\}, \quad F(1_A) = 1_{\{\star\}}, \quad G = 1_{\mathbf{Top}},
$$

where $\{\star\}$ denotes a singleton set regarded as a topological space in the obvious way. Then, the comma category $(F \downarrow G)$ is isomorphic to the category of pointed topological spaces $\mathbf{Top}_\star$. To this end, the objects of the corresponding comma category $(F \downarrow G)$ are triples (A, f, X), where $X \in \mathrm{Ob}\,\mathbf{Top}$ and $f \in \mathrm{Hom}_{\mathbf{Top}}(\{\star\}, X)$. The morphisms in $(F \downarrow G)$ between two objects (A, f, X) and (A, g, Y) are pairs (a, b), where $a \in \mathrm{Hom}_{\mathcal{A}}(A, A)$ and $b \in \mathrm{Hom}_{\mathbf{Top}}(X, Y)$ such that diagram (1.28) is commutative. Now since the only morphism in $\mathcal{A}$ is the identity on A, we get $(a, b) = (1_A, b)$, where $b \in \mathrm{Hom}_{\mathbf{Top}}(X, Y)$ fulfils the following compatibility:

$$
\begin{array}{ccc}
\{\star\} & \xrightarrow{\ 1_\star\ } & \{\star\} \\
{\scriptstyle f}\Big\downarrow & & \Big\downarrow{\scriptstyle g} \\
X & \xrightarrow[\ b\ ]{} & Y
\end{array}
\qquad \text{i.e.,}\quad b \circ f = g.
$$

$$\tag{1.29}$$

Therefore, the functor $H : (F \downarrow G) \to \mathbf{Top}_\star$ defined as follows for all (A, f, X), $(A, g, Y) \in \mathrm{Ob}(F \downarrow G)$ and all morphisms $(1_A, b)$ in $(F \downarrow G)$ between (A, f, X) and (A, g, Y):

$$
H(A, f, X) = \big(X, f(\star)\big), \qquad H(1_A, b) = b,
$$

is an isomorphism of categories. Indeed, note that the commutativity of (1.29) implies that b is indeed a morphism in $\mathbf{Top}_\star$ between $(X, f(\star))$ and $(Y, g(\star))$. Furthermore, since any morphism $f \in \mathrm{Hom}_{\mathbf{Top}}(\{\star\}, X)$ is uniquely determined by $f(\star)$ we can conclude that H is an isomorphism of categories. $\qquad\square$

In what follows we write down, for further use, two important special cases of comma categories:

Corollary 1.8.6 *Let $F : \mathcal{A} \to C$ and $G : \mathcal{B} \to C$ be two functors.*

(1) If $\mathcal{A}$ is the discrete category with only one object and $F : \mathcal{A} \to C$ is the constant functor at $C_0 \in \mathrm{Ob}\,C$ then the comma category $(F \downarrow G) = (C_0 \downarrow G)$ is isomorphic to the category defined as follows:

(i) the objects are pairs (f, B), where $B \in \mathrm{Ob}\,\mathcal{B}$, $f \in \mathrm{Hom}_C(C_0, G(B))$;

(ii) a morphism in $(C_0 \downarrow G)$ between two objects (f, B) and (f', B') is a morphism $b \in \mathrm{Hom}_{\mathcal{B}}(B, B')$ such that the following diagram is

commutative:

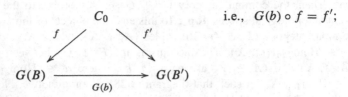

i.e., $G(b) \circ f = f'$;

(iii) *the composition of morphisms is given by the composition in* $\mathcal{B}$ *and the identities are* $1_{(f,\,B)} = 1_B$ *for all* $B \in \mathrm{Ob}\,\mathcal{B}$.

(2) *Similarly, if* $\mathcal{B}$ *is the discrete category with only one object and* $G : \mathcal{B} \to C$ *is the constant functor at* $C_0 \in \mathrm{Ob}\,C$, *then the comma category* $(F \downarrow G) = (F \downarrow C_0)$ *is defined as follows:*

  (i) *the objects are pairs* $(A,\,f)$, *where* $A \in \mathrm{Ob}\,\mathcal{A}$, $f \in \mathrm{Hom}_C(F(A),\,C_0)$;

  (ii) *a morphism in* $(F \downarrow C_0)$ *between two objects* $(A,\,f)$ *and* $(A',\,f')$ *is a morphism* $a \in \mathrm{Hom}_{\mathcal{A}}(A,\,A')$ *such that the following diagram is commutative:*

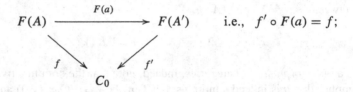

i.e., $f' \circ F(a) = f$;

  (iii) *the composition of morphisms is given by the composition in* $\mathcal{A}$ *and the identities are* $1_{(A,\,f)} = 1_A$ *for all* $A \in \mathrm{Ob}\,\mathcal{A}$.

If we specialize the previous result further by considering the non-trivial functor to be the identity then we obtain the so-called *slice* and *coslice categories*:

Corollary 1.8.7 *Let* C *be a category and* $C_0 \in \mathrm{Ob}\,C$.

(1) The category denoted by $(C_0 \downarrow C)$, *called the* category of objects under C_0 *or the* coslice category *with respect to* C_0, *is defined as follows:*

  (i) *the objects are pairs* $(f,\,C)$, *where* $C \in \mathrm{Ob}\,C$, $f \in \mathrm{Hom}_C(C_0,\,C)$;

  (ii) *a morphism in* $(C_0 \downarrow C)$ *between two objects* $(f,\,C)$ *and* $(f',\,C')$ *is a morphism* $h \in \mathrm{Hom}_C(C,\,C')$ *such that the following diagram is commutative:*

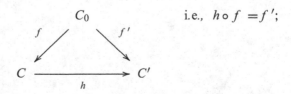

i.e., $h \circ f = f'$;

     (iii) the composition of morphisms is given by the composition in C and the identities are $1_{(f, C)} = 1_C$ *for all* $C \in \text{Ob} C$.

(2) *Similarly,* $(C \downarrow C_0)$ *is called the* category of objects over C_0 *or the* slice category *with respect to* C_0 *and is defined as follows:*

     (i) the objects are pairs (C, f), *where* $C \in \text{Ob} C$, $f \in \text{Hom}_C(C, C_0)$;
     (ii) a morphism in $(C \downarrow C_0)$ *between two objects* (C, f) *and* (C', f') *is a morphism* $h \in \text{Hom}_{\mathcal{A}}(C, C')$ *such that the following diagram is commutative:*

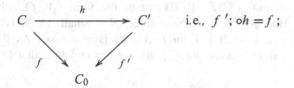

i.e., $f';oh = f$;

     (iii) the composition of morphisms is given by the composition in C and the identities are $1_{(C, f)} = 1_C$ *for all* $C \in \text{Ob} C$.

As we already mentioned, various properties of functors can be characterized in terms of certain related comma categories. One such property is representability:

Proposition 1.8.8 *A functor* $F : C \to \textbf{Set}$ *is representable if and only if the comma category* $(\{\star\} \downarrow F)$ *has an initial object, where* $\{\star\}$ *denotes a singleton set.*

Proof Assume first that (A, a) is a representing pair of F, where $A \in \text{Ob} C$ and $a \in F(A)$. We will show that the pair (f_a, A), where $f_a \in \text{Hom}_{\textbf{Set}}(\{\star\}, F(A))$ is defined by $f_a(\star) = a$, is the initial object of the comma category $(\{\star\} \downarrow F)$. Indeed, let $(g, B) \in \text{Ob}\left(\{\star\} \downarrow F\right)$, where $B \in \text{Ob} C$ and $g \in \text{Hom}_{\textbf{Set}}(\{\star\}, F(A))$, and consider $g(\star) = b \in F(B)$. By Proposition 1.7.7, there exists a unique $u \in \text{Hom}_{\textbf{Set}}(A, B)$ such that $F(u)(a) = b$. This is equivalent to u being the unique morphism such that $F(u) \circ f_a(\star) = g(\star)$, i.e., the following diagram is commutative:

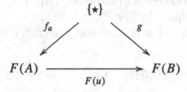

In other words, u is the unique morphism in $(\{\star\} \downarrow F)$ between (f_a, A) and (g, B), as desired.

     Conversely, let (f, A) be the initial object of $(\{\star\} \downarrow F)$, where $A \in \text{Ob} C$ and $f \in \text{Hom}_{\textbf{Set}}(\{\star\}, F(A))$, and consider $a = f(\star)$. Then, (A, a) is the representing pair of F. Indeed, note that if $B \in \text{Ob} C$ and $b \in F(B)$ then (f_b, B) is an object in $(\{\star\} \downarrow F)$, where $f_b \in \text{Hom}_{\textbf{Set}}(\{\star\}, F(A))$ is defined by $f_b(\star) = b$. As (f, A) is the initial object of $\left(\{\star\} \downarrow F\right)$, there exists a unique morphism $u \colon (f, A) \to (f_b, B)$ in $\left(\{\star\} \downarrow F\right)$ such that $F(u) \circ f_a = g$. To conclude, there exists a unique morphism $u \in \mathrm{Hom}_{\mathbf{Set}}(A, B)$ such that $F(u)(a) = b$ and therefore (A, a) is the representing pair of F by Proposition 1.7.7. $\square$

Next we discuss certain properties of comma categories which are of interest for the material developed in the sequel.

Lemma 1.8.9 *Let* $F \colon \mathcal{A} \to \mathcal{B}$ *be a functor and* $\mathcal{A}$ *a small category. Then, for all* $B \in \mathrm{Ob}\,\mathcal{B}$, *the comma-category* $(F \downarrow B)$ *is small.*

Proof The objects of $(F \downarrow B)$ are of the form (A, f), where $A \in \mathrm{Ob}\,\mathcal{A}$ and $f \in \mathrm{Hom}_{\mathcal{B}}(F(A), B)$. Now since $\mathcal{A}$ is small, $\mathrm{Ob}\,\mathcal{A}$ is a set and consequently $\bigcup_{A \in \mathrm{Ob}\,\mathcal{A}} \mathrm{Ob}\,\mathcal{A} \times \mathrm{Hom}(F(A), B)$ is also a set. As $\bigcup_{A \in \mathrm{Ob}\,\mathcal{A}} \left(\mathrm{Ob}\,\mathcal{A} \times \mathrm{Hom}(F(A), B)\right)$ contains $\mathrm{Ob}(F \downarrow B)$, we can conclude that $(F \downarrow B)$ is a small category. $\square$

Proposition 1.8.10 *Given two functors* $F \colon \mathcal{A} \to \mathcal{C}$ *and* $G \colon \mathcal{B} \to \mathcal{C}$, *we have an isomorphism of categories between* $\left(F \downarrow G\right)^{op}$ *and* $\left(G^{op} \downarrow F^{op}\right)$.

In particular, for any $C \in \mathrm{Ob}\,\mathcal{C}$ *we have an isomorphism of categories between* $\left(C \downarrow G\right)^{op}$ *and* $\left(G^{op} \downarrow C\right)$.

Proof Throughout the proof we will freely use the notation $\circ^{op}$ to denote the composition of morphisms in $\mathcal{C}^{op}$, $\mathcal{D}^{op}$ or $\left(F \downarrow G\right)^{op}$. It will, however, be obvious from the context which composition is used.

Note that the objects of the category $\left(G^{op} \downarrow F^{op}\right)$ are of the form $\left(B, f^{op}, A\right)$, where

$$B \in \mathrm{Ob}\,\mathcal{B}^{op} = \mathrm{Ob}\,\mathcal{B}, \quad A \in \mathrm{Ob}\,\mathcal{A}^{op} = \mathrm{Ob}\,\mathcal{A} \text{ and}$$

$$f^{op} \in \mathrm{Hom}_{\mathcal{C}^{op}}(G^{op}(B), F^{op}(A)) = \mathrm{Hom}_{\mathcal{C}^{op}}(G(B), F(A)).$$

This also shows that $f \in \mathrm{Hom}_{\mathcal{C}}(F(A), G(B))$. Therefore, we have $\left(B, f^{op}, A\right) \in \mathrm{Ob}\left(G^{op} \downarrow F^{op}\right)$ if and only if $\left(A, f, B\right) \in \mathrm{Ob}\left(F \downarrow G\right)^{op}$.

Furthermore, a morphism between two objects $\left(B_1, f_1^{op}, A_1\right)$ and $\left(B_2, f_2^{op}, A_2\right)$ in $\left(G^{op} \downarrow F^{op}\right)$ is a pair $\left(b^{op}, a^{op}\right)$, where $b^{op} \in \mathrm{Hom}_{\mathcal{B}^{op}}(B_1, B_2)$ and $a^{op} \in \mathrm{Hom}_{\mathcal{A}^{op}}(A_1, A_2)$ such that the following diagram is commutative:

$$
\begin{array}{ccc}
G^{op}(B_1) & \xrightarrow{\ G^{op}(b^{op})\ } & G^{op}(B_2) \\
{\scriptstyle f_1^{op}} \downarrow & & \downarrow {\scriptstyle f_2^{op}} \\
F^{op}(A_1) & \xrightarrow{\ F^{op}(a^{op})\ } & F^{op}(A_2)
\end{array}
\qquad \text{i.e.,} \quad F^{op}(a^{op}) \circ^{op} f_1^{op} = f_2^{op} \circ^{op} G^{op}(b^{op}).
$$

$$(1.30)$$

In particular, we have $a \in \mathrm{Hom}_{\mathcal{A}}(A_2, A_1)$ and $b \in \mathrm{Hom}_{\mathcal{B}}(B_2, B_1)$. Furthermore, (1.30) can be equivalently written as follows:

$$F^{op}(a^{op}) \circ^{op} f_1^{op} = f_2^{op} \circ^{op} G^{op}(b^{op})$$

$$\Leftrightarrow F(a)^{op} \circ^{op} f_1^{op} = f_2^{op} \circ^{op} G(b)^{op}$$

$$\Leftrightarrow \left(f_1 \circ F(a)\right)^{op} = \left(G(b) \circ f_2\right)^{op}$$

$$\Leftrightarrow f_1 \circ F(a) = G(b) \circ f_2.$$

This shows that (1.30) is commutative if and only if the following diagram is commutative:

$$
\begin{array}{ccc}
F(A_2) & \xrightarrow{\ F(a)\ } & F(A_1) \\
\Big\downarrow{\scriptstyle f_2} & & \Big\downarrow{\scriptstyle f_1} \\
G(B_2) & \xrightarrow[\ G(b)\ ]{} & G(B_1)
\end{array}
\qquad \text{i.e.,}\quad G(b) \circ f_2 = f_1 \circ F(a).
$$

$$(1.31)$$

Hence we have $(a, b) \in \mathrm{Hom}_{(F \downarrow G)}\big((A_2, f_2, B_2), (A_1, f_1, B_1)\big)$. In fact, all the above shows that $\left(b^{op}, a^{op}\right) \in \mathrm{Hom}_{\left(G^{op} \downarrow F^{op}\right)}\big((B_1, f_1^{op}, A_1), (B_2, f_2^{op}, A_2)\big)$ if and only if $(a, b) \in \mathrm{Hom}_{(F \downarrow G)}\big((A_2, f_2, B_2), (A_1, f_1, B_1)\big)$.

We can now define two functors $H \colon (F \downarrow G)^{op} \to \left(G^{op} \downarrow F^{op}\right)$ and $T \colon \left(G^{op} \downarrow F^{op}\right) \to (F \downarrow G)^{op}$ as follows:

$$H\big((A, f, B)\big) = \big(B, f^{op}, A\big),$$

$$(A, f, B) \in \mathrm{Ob}\,(F \downarrow G)^{op},$$

$$H\big((a, b)^{op}\big) = \big(b^{op}, a^{op}\big),$$

$$(a, b)^{op} \in \mathrm{Hom}_{(F \downarrow G)^{op}}\big((A_1, f_1, B_1), (A_2, f_2, B_2)\big),$$

$$T\big((B, f^{op}, A)\big) = (A, f, B),$$

$$(B, f^{op}, A) \in \mathrm{Ob}\,\big(G^{op} \downarrow F^{op}\big),$$

$$T\big((b^{op}, a^{op})\big) = (a, b)^{op},$$

$$(b^{op}, a^{op}) \in \mathrm{Hom}_{\left(G^{op} \downarrow F^{op}\right)}\big((B_1, f_1^{op}, A_1), (B_2, f_2^{op}, A_2)\big).$$

The above discussion shows that both H and T are well-defined. We will show that they are indeed functors. To start with, for all $A \in \mathrm{Ob}\,\mathcal{A}$ and $B \in \mathrm{Ob}\,\mathcal{B}$ we have

$$H\left(1^{op}_{(A,\,f,\,B)}\right) = H\left((1_A,\,1_B)^{op}\right) = \left(1^{op}_B,\,1^{op}_A\right) = 1_{\left(B,\,f^{op},\,A\right)},$$

$$T\left(1_{\left(B,\,f^{op},\,A\right)}\right) = T\left((1^{op}_B,\,1^{op}_A)\right) = (1_A,\,1_B)^{op} = 1^{op}_{(A,\,f,\,B)}.$$

Consider now two morphisms $(a,\,b)^{op} \in \mathrm{Hom}_{(F\downarrow G)^{op}}\left((A_1,\,f_1,\,B_1),\,(A_2,\,f_2,\,B_2)\right)$ and $(c,\,d)^{op} \in \mathrm{Hom}_{(F\downarrow G)^{op}}\left((A_2,\,f_2,\,B_2),\,(A_3,\,f_3,\,B_3)\right)$. We have

$$H\left((c,\,d)^{op}\,\circ^{op}\,(a,\,b)^{op}\right) = H\left(\left((a,\,b)\circ(c,\,d)\right)^{op}\right) = H\left(\left(a\circ c,\,b\circ d\right)^{op}\right)$$

$$= \left((b\circ d)^{op},\,(a\circ c)^{op}\right) = \left(d^{op}\,\circ^{op}\,b^{op},\,c^{op}\,\circ^{op}\,a^{op}\right) = \left(d^{op},\,c^{op}\right)\circ\left(b^{op},\,a^{op}\right)$$

$$= H\left((c,\,d)^{op}\right)\circ H\left((a,\,b)^{op}\right).$$

Finally, given two morphisms

$$\left(u^{op},\,v^{op}\right) \in \mathrm{Hom}_{\left(G^{op}\downarrow F^{op}\right)}\left((B_1,\,f_1^{op},\,A_1),\,(B_2,\,f_2^{op},\,A_2)\right)\text{ and }$$

$$\left(r^{op},\,s^{op}\right) \in \mathrm{Hom}_{\left(G^{op}\downarrow F^{op}\right)}\left((B_2,\,f_2^{op},\,A_2),\,(B_3,\,f_3^{op},\,A_3)\right)$$

we have

$$T\left((r^{op},\,s^{op})\circ(u^{op},\,v^{op})\right) = T\left((r^{op}\,\circ^{op}\,u^{op}),\,(s^{op}\,\circ^{op}\,v^{op})\right)$$

$$= T\left((u\circ r)^{op},\,(v\circ s)^{op}\right)$$

$$= \left(v\circ s,\,u\circ r\right)^{op} = \left((v,\,u)\circ(s,\,r)\right)^{op} = (s,\,r)^{op}\,\circ^{op}\,(v,\,u)^{op}$$

$$= T\left((r^{op},\,s^{op})\right)\circ^{op} T\left((u^{op},\,v^{op})\right).$$

The proof is now finished as H and T are obviously inverses to each other. □

1.9 Functor Categories

As the name suggests, *functor categories* have functors as objects and natural transformations as morphisms. The next result shows that this is indeed a category under certain conditions.

Proposition 1.9.1 *Let I and C be two categories. If I is a small category then the functors from I to C and the natural transformations between them as morphisms* *form a category, called a* functor category, *which we denote by Fun (I, C). If C is also small then Fun (I, C) is small.*

Proof The composition of natural transformations is given by the vertical composition as defined in Example 1.7.2, (5). This composition law is obviously associative and the identity at each functor F is just the identity natural transformation defined in Example 1.7.2, (4).

Note that $\bigcup_{i \in \mathrm{Ob}\, I} \mathrm{Hom}_C\big(F(i),\ G(i)\big)$ is a union of sets indexed by another set, namely Ob I. Hence $\bigcup_{i \in \mathrm{Ob}\, I} \mathrm{Hom}_C\big(F(i),\ G(i)\big)$ is a set as well. Finally, note that for any two functors $F,\ G \colon I \to C$, a natural transformation $\eta \colon F \to G$ is determined by a map

$$\eta \colon \mathrm{Ob}\, I \to \bigcup_{i \in \mathrm{Ob}\, I} \mathrm{Hom}_C\big(F(i),\ G(i)\big).$$

This shows that the class of all natural transformations between F and G is a subset of Ob I $\times$ $\bigcup_{i \in \mathrm{Ob}\, I} \mathrm{Hom}_C\big(F(i),\ G(i)\big)$. Therefore, the natural transformations between any two functors $F,\ G \colon I \to C$ form a set and the proof is now finished.

□

Examples 1.9.2

(1) For any small category C, a functor $F \colon C^{op} \to$ **Set** is called a (set-valued) *presheaf on* C and the functor category Fun(C^{op}, **Set**) is known as the *category of presheaves on* C.

(2) Let **2** denote the discrete category with two objects, say $*$ and $\diamond$. Then, for any category C the functor category Fun(**2**, C) is isomorphic to the product category $C \times C$. Indeed, first note that any functor $F \colon \mathbf{2} \to C$ is uniquely determined by two objects $C_F, D_F \in \mathrm{Ob}\, C$ such that

$$F(*) = C_F, \quad F(\diamond) = D_F, \quad F(1_*) = 1_{C_F}, \quad F(1_\diamond) = 1_{D_F}.$$

Consider now a natural transformation $\eta \colon F \to H$, where $F, H \colon \mathbf{2} \to C$ are functors. Then η is uniquely determined by two morphisms in C:

$$\eta_* \colon C_F \to C_H, \quad \eta_\diamond \colon D_F \to D_H.$$

Note that since there are no non-identity morphisms in **2** we have no non-trivial naturality diagram to impose additional conditions for the two morphisms η_* and $\eta_\diamond$. Therefore, the functor $V \colon \mathrm{Fun}(\mathbf{2}, C) \to C \times C$ defined below is an obvious isomorphism of categories:

$$V(F) = (C_F,\ D_F), \quad V(\eta) = (\eta_*,\ \eta_\diamond),$$

where $F, H \colon \mathbf{2} \to C$ are functors and $\eta \colon F \to H$ is a natural transformation.
□

The isomorphisms in a functor category can be easily characterized:

Proposition 1.9.3 *Let I be a small category, C an arbitrary category and F, $G \colon I \to C$ two functors. Then a natural transformation $\eta \colon F \to G$ is an isomorphism in $\mathrm{Fun}(I, C)$ if and only if η is a natural isomorphism.*

Proof Assume first that η is an isomorphism in $\mathrm{Fun}(I, C)$, i.e., there exists another natural transformation $\xi \colon G \to F$ such that $\eta \circ \xi = 1_G$ and $\xi \circ \eta = 1_F$. Hence, for any $i \in \mathrm{Ob}\, I$ we have $\eta_i \circ \xi_i = 1_{G(i)}$ and $\xi_i \circ \eta_i = 1_{F(i)}$. This proves that η_i is an isomorphism for all $i \in \mathrm{Ob}\, I$ and therefore η is a natural isomorphism.

Conversely, suppose now that η is a natural isomorphism. In particular, $\eta_i \colon F(i) \to G(i)$ is an isomorphism in C for all $i \in \mathrm{Ob}\, I$. We are left to show that $\eta^{-1} \colon G \to F$ assembled from the components η_i^{-1}, $i \in \mathrm{Ob}\, I$, is a natural transformation. To this end consider $f \in \mathrm{Hom}_I(i, j)$; the naturality of η yields the following commutative diagram:

$$
\begin{array}{ccc}
F(i) & \xrightarrow{\ \eta_i\ } & G(i) \\
{\scriptstyle F(f)} \downarrow & & \downarrow {\scriptstyle G(f)} \\
F(j) & \xrightarrow[\ \eta_j\ ]{} & G(j)
\end{array}
$$

i.e., $\eta_j \circ F(f) = G(f) \circ \eta_i$. Since each η_i is an isomorphism, we also have $F(f) \circ \eta_i^{-1} = \eta_j^{-1} \circ G(f)$. Therefore, the following diagram is commutative:

$$
\begin{array}{ccc}
G(i) & \xrightarrow{\ \eta_i^{-1}\ } & F(i) \\
{\scriptstyle G(f)} \downarrow & & \downarrow {\scriptstyle F(f)} \\
G(j) & \xrightarrow[\ \eta_j^{-1}\ ]{} & F(j)
\end{array}
$$

which proves that η^{-1} is indeed a natural transformation. $\square$

Some of the properties of a given category are inherited by all its corresponding functor categories, as can be seen in the following:

Proposition 1.9.4 *Let I be a small category and C a pointed category. Then the functor category $\mathrm{Fun}(I, C)$ is also pointed.*

Proof Let C_0 be the zero object of C. We will show that $\Delta_{C_0} \colon I \to C$, the constant functor at C_0 (see Example 1.5.3, (5)), is the zero object of the functor category $\mathrm{Fun}(I, C)$. To this end, consider another functor $F \colon I \to C$. We will show that there exists a unique natural transformation $\psi \colon \Delta_{C_0} \to F$ and a unique natural transformation $\varphi \colon F \to \Delta_{C_0}$.

Indeed, for all $i \in \mathrm{Ob}\, I$, we define ψ_i to be the unique morphism in C from C_0 to $F(i)$ (recall that C_0 is, in particular, an initial object in C). We are left to show that ψ as defined above is a natural transformation. Given any $f \in \mathrm{Hom}_I(i,\, j)$, both ψ_j and $F(f) \circ \psi_i$ are morphisms in C from C_0 to $F(j)$. As C_0 is an initial object in C we obtain $\psi_j = F(f) \circ \psi_i$. Hence, the following diagram is commutative:

$$
\begin{array}{ccc}
C_0 & \xrightarrow{\ \psi_i\ } & F(i) \\
{\scriptstyle 1_{C_0}}\Big\downarrow & & \Big\downarrow{\scriptstyle F(f)} \\
C_0 & \xrightarrow[\ \psi_j\ ]{} & F(j)
\end{array}
$$

which shows that ψ is indeed a natural transformation between Δ_{C_0} and F. Furthermore, ψ is the unique natural transformation between the aforementioned functors as for each $i \in \mathrm{Ob}\, I$ there exists a unique morphism between C_0 and $F(i)$.

Consider now φ_i to be the unique morphism in C from $F(i)$ to C_0, $i \in \mathrm{Ob}\, I$ (recall that C_0 is, in particular, a final object in C). Then, for any $f \in \mathrm{Hom}_I(i,\, j)$, both φ_i and $\varphi_j \circ F(f)$ are morphisms in C from $F(i)$ to C_0. As C_0 is the final object in C, we obtain $\varphi_i = \varphi_j \circ F(f)$. This shows that the following diagram commutes and therefore φ is a natural transformation between F and Δ_{C_0}:

$$
\begin{array}{ccc}
F(i) & \xrightarrow{\ \varphi_i\ } & C_0 \\
{\scriptstyle F(f)}\Big\downarrow & & \Big\downarrow{\scriptstyle 1_{C_0}} \\
F(j) & \xrightarrow[\ \varphi_j\ ]{} & C_0
\end{array}
$$

φ is the unique natural transformation which can be defined between the two functors above as for each $i \in \mathrm{Ob}\, I$ there exists a unique morphism between $F(i)$ and C_0. $\qquad\square$

Theorem 1.9.5 *Let $\mathcal{B}$, C be two categories with $\mathcal{B}$ small. Then we have an isomorphism of categories between* $\mathrm{Fun}(\mathcal{B}^{op},\, C^{op})$ *and* $\mathrm{Fun}(\mathcal{B},\, C)^{op}$.

Proof Throughout this proof, for any morphism $\alpha \colon F \to G$ in $\mathrm{Fun}(\mathcal{B}, C)$ (i.e., natural transformation) we denote by $\alpha^{\overline{op}} \colon G \to F$ the corresponding opposite morphism in $\mathrm{Fun}(\mathcal{B}, C)^{op}$ (Definition 1.4.1) while $\alpha^{op} \colon G^{op} \to F^{op}$ stands for the opposite natural transformation as defined in Proposition 1.8.3. Moreover, $\circ^{op}$ denotes the composition in $\mathrm{Fun}(\mathcal{B}, C)^{op}$ and $\circ_C$ (resp. $\circ_C^{op}$) stands for the composition in the category C (resp. C^{op}). Finally, we use $\bullet$ to denote the composition in $\mathrm{Fun}(\mathcal{B}^{op}, C^{op})$ and the unadorned $\circ$ for the composition in $\mathrm{Fun}(\mathcal{B}, C)$.

Define a functor $T: \mathrm{Fun}(\mathcal{B},\,C)^{op} \to \mathrm{Fun}(\mathcal{B}^{op},\,C^{op})$ as follows:

$$T(F) = F^{op}, \quad T(\alpha^{\overline{op}}) = \alpha^{op} \tag{1.32}$$

for all functors $F,\,G: \mathcal{B} \to C$ and natural transformations $\alpha: F \to G$. Note that T is well-defined as $\alpha^{\overline{op}}: G \to F$ in $\mathrm{Fun}(\mathcal{B},\,C)^{op}$ and $\alpha^{op}: G^{op} \to F^{op}$. Furthermore, consider three functors $F,\,G,\,H: \mathcal{B} \to C$ and let $\alpha: F \to G,\ \beta: G \to H$ be natural transformations. Then, for all $B \in \mathrm{Ob}\,\mathcal{B}$ we have

$$\left((\beta \circ \alpha)^{op}\right)_B \overset{(1.26)}{=} \left((\beta \circ \alpha)_B\right)^{op} = \left(\beta_B \circ_C \alpha_B\right)^{op} = (\alpha_B)^{op} \circ_C^{op} (\beta_B)^{op}$$

$$\overset{(1.26)}{=} \left(\alpha^{op}\right)_B \circ_C^{op} \left(\beta^{op}\right)_B \overset{(1.17)}{=} (\alpha^{op} \bullet \beta^{op})_B,$$

which leads to:

$$(\beta \circ \alpha)^{op} = \alpha^{op} \bullet \beta^{op}. \tag{1.33}$$

We obtain:

$$T\left(\alpha^{\overline{op}} \circ^{op} \beta^{\overline{op}}\right) = T\left((\beta \circ \alpha)^{\overline{op}}\right) = (\beta \circ \alpha)^{op} \overset{(1.33)}{=} \alpha^{op} \bullet \beta^{op} = T\left(\alpha^{\overline{op}}\right) \bullet T\left(\beta^{\overline{op}}\right)$$

and therefore T is indeed a functor. We are left to construct the inverse functor of T. To this end, consider $S: \mathrm{Fun}(\mathcal{B}^{op},\,C^{op}) \to \mathrm{Fun}(\mathcal{B},\,C)^{op}$ defined by

$$S(U) = U^{op}, \quad S(\beta) = (\beta^{op})^{\overline{op}} \tag{1.34}$$

for all functors $U,\,V: \mathcal{B}^{op} \to C^{op}$ and all natural transformations $\beta: U \to V$. Note that $\beta^{op}: V^{op} \to U^{op}$ is a natural transformation between the functors U^{op}, $V^{op}: \mathcal{B} \to C$ and therefore $(\beta^{op})^{\overline{op}}$ is a morphism in $\mathrm{Fun}(\mathcal{B},\,C)^{op}$. Consider now three functors $U,\,V,\,W: \mathcal{B}^{op} \to C^{op}$ and natural transformations $\alpha: U \to V$, $\beta: V \to W$.

We aim to show that $S(\beta \bullet \alpha) = S(\beta) \circ^{op} S(\alpha)$. To start with, note that for any $B \in \mathrm{Ob}\,\mathcal{B}$ we have $\alpha_B \in \mathrm{Hom}_{C^{op}}(U(B),\,V(B))$ and therefore $\alpha_B = \overline{\alpha}_B^{op}$ for some morphism $\overline{\alpha}_B \in \mathrm{Hom}_C(V(B),\,U(B))$. Similarly, $\beta_B = \overline{\beta}_B^{op}$ for some morphism $\overline{\beta} \in \mathrm{Hom}_C(W(B),\,V(B))$. This leads to the following:

$$\left((\beta \bullet \alpha)^{op}\right)_B = \left((\beta \bullet \alpha)_B\right)^{op} = \left(\beta_B \circ_C^{op} \alpha_B\right)^{op} = \left(\overline{\beta}_B^{op} \circ_C^{op} \overline{\alpha}_B^{op}\right)^{op}$$

$$= \left((\overline{\alpha}_B \circ_C \overline{\beta}_B)^{op}\right)^{op} = \overline{\alpha}_B \circ_C \overline{\beta}_B = (\overline{\alpha}_B^{op})^{op} \circ_C (\overline{\beta}_B^{op})^{op}$$

$$= \alpha_B^{op} \circ_C \beta_B^{op} = (\alpha^{op})_B \circ_C (\beta^{op})_B = \left(\alpha^{op} \circ \beta^{op}\right)_B.$$

Hence, we have:

$$(\beta \bullet \alpha)^{op} = \alpha^{op} \circ \beta^{op}. \tag{1.35}$$

Putting everything together, we obtain:

$$S(\beta \bullet \alpha) = \left((\beta \bullet \alpha)^{op}\right)^{\overline{op}} \overset{(1.35)}{=} \left(\alpha^{op} \circ \beta^{op}\right)^{\overline{op}} = \left(\beta^{op}\right)^{\overline{op}} \circ^{op} \left(\alpha^{op}\right)^{\overline{op}} = S(\beta) \circ^{op} S(\alpha),$$

as desired. Finally, we will show that T and S are inverses to each other. To this end, given two functors $F, G \colon \mathcal{B} \to \mathcal{C}$ and a natural transformation $\alpha \colon F \to G$ we have

$$S(T(F)) \overset{(1.32)}{=} S(F^{op}) \overset{(1.34)}{=} (F^{op})^{op} = F,$$

$$S(T(\alpha^{\overline{op}})) \overset{(1.32)}{=} S(\alpha^{op}) \overset{(1.34)}{=} ((\alpha^{op})^{op})^{\overline{op}} = \alpha^{\overline{op}}.$$

Similarly, for all functors $U, V \colon \mathcal{B}^{op} \to \mathcal{C}^{op}$ and natural transformations $\beta \colon U \to V$ we have

$$T(S(U)) \overset{(1.34)}{=} T(U^{op}) \overset{(1.32)}{=} (U^{op})^{op} = U,$$

$$T(S(\beta)) \overset{(1.34)}{=} T((\beta^{op})^{\overline{op}}) \overset{(1.32)}{=} (\beta^{op})^{op} = \beta$$

and the proof is now finished.

$\square$

Lemma 1.9.6 *Let I be a small category and C an arbitrary category. If F, $G \colon I \to C$ are two functors and $\psi \colon F \to G$ is a natural transformation then ψ is a monomorphism in $\mathrm{Fun}(I, C)$ if and only if the dual natural transformation $\psi^{op} \colon F^{op} \to G^{op}$ is an epimorphism in $\mathrm{Fun}(I^{op}, C^{op})$.*

Proof Indeed, using the duality principle, ψ is a monomorphism in $\mathrm{Fun}(I, C)$ if and only if $\psi^{\overline{op}}$ is an epimorphism in $\mathrm{Fun}(I, C)^{op}$. Furthermore, by Theorem 1.9.5 this is equivalent to $T\left(\psi^{\overline{op}}\right) = \psi^{op}$ being an epimorphism in $\mathrm{Fun}(I^{op}, C^{op})$ and the proof is now finished.

$\square$

There are various methods of constructing functors between functor categories. For instance, if I is a small category and $C, \mathcal{D}$ are arbitrary categories then any functor $F \colon C \to \mathcal{D}$ induces a functor between the corresponding functor categories as follows:

$$F_\star \colon \mathrm{Fun}(I, C) \to \mathrm{Fun}(I, \mathcal{D}), \quad F_\star(G) = FG, \quad F_\star(\psi)_i = F(\psi_i) \tag{1.36}$$

for all functors $G, G' \colon I \to C$, all natural transformations $\psi \colon G \to G'$ and all $i \in \mathrm{Ob}\, I$.

Similarly, if $F: I \to J$ is a functor between small categories and C is an arbitrary category, we can define another induced functor F^* between the corresponding functor categories as follows:

$$F^*: \mathrm{Fun}(J, C) \to \mathrm{Fun}(I, C), \quad F^*(G) = GF, \quad F^*(\psi)_i = \psi_{F(i)} \tag{1.37}$$

for all functors $G, G': J \to C$, all natural transformations $\psi: G \to G'$ and all $i \in \mathrm{Ob}\, I$.

Finally, a functor $F: I \to J$ between small categories induces by *precomposition* with F a new functor $- \circ F: \mathrm{Fun}(J, C) \to \mathrm{Fun}(I, C)$ defined for all functors $G, G': J \to C$ and all natural transformations $\psi: G \to G'$ by $\psi \circ F = \psi_F$.

Furthermore, for any functor category we can define an *evaluation bifunctor* as follows:

Proposition 1.9.7 *Given a small category I and an arbitrary category C, there exists a bifunctor* $\mathrm{Ev}: \mathrm{Fun}(I, C) \times I \to C$, *called the* evaluation bifunctor, *defined as follows for all $i, j \in \mathrm{Ob}\, I$, $f \in \mathrm{Hom}_I(i, j)$, all functors $F, G: I \to C$ and all natural transformations $\eta: F \to G$:*

$$\mathrm{Ev}(F, i) = F(i), \qquad \mathrm{Ev}(\eta, f) = G(f) \circ \eta_i.$$

Proof To start with, note that as $\eta: F \to G$ is a natural transformation, the following diagram commutes:

$$
\begin{array}{ccc}
F(i) & \xrightarrow{\eta_i} & G(i) \\
{\scriptstyle F(f)}\Big\downarrow & & \Big\downarrow{\scriptstyle G(f)} \\
F(j) & \xrightarrow[\eta_j]{} & G(j)
\end{array}
\qquad \text{i.e.,} \quad \eta_j \circ F(f) = G(f) \circ \eta_i.
$$

$$\tag{1.38}$$

Hence, we obtain $\mathrm{Ev}(\eta, f) = G(f) \circ \eta_i = \eta_j \circ F(f)$, which shows that Ev is well-defined.

Moreover, for all $i \in \mathrm{Ob}\, I$ and all functors $F: I \to C$ we have

$$\mathrm{Ev}(1_F, 1_i) = F(1_i) \circ 1_{F(i)} = 1_{F(i)} \circ 1_{F(i)} = 1_{F(i)} = 1_{\mathrm{Ev}(F, i)}.$$

We have proved that Ev preserves identities. Consider now $f \in \mathrm{Hom}_I(i, j)$, $g \in \mathrm{Hom}_I(j, k)$ and $\eta: F \to G$, $\xi: G \to H$, natural transformations, where $F, G, H: I \to C$. We obtain:

$$\mathrm{Ev}\big((\xi, g) \circ (\eta, f)\big) = \mathrm{Ev}(\xi \circ \eta, g \circ f) = H(g \circ f) \circ (\xi \circ \eta)_i$$

$$= H(g) \circ \underline{H(f) \circ \xi_i} \circ \eta_i = H(g) \circ \xi_j \circ G(f) \circ \eta_i = \mathrm{Ev}(\xi, g) \circ \mathrm{Ev}(\eta, f),$$

where in the fourth equality we used the naturality of ξ. Hence Ev preserves compositions and is indeed a functor. $\square$

In light of Proposition 1.5.5, we have two other functors induced by the evaluation bifunctor Ev. Indeed, given $i \in \mathrm{Ob}\,I$, the left associated functor at i denoted by $\mathrm{Ev}(-, i)\colon \mathrm{Fun}(I, C) \to C$ will be called the *evaluation functor at i* and is defined as follows:

$$\mathrm{Ev}(F, i) = F(i) \quad \text{and} \quad \mathrm{Ev}(\eta, 1_i) = G(1_i) \circ \eta_i = \eta_i \qquad (1.39)$$

for all functors $F, G\colon I \to C$ and all natural transformations $\eta\colon F \to G$.

On the other hand, for any functor $F\colon I \to C$, the right associated functor $\mathrm{Ev}(F, -)\colon I \to C$ coincides with F. Indeed, for all $i, j \in \mathrm{Ob}\,I$ and all $f \in \mathrm{Hom}_I(i, j)$ we have

$$\mathrm{Ev}(F, -)(i) = \mathrm{Ev}(F, i) = F(i) \text{ and}$$

$$\mathrm{Ev}(F, -)(f) = \mathrm{Ev}(1_F, f) = F(f) \circ 1_{F(i)} = F(f).$$

Another functor which will later turn out to be important is the *diagonal functor* defined as follows:

Proposition 1.9.8 *Let I be a small category and C an arbitrary category. There exists a functor $\Delta\colon C \to \mathrm{Fun}(I, C)$, called the* diagonal functor, *defined as follows for all $C, D \in \mathrm{Ob}\,C$ and all $f \in \mathrm{Hom}_C(C, D)$:*

$$\Delta(C) = \Delta_C, \qquad \Delta(f) = \eta\colon \Delta_C \to \Delta_D,$$

where $\Delta_C\colon I \to C$ is the constant functor at C and η is the natural transformation given by $\eta_i = f$ for all $i \in \mathrm{Ob}\,I$.

Proof Recall from Example 1.5.3, (5) that $\Delta_C\colon I \to C$ is the functor that sends each object of the category I to C and each morphism of I to the identity 1_C. To start with, it is straightforward to see that for any $t \in \mathrm{Hom}_I(i, j)$ the following diagram

$$
\begin{array}{ccc}
\Delta_C(i) & \xrightarrow{\ \eta_i\ } & \Delta_D(i) \\
{\scriptstyle \Delta_C(t)}\big\downarrow & & \big\downarrow{\scriptstyle \Delta_D(t)} \\
\Delta_C(j) & \xrightarrow[\ \eta_j\ ]{} & \Delta_D(j)
\end{array}
$$

comes down to

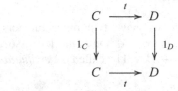

which is obviously commutative. This shows that η defined by $\eta_i = t$ for all $i \in$ Ob I is indeed a natural transformation and therefore Δ is well-defined.

Clearly, for any $C \in$ ObC, we have $\Delta(1_C) = \xi$, where $\xi \colon \Delta_C \to \Delta_C$ is the natural transformation defined by $\xi_i = 1_C$ for all $i \in$ Ob I. Hence ξ is the identity natural transformation 1_{Δ_C} and we have proved that Δ respects identities.

Consider now $f \in \mathrm{Hom}_C(C, D)$ and $g \in \mathrm{Hom}_C(D, E)$ and let $\Delta(f) = \mu$, $\Delta(g) = \nu$ where $\mu \colon \Delta_C \to \Delta_D$ and $\nu \colon \Delta_D \to \Delta_E$ are natural transformations defined by $\mu_i = f$ and $\nu_i = g$ for all $i \in$ Ob I. Furthermore, we have $\Delta(g \circ f) = \mu$, where $\eta \colon \Delta_C \to \Delta_E$ is the natural transformation defined by $\eta_i = g \circ f$ for all $i \in$ Ob I. Therefore, $\eta_i = \nu_i \circ \mu_i$ for all $i \in$ Ob I, which implies that the natural transformations η and $\nu \circ \mu$ coincide. Consequently, $\Delta(g \circ f) = \Delta(g) \circ \Delta(f)$ and we have proved that Δ is indeed a functor.   □

1.10 Yoneda's Lemma

Yoneda's lemma is arguably one of the most important results contained in this book. It allows us to *embed* any category into its category of presheaves by means of a fully faithful functor and this approach opens the way to a plethora of applications. Some important consequences of Yoneda's lemma, which go beyond the scope of this book, are mentioned at the end of the section.

Theorem 1.10.1 (Yoneda's Lemma) *Let $F \colon C \to$ **Set** be a functor and $C \in$ ObC. Then the natural transformations from $\mathrm{Hom}_C(C, -)$ to F are in bijection with the elements of the set $F(C)$ and the bijection is given for any natural transformation $\varphi \colon \mathrm{Hom}_C(C, -) \to F$ as follows:*

$$\pi_{C, F} \colon \mathrm{Nat}\big(\mathrm{Hom}_C(C, -), F\big) \to F(C), \quad \pi_{C, F}(\varphi) = \varphi_C(1_C) \in F(C). \tag{1.40}$$

In particular, $\mathrm{Nat}\big(\mathrm{Hom}_C(C, -), F\big)$ *is a set.*

Proof Consider $\tau_{C,F}: F(C) \rightarrow \mathrm{Nat}\big(\mathrm{Hom}_C(C, -),\, F\big)$, defined for every $x \in F(C)$ by:

$$\tau_{C,F}(x) = h^x \text{ where } \big(h^x\big)_D(f) = F(f)(x) \tag{1.41}$$

for every $D \in \mathrm{Ob}\, C$ and $f \in \mathrm{Hom}_C(C, D)$. First we have to check that h^x is a natural transformation. This comes down to proving that given a morphism $f \in \mathrm{Hom}_C(A, B)$ the following diagram is commutative:

$$
\begin{array}{ccc}
\mathrm{Hom}_C(C, A) & \xrightarrow{\;\;(h^x)_A\;\;} & F(A) \\
{\scriptstyle \mathrm{Hom}_C(C, f)}\Big\downarrow & & \Big\downarrow{\scriptstyle F(f)} \\
\mathrm{Hom}_C(C, B) & \xrightarrow[\;\;(h^x)_B\;\;]{} & F(B)
\end{array}
$$

Indeed, for any $g \in \mathrm{Hom}_C(C, A)$ we have

$$
\begin{aligned}
(h^x)_B\big(\mathrm{Hom}_C(C, f)\big)(g) &= (h^x)_B(f \circ g) \\
&= F(f \circ g)(x) = F(f) \circ F(g)(x) \\
&= F(f)(h^x)_A(g).
\end{aligned}
$$

Thus h^x is a natural transformation. The proof will be finished once we show that $\pi_{C,F}$ and $\tau_{C,F}$ are inverses to each other. To start with, consider $x \in F(C)$. Then we have

$$
\begin{aligned}
(\pi_{C,F} \circ \tau_{C,F})(x) &= \pi_{C,F}\big(\tau_{C,F}(x)\big) = \pi_{C,F}\big(h^x\big) \\
&= \big(h^x\big)_C(1_C) = F(1_C)(x) \\
&= 1_{F(C)}(x) = x.
\end{aligned}
$$

Thus $\pi_{C,F} \circ \tau_{C,F} = 1_{F(C)}$. Consider now a natural transformation $\varphi : \mathrm{Hom}_C(C, -) \rightarrow F$. We want to prove that $(\tau_{C,F} \circ \pi_{C,F})(\varphi) = \varphi$. Indeed, as φ is a natural transformation, for every $D \in \mathrm{Ob}\, C$ and every $f \in \mathrm{Hom}_C(C, D)$ the following diagram is commutative:

$$
\begin{array}{ccc}
\mathrm{Hom}_C(C, C) & \xrightarrow{\;\;\varphi_C\;\;} & F(C) \\
{\scriptstyle \mathrm{Hom}_C(C, f)}\Big\downarrow & & \Big\downarrow{\scriptstyle F(f)} \\
\mathrm{Hom}_C(C, D) & \xrightarrow[\;\;\varphi_D\;\;]{} & F(D)
\end{array}
$$

In particular, by evaluating the above diagram at $1_C \in \mathrm{Hom}_C(C, C)$ we obtain:

$$F(f)\big(\varphi_C(1_C)\big) = \varphi_D(f). \tag{1.42}$$

Hence for every $D \in \mathrm{Ob}\,C$ and every $f \in \mathrm{Hom}_C(C, D)$ we have

$$\tau_{C,F}\big(\pi_{C,F}(\varphi)\big)_D(f) = \tau_{C,F}\big(\varphi_C(1_C)\big)_D(f)$$
$$= F(f)\big(\varphi_C(1_C)\big) \overset{(1.42)}{=} \varphi_D(f),$$

which implies $\tau_{C,F} \circ \pi_{C,F}(\varphi) = \varphi$ and the proof is now complete. $\square$

It turns out that the bijections $\pi_{C,F}$ defined in (1.40) form a natural transformation in the variable C. Furthermore, if C is a small category, then the bijections $\pi_{C,F}$ form a natural transformation in the variable F as well. The precise statement is the following:

Theorem 1.10.2 *Let C be a category, $C \in \mathrm{Ob}\,C$ and $F : C \to \mathbf{Set}$ a functor.*

(1) If $G : C \to \mathbf{Set}$ is the functor defined as follows for all $C \in \mathrm{Ob}\,C$ and $f \in \mathrm{Hom}_C(C, C')$:

$$G(C) = \mathrm{Nat}\big(\mathrm{Hom}_C(C, -), F\big),$$

$$G(f) : \mathrm{Nat}\big(\mathrm{Hom}_C(C, -), F\big) \to \mathrm{Nat}\big(\mathrm{Hom}_C(C', -), F\big),$$

$$G(f)(\psi) = \psi \circ \mathrm{Hom}_C(f, -), \quad \psi \in \mathrm{Nat}\big(\mathrm{Hom}_C(C, -), F\big)$$

then $\pi_{-,F} : G \to F$ defined in (1.40) is a natural transformation.

(2) Assume C is a small category. If $H : \mathrm{Fun}(C, \mathbf{Set}) \to \mathbf{Set}$ is the functor defined as follows for all functors $F : C \to \mathbf{Set}$ and natural transformations $\psi : F \to F'$:

$$H(F) = \mathrm{Nat}\big(\mathrm{Hom}_C(C, -), F\big),$$

$$H(\psi) : \mathrm{Nat}\big(\mathrm{Hom}_C(C, -), F\big) \to \mathrm{Nat}\big(\mathrm{Hom}_C(C, -), F'\big),$$

$$H(\psi)(\varphi) = \psi \circ \varphi, \quad \varphi \in \mathrm{Nat}\big(\mathrm{Hom}_C(C, -), F\big)$$

then $\pi_{C,-} : H \to \mathrm{Ev}_C$ defined in (1.40) is a natural transformation, where Ev_C is the evaluation functor at C.

Proof

(1) Let $f \in \mathrm{Hom}_C(C, C')$; we need to prove the commutativity of the following diagram:

$$
\begin{array}{ccc}
G(C) & \xrightarrow{\ \pi_{C,F}\ } & F(C) \\
{\scriptstyle G(f)}\downarrow & & \downarrow{\scriptstyle F(f)} \\
G(C') & \xrightarrow[\ \pi_{C',F}\ ]{} & F(C')
\end{array}
\qquad \text{i.e.,}\quad F(f) \circ \pi_{C,F} = \pi_{C',F} \circ G(f).
$$

Indeed, for all $\varphi \in G(C) = \mathrm{Nat}\big(\mathrm{Hom}_C(C, -), F\big)$ we have

$$
\big(F(f) \circ \pi_{C,F}\big)(\varphi) \overset{(1.40)}{=} F(f)\big(\varphi_C(1_C)\big) = \varphi_{C'}(f) = \varphi_{C'}\big(\mathrm{Hom}_C(f, C')(1_{C'})\big)
$$

$$
= \big(\varphi \circ \mathrm{Hom}_C(f, -)\big)_{C'}(1_{C'}) = \pi_{C',F}\big(\varphi \circ \mathrm{Hom}_C(f, -)\big) = \big(\pi_{C',F} \circ G(f)\big)(\varphi),
$$

where the second equality holds because $\varphi \colon \mathrm{Hom}_C(C, -) \to F$ is a natural transformation.

(2) Consider two functors $F, F' \colon C \to \mathbf{Set}$ and a natural transformation $\psi \colon F \to F'$. The proof will be finished once we show the commutativity of the following diagram:

$$
\begin{array}{ccc}
H(F) & \xrightarrow{\ \pi_{C,F}\ } & \mathrm{Ev}_C(F) \\
{\scriptstyle H(\psi)}\downarrow & & \downarrow{\scriptstyle \mathrm{Ev}_C(\psi)} \\
H(F') & \xrightarrow[\ \pi_{C,F'}\ ]{} & \mathrm{Ev}_C(F')
\end{array}
\qquad \text{i.e.,}\quad \mathrm{Ev}_C(\psi) \circ \pi_{C,F} = \pi_{C,F'} \circ H(\psi).
$$

To this end, let $\varphi \in H(F) = \mathrm{Nat}\big(\mathrm{Hom}_C(C, -), F\big)$; we obtain

$$
\big(\mathrm{Ev}_C(\psi) \circ \pi_{C,F}\big)(\varphi) = \mathrm{Ev}_C(\psi)\big(\pi_{C,F}(\varphi)\big) \overset{(1.40)}{=} \mathrm{Ev}_C(\psi)\big(\varphi_C(1_C)\big)
$$

$$
\overset{(1.39)}{=} \psi_C\big(\varphi_C(1_C)\big) = (\psi \circ \varphi)_C(1_C) = \pi_{C,F'}(\psi \circ \varphi)
$$

$$
= \pi_{C,F'}\big(H(\psi)(\varphi)\big) = \big(\pi_{C,F'} \circ H(\psi)\big)(\varphi),
$$

as desired.

$\square$

Remark 1.10.3 The contravariant version of Yoneda's lemma can be easily obtained by replacing the category C with $\mathcal{D}^{op}$ in Theorem 1.10.1 and noticing that the functors $\mathrm{Hom}_{\mathcal{D}^{op}}(C, -) \colon \mathcal{D}^{op} \to \mathbf{Set}$ and $\mathrm{Hom}_{\mathcal{D}}(-, C) \colon \mathcal{D}^{op} \to \mathbf{Set}$ coincide.

As a first consequence, we can describe all natural transformations between hom functors.

Corollary 1.10.4 *Let C be a category and $C, D \in \mathrm{Ob}\, C$. Then:*

(1) $\alpha: \mathrm{Hom}_C(C, -) \to \mathrm{Hom}_C(D, -)$ is a natural transformation if and only if there exists a unique $\phi \in \mathrm{Hom}_C(D, C)$ such that $\alpha = h^\phi$, where $h^\phi: \mathrm{Hom}_C(C, -) \to \mathrm{Hom}_C(D, -)$ is the natural transformation defined as follows:

$$\left(h^\phi\right)_X(f) = f \circ \phi, \quad \text{for all } X \in \mathrm{Ob}\, C \text{ and } f \in \mathrm{Hom}_C(C, X). \tag{1.43}$$

(2) $\beta: \mathrm{Hom}_C(-, C) \to \mathrm{Hom}_C(-, D)$ is a natural transformation if and only if there exists a unique $\delta \in \mathrm{Hom}_C(C, D)$ such that $\beta = t^\delta$, where $t^\delta: \mathrm{Hom}_C(-, C) \to \mathrm{Hom}_C(-, D)$ is the natural transformation defined as follows:

$$\left(t^\delta\right)_X(g) = \delta \circ g, \quad \text{for all } X \in \mathrm{Ob}\, C \text{ and } g \in \mathrm{Hom}_C(X, C). \tag{1.44}$$

Proof

(1) We apply the Yoneda lemma for $F = \mathrm{Hom}_C(D, -)$. This gives a set bijection $\tau: \mathrm{Hom}_C(D, C) \to \mathrm{Nat}\big(\mathrm{Hom}_C(C, -), \mathrm{Hom}_C(D, -)\big)$ defined for every $\phi \in \mathrm{Hom}_C(D, C)$ by:

$$\tau(\phi) = h^\phi, \quad \text{where } \left(h^\phi\right)_X(f) = \mathrm{Hom}_C(D, f)(\phi) = f \circ \phi$$

for every $X \in \mathrm{Ob}\, C$ and $f \in \mathrm{Hom}_C(C, X)$. Hence, any natural transformation between $\mathrm{Hom}_C(C, -)$ and $\mathrm{Hom}_C(D, -)$ is of the form h^ϕ for a unique $\phi \in \mathrm{Hom}_C(D, C)$, as desired.

(1) Follows from the contravariant version of Yoneda's lemma.

$\square$

Proposition 1.10.5 *Let C be a category and $C, D \in \mathrm{Ob}\, C$. Then C and D are isomorphic if and only if the functors $\mathrm{Hom}_C(C, -)$ and $\mathrm{Hom}_C(D, -)$ are naturally isomorphic.*

Proof Suppose first that C and D are isomorphic objects in C and let $\phi \in \mathrm{Hom}_C(D, C)$ be an isomorphism. As proved in Corollary 1.10.4, (1) we have a natural transformation defined by $\tau(\phi) = h^\phi : \mathrm{Hom}_C(C, -) \to \mathrm{Hom}_C(D, -)$, where $\left(h^\phi\right)_X(f) = \mathrm{Hom}_C(D, f)(\phi) = f \circ \phi$ for all $X \in \mathrm{Ob}\, C$ and $f \in \mathrm{Hom}_C(C, X)$. We will prove that $\left(h^\phi\right)_X$ is a set bijection for every $X \in \mathrm{Ob}\, C$. To this end, we define $\left(\mu^\phi\right)_X : \mathrm{Hom}_C(D, X) \to \mathrm{Hom}_C(C, X)$ by $\left(\mu^\phi\right)_X(g) = g \circ \phi^{-1}$ for any $g \in \mathrm{Hom}_C(D, X)$. We will see that $\left(\mu^\phi\right)_X$ is the inverse of $\left(h^\phi\right)_X$. Indeed, for any $g \in \mathrm{Hom}_C(D, X)$ and any $f \in \mathrm{Hom}_C(C, X)$ we have

$$\left(h^\phi\right)_X \circ \left(\mu^\phi\right)_X (g) = \left(h^\phi\right)_X \left(g \circ \phi^{-1}\right) = g \circ \phi^{-1} \circ \phi = g,$$

$$\left(\mu^\phi\right)_X \circ \left(h^\phi\right)_X (f) = \left(\mu^\phi\right)_X \left(f \circ \phi\right) = f \circ \phi \circ \phi^{-1} = f.$$

Hence, $\tau(\phi)\colon \mathrm{Hom}_C(C, -) \to \mathrm{Hom}_C(D, -)$ is a natural isomorphism, as desired.

Conversely, let $\alpha\colon \mathrm{Hom}_C(C, -) \to \mathrm{Hom}_C(D, -)$ be a natural isomorphism. Denote $\alpha_C(1_C) \in \mathrm{Hom}_C(D, C)$ by u; we will prove that u is an isomorphism. We start by pointing out that $\alpha_D\colon \mathrm{Hom}_C(C, D) \to \mathrm{Hom}_C(D, D)$ is a bijection and since $1_D \in \mathrm{Hom}_C(D, D)$ we can find $v \in \mathrm{Hom}_C(C, D)$ such that $\alpha_D(v) = 1_D$.

Now as α is a natural transformation, the following diagram is commutative:

$$
\begin{array}{ccc}
\mathrm{Hom}_C(C, D) & \xrightarrow{\ \alpha_D\ } & \mathrm{Hom}_C(D, D) \\
{\scriptstyle \mathrm{Hom}_C(C, u)} \downarrow & & \downarrow {\scriptstyle \mathrm{Hom}_C(D, u)} \\
\mathrm{Hom}_C(C, C) & \xrightarrow[\ \alpha_C\ ]{} & \mathrm{Hom}_C(D, C)
\end{array}
$$

i.e., $\mathrm{Hom}_C(D, u) \circ \alpha_D = \alpha_C \circ \mathrm{Hom}_C(C, u)$. By evaluating this diagram at $v \in \mathrm{Hom}_C(C, D)$ and using $\alpha_D(v) = 1_D$ we obtain

$$\mathrm{Hom}_C(D, u) \circ \alpha_D \circ v = \alpha_C \circ \mathrm{Hom}_C(C, u) \circ v$$

$$\Leftrightarrow u \circ \alpha_D(v) = \alpha_C(u \circ v)$$

$$\Leftrightarrow u = \alpha_C(u \circ v).$$

Since we also have $u = \alpha_C(1_C)$ and α_C is a set bijection, we get $u \circ v = 1_C$.

Using again the fact that α is a natural transformation the following diagram is commutative:

$$
\begin{array}{ccc}
\mathrm{Hom}_C(C, C) & \xrightarrow{\ \alpha_C\ } & \mathrm{Hom}_C(D, C) \\
{\scriptstyle \mathrm{Hom}_C(C, v)} \downarrow & & \downarrow {\scriptstyle \mathrm{Hom}_C(D, v)} \\
\mathrm{Hom}_C(C, D) & \xrightarrow[\ \alpha_D\ ]{} & \mathrm{Hom}_C(D, D)
\end{array}
$$

Evaluating the above diagram at $1_C \in \mathrm{Hom}_C(C, C)$ and using $\alpha_D(v) = 1_D$ we obtain

$$\mathrm{Hom}_C(D, v) \circ \alpha_C \circ 1_C = \alpha_D \circ \mathrm{Hom}_C(C, v) \circ 1_C$$

$$\Leftrightarrow v \circ \alpha_C(1_C) = \alpha_D(v \circ 1_C)$$

$$\Leftrightarrow v \circ u = 1_D,$$

therefore v is the inverse of u and the proof is now finished.   □

In light of the above result we obtain:

Corollary 1.10.6 *Any two representing objects of a given representable functor* $F: C \to$ **Set** *are isomorphic.*

Proof Let $F: C \to$ **Set** be a functor and let $C, D \in \mathrm{Ob}\,C$ be representing objects of F. Therefore we have two natural isomorphisms $\alpha: F \to \mathrm{Hom}_C(C, -)$ and $\beta: F \to \mathrm{Hom}_C(D, -)$. This shows that $\beta \circ \alpha^{-1}: \mathrm{Hom}_C(C, -) \to \mathrm{Hom}_C(D, -)$ is a natural isomorphism, where α^{-1} is the natural transformation defined in Example 1.7.2, (6). Now Proposition 1.10.5 implies that C and D are isomorphic objects in C.   □

Definition 1.10.7 Given a small category C, the functor $Y: C \to \mathrm{Fun}(C^{op}, \mathbf{Set})$ which sends any $C \in \mathrm{Ob}\,C$ to the hom functor $\mathrm{Hom}_C(-, C)^{22}$ and any $f \in \mathrm{Hom}_C(C, D)$ to the natural transformation t^f defined in (1.44) is called the *Yoneda embedding*.

Corollary 1.10.8 *For any small category C, the Yoneda embedding functor* $Y: C \to \mathrm{Fun}(C^{op}, \mathbf{Set})$ *is fully faithful.*

Proof Let $C, D \in \mathrm{Ob}\,C$; we need to prove that the mapping defined below is bijective:

$$\mathcal{Y}_{C,D}: \mathrm{Hom}_C(C, D) \to \mathrm{Nat}\big(\mathrm{Hom}_C(-, C), \mathrm{Hom}_C(-, D)\big),$$

$$\mathcal{Y}_{C,D}(f) = Y(f) = t^f, \quad f \in \mathrm{Hom}_C(C, D).$$

This follows trivially from Corollary 1.10.4, (2).   □

Example 1.10.9 Cayley's theorem is a well-known result in group theory. It states that any group G is isomorphic to a subgroup of the symmetric group on G (i.e., the group of all permutations on the set G). In what follows, we show that Cayley's theorem can be obtained as an easy consequence of Yoneda's lemma. Indeed, recall that any group $(G, \cdot)$ can be made into a category $\mathcal{G}$ with one object, say $\star$, and $\mathrm{Hom}_{\mathcal{G}}(\star, \star) = G$ as in Example 1.2.2, (3). Now Corollary 1.10.4, (2) of Yoneda's lemma gives a set bijection:

$$\mathcal{Y}_{\star,\star}: \mathrm{Hom}_{\mathcal{G}}(\star, \star) \to \mathrm{Nat}\big(\mathrm{Hom}_{\mathcal{G}}(-, \star), \mathrm{Hom}_{\mathcal{G}}(-, \star)\big)$$

$$g \mapsto t^g, \text{ where } (t^g)_\star(x) = g \cdot x \text{ for all } g, x \in G.$$

Note that $(t^g)_\star: G \to G$ is a bijective map or, equivalently, a permutation on the set G. Furthermore, we have a group structure on the natural transformations $\{t^g \mid g \in$

---

[22] Note that $\mathrm{Hom}_C(-, C): C^{op} \to$ **Set** is a covariant functor and therefore it is an object of the category of presheaves on C.

$G\}$ whose multiplication is given as follows: $t^{g_1} \circ t^{g_2} = t^{g_1 \cdot g_2}$ for all $g_1, g_2 \in G$. With this multiplication, the set bijection $\mathcal{Y}_{*,*}$ becomes a group homomorphism. Indeed, for all g_1, g_2 we have

$$\mathcal{Y}_{*,*}(g_1 \cdot g_2) = t^{g_1 \cdot g_2} = t^{g_1} \circ t^{g_2},$$

as desired. Therefore Yoneda's lemma gives a bijection between G and some subgroup of permutations on G, as stated in Cayley's theorem. $\square$

The influence of Yoneda's lemma in various areas of mathematics can hardly be exaggerated. As we cannot possibly do justice to the vast and far reaching applications of Yoneda's lemma without exceeding the scope of this introductory book, we only provide here a few scattered examples with pointers to the literature for details. Among the areas where Yoneda's lemma is a key ingredient in proving fundamental results we mention algebraic geometry (e.g., in showing that the category of all solution functors of a system of polynomial equations is equivalent to the category of relative schemes that are of finite presentation over the base field [11, Part B]), the theory of locally presentable and accesible categories (e.g., in showing that given a small category C, every hom-functor can be seen as a finitely presentable object in the corresponding category of presheaves [1]), topos theory (e.g., in showing the existence of a subobject classifier for an arbitrary category of presheaves [36, Chapter I, §4]). Further applications of Yoneda's lemma will be highlighted in Example 2.7.4 in connection to the (co)completeness of the category of presheaves on an arbitrary small category.

1.11 Exercises

1.1 Is **Grp** a subcategory of **Set**?

1.2 If $A \in \mathrm{Ob}\,C$ is an initial object and $B \in \mathrm{Ob}\,C$ is such that A and B are isomorphic then B is also an initial object in C.

1.3 Consider the category $\mathcal{P}$ whose objects are triples (X, x, f), where $X \in \mathrm{Ob}\,\mathbf{Set}$, $x \in X$, $f: X \to X$ is a function, and a morphism in $\mathcal{P}$ between two objects (X, x, f) and (Y, y, g) is a function $\theta: X \to Y$ such that $\theta(x) = y$ which renders the following diagram commutative:

$$
\begin{array}{ccc}
X & \xrightarrow{\theta} & Y \\
f \downarrow & & \downarrow g \\
X & \xrightarrow{\theta} & Y
\end{array}
\qquad g \circ \theta = \theta \circ f.
$$

Show that $\mathcal{P}$ has an initial object.

1.4 Let X be an object in C.

    (a) If X is an initial object then any monomorphism $m \in \text{Hom}_C(C, X)$ is an isomorphism.

    (b) If X is a final object then any epimorphism $m \in \text{Hom}_C(X, C)$ is an isomorphism.

1.5 Let C be a category and $f \in \text{Hom}_C(A, B)$, $g \in \text{Hom}_C(B, C)$. Prove that:

    (a) if both f and g are monomorphisms (epimorphisms) then $g \circ f$ is a monomorphism (epimorphism);

    (b) if the composition $g \circ f$ is a monomorphism then f is a monomorphism;

    (c) if the composition $g \circ f$ is an epimorphism then g is an epimorphism.

1.6 A morphism $f \in \text{Hom}_C(A, B)$ is called a *split monomorphism* if there exists some $t \in \text{Hom}_C(B, A)$ such that $t \circ f = 1_A$ (i.e., f is left invertible). Dually, a morphism f in C is called a *split epimorphism* if f^{op} is a split monomorphism in C^{op}.

    (a) Show that any split monomorphism (resp. split epimorphism) is a monomorphism (resp. epimorphism) and prove by a counterexample that the converse does not hold.

    (b) Show that f is both an epimorphism and a split monomorphism if and only if f is an isomorphism. Dually, f is both a monomorphism and a split epimorphism if and only if f is an isomorphism.

1.7 An epimorphism $f \in \text{Hom}_C(A, B)$ is called a *strong epimorphism* if for any commutative square:

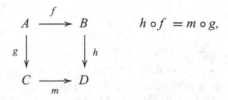

where $m \in \text{Hom}_C(C, D)$ is a monomorphism, there exists a unique $v \in \text{Hom}_C(B, C)$ which gives rise to two commutative triangles, i.e., $m \circ v = h$ and $v \circ f = g$. Dually, a morphism f in C is called a *strong monomorphism* if f^{op} is a strong epimorphism in C^{op}. Show that the following are equivalent:

    (a) f is an isomorphism;

    (b) f a strong epimorphism and a monomorphism;

    (c) f a strong monomorphism and an epimorphism.

1.8 Prove that any group regarded as a one-object category is isomorphic to its opposite. Is the assertion still true for monoids?

1.9 Give examples to show that:

(a) arbitrary functors do not necessarily preserve or reflect monomorphisms and epimorphisms;

(b) functors which reflect isomorphisms are not necessarily fully faithful.

1.10 Construct an example of:

(a) a full functor which is not surjective on objects/morphisms;

(b) a faithful functor which is not injective on objects/morphisms.

1.11 Is there a functor $F:$ **Grp** $\to$ **Grp** such that $F(G)$ is equal to the center of G for all groups G?

1.12 Is there a functor $F:$ **Top** $\to$ **Set** such that $F(X)$ is equal to the set of connected components of X for all topological spaces X?

1.13 Let $F: C \to \mathcal{D}$ and $G: \mathcal{D} \to C$ be functors such that GF is naturally isomorphic to 1_C. Prove that F is faithful.

1.14 Let $F: C \to \mathcal{D}$ and $G: \mathcal{D} \to C$ be two functors.

(a) If GF is faithful then F is faithful.

(b) If GF is full and G is faithful then F is full.

1.15 Let I be a small category and $F: C \to \mathcal{D}$ a fully faithful functor. If $F_\star:$ Fun(I, C) $\to$ Fun(I, $\mathcal{D}$) is the induced functor defined in (1.36), show that:

(a) $F_\star$ is fully faithful;

(b) if $G, H: I \to C$ are two functors such that $F_\star(G)$ and $F_\star(H)$ are naturally isomorphic, then G and H are naturally isomorphic.

1.16 Let $U:$ **Mon** $\to$ **Set** be the forgetful functor and define $U \times U:$ **Mon** $\to$ **Set** as follows:

$(U \times U)(M) = U(M) \times U(M)$ for all $M \in$ Ob **Mon**;

$(U \times U)(f)(m, m') = (f(m), f(m'))$ for all $f \in$ Hom$_{\mathbf{Mon}}(M, N)$, m, $m' \in U(M)$.

Prove that $\gamma: U \times U \to U$, defined by $\gamma_M = m_M$ for all $M \in$ Ob **Mon**, is a natural transformation, where m_M denotes the multiplication of the monoid M.

1.17 Prove the *interchange law between vertical and horizontal composition of natural transformations*, i.e., $(\delta * \gamma) \circ (\beta * \alpha) = (\delta \circ \beta) * (\gamma \circ \alpha)$, where the natural transformations involved fit into the following picture:

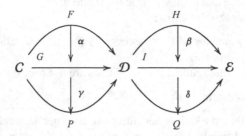

1.18 Let G be a group and let $\mathcal{G}$ be the corresponding category. Describe $\mathrm{Nat}(1_{\mathcal{G}},\, 1_{\mathcal{G}})$.

1.19 Let $(X,\, \leqslant)$ and $(Y,\, \ll)$ be two pre-ordered sets and $\mathrm{PO}(X,\, \leqslant)$, respectively $\mathrm{PO}(Y,\, \ll)$, the associated categories.

   (a) Describe the functors between these two categories. If $F,\, G\colon \mathrm{PO}(X,\, \leqslant) \to \mathrm{PO}(Y,\, \ll)$ are two such functors, describe $\mathrm{Nat}(F,\, G)$.
   (b) Show that $\mathrm{Fun}\big(\mathrm{PO}(X,\, \leqslant),\, \mathrm{PO}(Y,\, \ll)\big)$ is isomorphic to a category associated to a pre-ordered set.

1.20 Let V be a finite-dimensional vector space. Show that the functors $\mathrm{Hom}_{K}\mathcal{M}(V,\, -)$, $V^{*} \otimes - \colon {}_{K}\mathcal{M} \to {}_{K}\mathcal{M}$, are naturally isomorphic.

1.21 Prove that for any category C there exists a functor $F\colon C \to \mathbf{Set}$ which is not representable.

1.22 Prove that the contravariant power-set functor $P^{c}\colon \mathbf{Set} \to \mathbf{Set}$ is representable.

1.23 Let $G \in \mathrm{Ob}\,\mathbf{Grp}$. Prove that the functor $F_{G}\colon \mathbf{Grp} \to \mathbf{Set}$ defined as follows for any $H \in \mathrm{Ob}\,\mathbf{Grp}$, $f \in \mathrm{Hom}_{\mathbf{Grp}}(G,\, H)$:

$$F_{G}(H) = \{f \in \mathrm{Hom}_{\mathbf{Set}}(G,\, H) \mid f \text{ is an antimorphism of groups}\},$$

$$F(f)(g) = f \circ g,\ g \in F_{G}(H)$$

is representable.

1.24 An object I of a category C is called *injective* if for any $u \in \mathrm{Hom}_{C}(A,\, I)$ and any monomorphism $m \in \mathrm{Hom}_{C}(A,\, B)$ there exists a morphism $v \in \mathrm{Hom}_{C}(B,\, I)$ such that the following diagram is commutative:

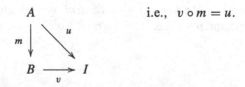

i.e., $v \circ m = u$.

Dually, an object P of C is called *projective* if it is an injective object in C^{op}.
Describe the injective (resp. projective) objects in **Ab** and $_R\mathcal{M}$.

1.25 Let I, J be two small categories and C an arbitrary category. Prove that the
functor categories $\mathrm{Fun}\big(I, \mathrm{Fun}(J, C)\big)$ and $\mathrm{Fun}(I \times J, C)$ are isomorphic.

Chapter 2
Limits and Colimits

Limits and colimits are fundamental and unifying concepts in category theory. Many seemingly unrelated constructions from different fields of mathematics such as the free product of groups (with amalgamation), the tensor product of (co)algebras or the direct sum of modules are in fact special instances of these very general concepts.

2.1 (Co)products, (Co)equalizers, Pullbacks and Pushouts

In order to achieve a better understanding of (co)limits we will introduce them gradually, starting with some generic special cases, namely (co)products, (co)equalizers, pullbacks and pushouts.

Definition 2.1.1 Let I be a set and $(P_i)_{i \in I}$ a family of objects in a category $\mathcal{C}$. A *product* of the family $(P_i)_{i \in I}$ is a pair $(\prod_{i \in I} P_i, (p_i)_{i \in I})$, where

(1) $\prod_{i \in I} P_i \in \mathrm{Ob}\,\mathcal{C}$;
(2) $p_j \colon \prod_{i \in I} P_i \to P_j$ are morphisms in $\mathcal{C}$ for all $j \in I$,

and for any other pair $(P, (f_i)_{i \in I})$ where

(1) $P \in \mathrm{Ob}\,\mathcal{C}$;
(2) $f_j \colon P \to P_j$ are morphisms in $\mathcal{C}$ for all $j \in I$,

© The Author(s), under exclusive license to Springer Nature Switzerland AG 2023
A. Agore, *A First Course in Category Theory*, Universitext,
https://doi.org/10.1007/978-3-031-42899-9_2

there exists a unique morphism $f : P \to \prod_{i \in I} P_i$ in C such that the following diagram commutes for all $j \in I$:

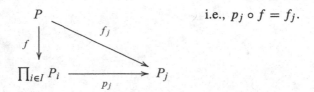

We say that C is a *category with (finite) products* or that C *has (finite) products* if there exists a product in C for any (finite) family of objects. If I is a finite set, then we also write $P_1 \times \ldots \times P_n$ instead of $\prod_{i=\overline{1,n}} P_i$.

Coproducts are the dual notion of products; that is, the coproduct of a family $(X_i)_{i \in I}$ of objects of a category C is defined to be the product of the same family of objects in the dual category C^{op}. This comes down to the following:

Definition 2.1.2 Let I be a set and $(Q_i)_{i \in I}$ a family of objects in a category C. A *coproduct* of the family $(Q_i)_{i \in I}$ is a pair $\left(\coprod_{i \in I} Q_i, (q_i)_{i \in I} \right)$, where

(1) $\coprod_{i \in I} Q_i \in \mathrm{Ob}\, C$;
(2) $q_j : Q_j \to \coprod_{i \in I} Q_i$ are morphisms in C for all $j \in I$,

and for any other pair $\left(Q, (f_i)_{i \in I} \right)$, where

(1) $Q \in \mathrm{Ob}\, C$;
(2) $f_j : Q_j \to Q$ are morphisms in C for all $j \in I$,

there exists a unique morphism $f : \coprod_{i \in I} Q_i \to Q$ in C such that the following diagram commutes for all $j \in I$:

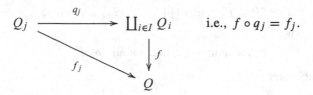

We say that C is a *category with (finite) coproducts* or that C *has (finite) coproducts* if there is a coproduct in C for any (finite) family of objects.

Proposition 2.1.3 *When it exists, the (co)product of a family of objects is unique up to isomorphism.*

Proof We start by proving the uniqueness up to isomorphism of the product. Let C be a category and consider two products $(P, (p_i)_{i \in I})$, respectively $(\overline{P}, (\overline{p}_i)_{i \in I})$, in C of the same family of objects $(P_i)_{i \in I}$. Since $(\overline{P}, (\overline{p}_i)_{i \in I})$ is assumed to be a product, there exists a unique $f \in \mathrm{Hom}_C(P, \overline{P})$ such that for any $j \in I$ we

have

$$\overline{p}_j \circ f = p_j. \tag{2.1}$$

Similarly, as $\big(P, (p_i)_{i \in I}\big)$ is also a product of the same family of objects, we obtain a unique $g \in \mathrm{Hom}_{\mathcal{C}}(\overline{P}, P)$ such that for any $j \in I$ we have

$$p_j \circ g = \overline{p}_j. \tag{2.2}$$

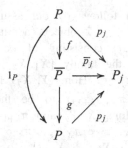

By composing the equality in (2.2) with f on the right and using (2.1) we obtain

$$p_j \circ (g \circ f) = \overline{p}_j \circ f \overset{(2.1)}{=} p_j \tag{2.3}$$

for any $j \in I$. Applying Definition 2.1.1 to the pair $\big(P, (p_i)_{i \in I}\big)$, seen both as a product and as the other pair, yields a unique $h \in \mathrm{Hom}_{\mathcal{C}}(P, P)$ such that $p_j \circ h = p_j$ for any $j \in I$. By the uniqueness of h we must have $h = 1_P$. Moreover, since by (2.3) the map $g \circ f$ also fulfills the above identity we obtain $g \circ f = 1_P$. In the same manner we obtain $f \circ g = 1_{\overline{P}}$ and therefore P and $\overline{P}$ are isomorphic.

The assertion regarding coproducts follows by applying the duality principle. Indeed, assume now that $\big(Q, (q_i)_{i \in I}\big)$ and $\big(\overline{Q}, (\overline{q}_i)_{i \in I}\big)$ are coproducts in $\mathcal{C}$ of the same family of objects $\big(Q_i\big)_{i \in I}$. Then $\big(Q, (q_i^{op})_{i \in I}\big)$ and $\big(\overline{Q}, (\overline{q}_i^{op})_{i \in I}\big)$ are both products in $\mathcal{C}^{op}$ of the family of objects $\big(Q_i\big)_{i \in I}$. According to the above proof, there exists an isomorphism $f^{op} \in \mathrm{Hom}_{\mathcal{C}^{op}}(Q, \overline{Q})$. Therefore, we have an isomorphism $f \in \mathrm{Hom}_{\mathcal{C}}(\overline{Q}, Q)$. $\square$

Proposition 2.1.4 *Let $\mathcal{C}$ be a category such that any two objects admit a (co)product. Then, any non-empty finite family of objects in $\mathcal{C}$ admits a (co)product.*

Proof Let $\{X_1, X_2, \ldots, X_n\}$ be a non-empty family of objects in $\mathcal{C}$. We use induction on n to construct the product of this family. If $n = 1$ then the family $\{X\}$ has a product given by the pair $(X, 1_X)$. Indeed, if $\big(Y, f \in \mathrm{Hom}_{\mathcal{C}}(Y, X)\big)$

then the unique morphism which makes the following diagram commutative is precisely f:

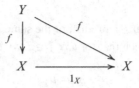

For $n = 2$ the conclusion follows by our assumption. Assume now that any family with at most $k \geq 2$ objects admits a product and consider the family $\{X_1, X_2, \ldots, X_{k+1}\}$ consisting of $(k + 1)$ objects. According to the inductive hypothesis, the family $\{X_1, X_2, \ldots, X_k\}$ admits a product, say $\left(Y, (\pi_j)_{j=\overline{1,k}}\right)$, where $\pi_j \in \operatorname{Hom}_{\mathcal{C}}(Y, X_j)$ for all $j = 1, 2, \ldots, k$. Furthermore, consider $\left(X, (\pi_X, \pi_{k+1})\right)$ to be the product of the family $\{Y, X_{k+1}\}$, where $\pi_X \in \operatorname{Hom}_{\mathcal{C}}(X, Y)$ and $\pi_{k+1} \in \operatorname{Hom}_{\mathcal{C}}(X, X_{k+1})$. We will show that $\left(X, (\pi_1 \circ \pi_X, \pi_2 \circ \pi_X, \ldots, \pi_k \circ \pi_X, \pi_{k+1})\right)$ is the product of the family of objects $\{X_1, X_2, \ldots, X_{k+1}\}$. To this end, let $Z \in \operatorname{Ob}\mathcal{C}$ and $f_j \in \operatorname{Hom}_{\mathcal{C}}(Z, X_j)$, $j = 1, 2, \ldots, (k + 1)$. As $\left(Y, (\pi_j)_{j=\overline{1,k}}\right)$ is the product of the family of objects $\{X_1, X_2, \ldots, X_k\}$, there exists a unique $f \in \operatorname{Hom}_{\mathcal{C}}(Z, Y)$ such that the following diagram is commutative for all $i = 1, 2, \ldots, k$:

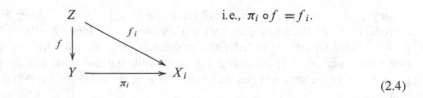

$$\text{i.e., } \pi_i \circ f = f_i.$$

$$(2.4)$$

Moreover, as $\left(X, (\pi_X, \pi_{k+1})\right)$ is the product of the family $\{Y, X_{k+1}\}$, there exists a unique $g \in \operatorname{Hom}_{\mathcal{C}}(Z, X)$ such that the following diagram is commutative:

$$Y \xleftarrow{\pi_X} X \xrightarrow{\pi_{k+1}} X_{k+1} \qquad \text{i.e., } \pi_X \circ g = f \text{ and } \pi_{k+1} \circ g = f_{k+1}.$$

(with $g: Z \to X$, $f: Z \to Y$, $f_{k+1}: Z \to X_{k+1}$)

$$(2.5)$$

Note that $g \in \operatorname{Hom}_{\mathcal{C}}(Z, X)$ is a morphism such that $\pi_{k+1} \circ g = f_{k+1}$ and $(\pi_j \circ \pi_X) \circ g = \pi_j \circ (\pi_X \circ g) \overset{(2.5)}{=} \pi_j \circ f \overset{(2.4)}{=} f_j$ for all $j = 1, 2, \ldots, k$. The proof will be finished once we show that g is the unique morphism with these properties. Indeed, assume there exists a $g' \in \text{Hom}_{\mathcal{C}}(Z, X)$ such that

$$\pi_{k+1} \circ g' = f_{k+1} \quad \text{and} \quad (\pi_j \circ \pi_X) \circ g' = f_j, \text{ for all } j = 1, 2, \ldots, k.$$

In particular, we have $\pi_j \circ (\pi_X \circ g') = f_j$ for all $j = 1, 2, \ldots, k$ and by the uniqueness of the morphism which makes (2.4) commutative we obtain $f = \pi_X \circ g'$. Putting everything together, we obtain $\pi_{k+1} \circ g' = f_{k+1}$ and $\pi_X \circ g' = f$. As g is the unique morphism for which (2.5) holds it follows that $g = g'$, as desired.

The claim concerning coproducts follows by duality. Indeed, assume that $\mathcal{C}$ is a category such that any two objects admit a coproduct. Then, in $\mathcal{C}^{op}$ any two objects admit a product and, according to the above proof, any non-empty finite family of objects in $\mathcal{C}^{op}$ admits a product. Therefore, any non-empty finite family of objects in $\mathcal{C}$ admits a coproduct, as desired. □

Examples 2.1.5

(1) The product of any family $(X_i)_{i \in I}$ of objects in **Set**, where I is a set, is given by the corresponding cartesian product $\prod_{i \in I} X_i$ together with the canonical projections $(\pi_j \colon \prod_{i \in I} X_i \to X_j)_{j \in I}$. More precisely, we have

$$\prod_{i \in I} X_i = \{(x_i)_{i \in I} \mid x_i \in X_i\}, \quad \pi_j \colon \prod_{i \in I} X_i \to X_j,$$

$$\pi_j((x_i)_{i \in I}) = x_j \text{ for all } j \in I.$$

Indeed, consider $X \in \text{Ob Set}$ and a family of functions $p_i \colon X \to X_i, i \in I$. We will show that there exists a unique map $f \colon X \to \prod_{i \in I} X_i$ such that $\pi_i \circ f = p_i$ for all $i \in I$. To this end, define $f \colon X \to \prod_{i \in I} X_i$ by $f(x) = (p_i(x))_{i \in I}$ for all $x \in X$. Then, for all $i \in I$ and $x \in X$ we obviously have $\pi_i \circ f(x) = p_i(x)$. Consider now $g \colon X \to \prod_{i \in I} X_i$ to be another map such that $\pi_i \circ g = p_i$ for all $i \in I$. As π_i is just the projection on the i-th component, and $\pi_i \circ g(x) = p_i(x)$ for all $i \in I$ and all $x \in X$, we obtain $f = g$, as desired.

(2) The product of any family of objects in **Grp**, **Ab**, $_R\mathcal{M}$, **Ring** is given by the cartesian product of the underlying sets endowed with componentwise operations together with the canonical projections. For instance, in the category **Grp** for any family $(G_i)_{i \in I}$ of objects, the group structure on the cartesian product $\prod_{i \in I} G_i$ is defined as follows:

$$((g_i)_{i \in I}) \cdot ((h_i)_{i \in I}) = ((g_i \cdot_i h_i)_{i \in I}),$$

where $\cdot_i$ denotes the group multiplication in G_i. It can now be easily checked as in the previous example that this is indeed the product in **Grp** of the family $(G_i)_{i \in I}$.

(3) The product of any family of objects in **Top** is given by the cartesian product of the underlying sets endowed with the product topology together with the canonical projections. Let I be a set, $(X_i)_{i \in I}$ a family of topological spaces and $\left(\prod_{i \in I} X_i, (\pi_i)_{i \in I}\right)$ the product of the underlying sets. Then, each π_j is obviously continuous as the product topology is the coarsest topology on $\prod_{i \in I} X_i$ for which all projections $\pi_j : \prod_{i \in I} X_i \to X_j$ are continuous. Assume now that $X \in \mathrm{Ob}\,\mathbf{Top}$ and $p_j \in \mathrm{Hom}_{\mathbf{Top}}(X, X_j)$, $j \in I$, is a family of continuous maps. It can be easily seen that the unique map $f : X \to \prod_{i \in I} X_i$ defined in Example 2.1.5, (1) is continuous too. Indeed, recall that the product topology is generated by the sub-basis $\mathcal{S} = \bigcup_{j \in I} \mathcal{S}_j$, where $\mathcal{S}_j = \{\pi_j^{-1}(U_j) \mid U_j \text{ open in } X_j\}$ for all $j \in J$, and therefore in order to prove continuity of f it will suffice to show that the inverse image of each sub-basis element of $\prod_{i \in I} X_i$ is open [39, p. 103]. To this end, for all $j \in I$ we have $f^{-1}\!\left(\pi_j^{-1}(U_j)\right) = (\pi_j \circ f)^{-1}(U_j) = p_j^{-1}(U_j)$ and since p_j is continuous, the latter term is an open set, as desired.

(4) Products in **Haus** and **KHaus**, the categories of Hausdorff and compact Hausdorff spaces respectively, are constructed as in **Top**. Indeed, it will suffice to show that given a family of Hausdorff (resp. compact Hausdorff) spaces $(X_i)_{i \in I}$, where I is a set, the product of the underlying spaces together with the product topology defined in the previous examples is a Hausdorff (resp. compact Hausdorff) space. To this end, let $x = (x_i)_{i \in I}$, $y = (y_i)_{i \in I} \in \prod_{i \in I} X_i$ such that $x \neq y$. Thus, there exists a $t \in I$ such that $x_t \neq y_t$ and since X_t is a Hausdorff space we can find two disjoint open spaces in X_t such that $x_t \in U$ and $y_t \in V$. As $\pi_t : \prod_{i \in I} X_i \to X_t$ is a continuous map, both $\pi_t^{-1}(U)$ and $\pi_t^{-1}(V)$ are open subsets in $\prod_{i \in I} X_i$ such that $x \in \pi_t^{-1}(U)$ and $y \in \pi_t^{-1}(V)$. Moreover, we have $\pi_t^{-1}(U) \cap \pi_t^{-1}(V) = \pi_t^{-1}(U \cap V) = \emptyset$ ([37, Exercise 1.5]). To conclude, we have found two disjoint neighborhoods $\pi_t^{-1}(U)$ and $\pi_t^{-1}(V)$ of x and y respectively, which shows that $\prod_{i \in I} X_i$ is indeed a Hausdorff space.

Furthermore, if $(X_i)_{i \in I}$ is a family of compact Hausdorff spaces, Tychonoff's theorem shows that the product topology $\prod_{i \in I} X_i$ is a compact space as well [39, Theorem 37.3].

(5) Let $\mathrm{PO}(X, \leqslant)$ be the category corresponding to the pre-ordered set $(X, \leqslant)$ (as defined in Example 1.2.2, (2)) and $(x_i)_{i \in I}$ a family of objects in $\mathrm{PO}(X, \leqslant)$ indexed by a set I, i.e., $x_i \in X$ for any $i \in I$. If it exists, the product of this family is a pair $\left(p, (\pi_i)_{i \in I}\right)$, where $p \in X$ and $\pi_i : p \to x_i$ are morphisms in $\mathrm{PO}(X, \leqslant)$. This comes down to $p \leqslant x_i$ for any $i \in I$. Moreover, for any other pair $\left(q, (u_i)_{i \in I}\right)$, where $q \in X$ and $u_i : q \to x_i$ are morphisms in $\mathrm{PO}(X, \leqslant)$, there exists a morphism $f : q \to p$. In other words, for any $q \in X$ such that $q \leqslant x_i$ for any $i \in I$, we also have $q \leqslant p$. Therefore, p is precisely the infimum (if it exists) of the family $(x_i)_{i \in I}$.

(6) The product of categories as defined in Definition 1.4.4 together with the corresponding projection functors (Example 1.5.3, (6)) is the product in the category **Cat** of small categories (Proposition 1.5.4). Indeed, if $(\mathcal{C}_i)_{i \in I}$ is a family of small categories indexed by the set I, then $\left(\prod_{i \in I} \mathcal{C}_i, (p_{\mathcal{C}_i})_{i \in I}\right)$ is the

product of this family in **Cat**. It can be easily seen that for any small category $\mathcal{D}$ and any family of functors $U_{\mathcal{C}_i} : \mathcal{D} \to \mathcal{C}_i$ there exists a unique functor $F : \mathcal{D} \to \prod_{i \in I} \mathcal{C}_i$ such that $p_{\mathcal{C}_i} \circ F = U_{\mathcal{C}_i}$ for all $i \in I$; more precisely, F is defined as follows for all $D \in \mathrm{Ob}\,\mathcal{D}$ and $f \in \mathrm{Hom}_{\mathcal{D}}(D, D')$:

$$F(D) = \left(U_{\mathcal{C}_i}(D)\right)_{i \in I}, \quad F(f) = \left(U_{\mathcal{C}_i}(f)\right)_{i \in I}.$$

(7) The category **SiGrp** of simple groups does not admit products or coproducts. Indeed, suppose this category admits products and let H, K be simple groups. Let $\left(X, (p, q)\right)$ be the product in **SiGrp** of H and K. In particular, X is a simple group and $p : X \to H, q : X \to K$ are group homomorphisms. Consider now the pair $\left(H, (\mathrm{Id}_H, 0_K)\right)$ where Id_H is the identity homomorphism on H while $0_K : H \to K$ denotes the group homomorphism defined by $0_K(h) = 1_K$ for all $h \in H$. By Definition 2.1.1 there exists a unique homomorphism of groups $f : H \to X$ such that the following diagram is commutative:

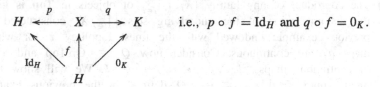

i.e., $p \circ f = \mathrm{Id}_H$ and $q \circ f = 0_K$.

From $p \circ f = \mathrm{Id}_H$ it follows that f is injective and $q \circ f = 0_K$ implies that $\mathrm{Im}(f) \subseteq \ker(q) \subseteq X$. Putting all this together we have $\{1\} \neq H \cong \mathrm{Im}(f) \subseteq \ker(q) \trianglelefteq X$ and since $\ker(q)$ is a normal subgroup of the simple group X, we must have $\ker(q) = X$, i.e., $q = 0_K$. Next, we consider the pair $\left(K, (0_H, \mathrm{Id}_K)\right)$ where Id_K is the identity homomorphism on K while $0_H : K \to H$ denotes the group homomorphism defined by $0_H(k) = 1_H$ for all $k \in K$. Using again Definition 2.1.1 yields a unique homomorphism of groups $g : K \to X$ such that the following diagram is commutative:

$$H \xleftarrow{\ p\ } X \xrightarrow{\ q\ } K \qquad \text{i.e., } p \circ g = 0_H \text{ and } q \circ g = \mathrm{Id}_K.$$

$$\begin{array}{c} H \xleftarrow{\ p\ } X \xrightarrow{\ q\ } K \\[2pt] {}_{0_H}\nwarrow \ \uparrow{\scriptstyle g}\ \nearrow{}_{\mathrm{Id}_K} \\[2pt] K \end{array}$$

Since $q = 0_K$ the last equality gives $K = \{1\}$, which is a contradiction. A similar argument shows that **SiGrp** does not have coproducts either.

(8) The category **Field** does not have products or coproducts. Indeed, consider the fields $\mathbb{Z}_2$ and $\mathbb{Z}_3$ of integers modulo 2, respectively 3 and let $(K, (i, j))$ be their coproduct in **Field**, where $i : \mathbb{Z}_2 \to K$ and $j : \mathbb{Z}_3 \to K$ are morphisms of fields. Thus, in K we have both $1 + 1 = 0$ and $1 + 1 + 1 = 0$, which yields $1 = 0$, an obvious contradiction. Similarly one can prove that **Field** does not have products either.

(9) In **Set**, the coproduct of a family $(X_i)_{i \in I}$ is just its disjoint union, i.e., the union of the sets $X_i' = X_i \times \{i\}$. Thus, the coproduct of the family $(X_i)_{i \in I}$ is the pair $\left(\coprod_{i \in I} X_i, (q_i)_{i \in I} \right)$, where $\coprod_{i \in I} X_i = \{(x, i) \mid i \in I, x \in X_i\}$ and $q_j \colon X_j \to \coprod_{i \in I} X_i$, $q_j(x) = (x, j)$ for all $j \in I$. Indeed, given a set Q together with a collection of maps $f_j \colon X_j \to Q$, $j \in I$, define $f \colon \coprod_{i \in I} X_i \to Q$ by considering $f((x, j)) = f_j(x)$ for all $(x, j) \in X_j' \subset \coprod_{i \in I} X_i$. Since each (x, j) lies inside a unique copy of X_j', the following map is well-defined:

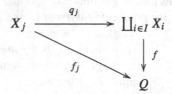

(10) The coproduct of any family $(X_i, \tau_i)_{i \in I}$ of objects in **Top** is given by the disjoint union of the underlying sets $\coprod_{i \in I} X_i$ constructed in the previous example endowed with the finest topology τ for which all maps q_j are continuous. Consider now $Q \in \mathrm{Ob}\,\mathbf{Top}$ and a family of continuous maps $f_j \colon X_j \to Q$, $j \in I$. We will show that the unique map $f \colon \coprod_{i \in I} X_i \to Q$ defined in the previous example is continuous. To this end, let $U \subseteq Q$ be an open set. Note that since τ is the finest topology for which all q_j are continuous, in order to show that $f^{-1}(U)$ is an open set it will suffice to prove that $q_j^{-1}(f^{-1}(U))$ is open for all $j \in I$. Indeed, we have $q_j^{-1}(f^{-1}(U)) = (f \circ q_j)^{-1}(U) = f_j^{-1}(U)$ and since each f_j is continuous, the desired conclusion follows.

(11) For certain categories, such as **Grp**, the coproducts are more complicated than the products and the constructions do not rely on the ones performed in **Set**. This is basically because unions do not usually preserve operations (for instance, the union of an arbitrary family of groups is not necessarily a group). In group theory, the construction which gives the coproducts is called the *free product of groups* (see [50, Chapter 11, p. 388]). Let $(G_i)_{i \in I}$ be a family of groups, where I is a set, and denote by $\left(*_{i \in I} G_i, (j_k \colon G_k \to *_{i \in I} G_i,)_{k \in I} \right)$ the corresponding free product. Consider now $G \in \mathrm{Ob}\,\mathbf{Grp}$ and a family of group homomorphisms $f_k \colon G_k \to G$, $k \in I$. The definition of the free product yields a unique group homomorphism $f \colon *_{i \in I} G_i \to G$ such that $f \circ j_k = f_k$ for all $k \in I$. Therefore, $\left(*_{i \in I} G_i, (j_k \colon G_k \to *_{i \in I} G_i,)_{k \in I} \right)$ is the coproduct of the family $(G_i)_{i \in I}$ in **Grp**.

(12) In the category **Ab** of abelian groups the coproduct is given by the direct sum with componentwise multiplication law. More precisely, for any family

$(A_i)_{i \in I}$ of abelian groups we have

$$\coprod_{i \in I} A_i = \{(a_i)_{i \in I} \mid a_i \in A_i, \ \{i \mid a_i \neq 0\} \text{ is finite}\},$$

$$q_{i_0} : A_{i_0} \to \coprod_{i \in I} A_i, \quad q_{i_0}(a) = (a_i)_{i \in I},$$

where $a_{i_0} = a$ and $a_j = 0$ for all $j \neq i_0$. Indeed, consider the pair $\big(H, (f_i)_{i \in I}\big)$, where H is an abelian group and $f_j : A_j \to H$ are group homomorphisms for all $j \in I$. Then, the unique homomorphism of groups $f : \coprod_{i \in I} A_i \to H$ which makes the following diagram commutative for all $j \in I$:

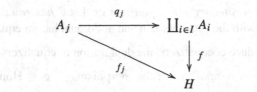

is given by $f\big((a_i)_{i \in I}\big) = \sum_{i \in I} f_i(a_i)$.[1] Suppose now that $g : \coprod_{i \in I} A_i \to H$ is another group homomorphism such that $g \circ q_j = f_j$ for all $j \in I$. Then $(f - g) \circ q_j$ is the zero map from A_j to H and thus the image of q_j is contained in $\ker(f - g)$ for all $j \in I$. Now observe that any element in $\coprod_{i \in I} A_i$ is a sum of finitely many elements of the form $q_j(a_j)$. Therefore, since $\ker(f - g)$ is a subgroup in $\coprod_{i \in I} A_i$, we obtain that $\ker(f - g) = \coprod_{i \in I} A_i$, i.e., $f = g$ and the proof is now finished.

An analogous description of coproducts holds for the category ${}_R \mathcal{M}$ of modules over a ring R.

(13) Let $\mathrm{PO}(X, \leqslant)$ be the category corresponding to the pre-ordered set $(X, \leqslant)$. Then, the coproduct of a family $(x_i)_{i \in I}$, if it exists, is just its supremum. $\square$

Another important example of a limit is an equalizer.

Definition 2.1.6 An *equalizer* of the morphisms $f, g \in \mathrm{Hom}_{\mathcal{C}}(X, Y)$ is a pair (E, p), where

(1) $E \in \mathrm{Ob}\,\mathcal{C}$;
(2) $p \in \mathrm{Hom}_{\mathcal{C}}(E, X)$ such that $f \circ p = g \circ p$,

---

[1] Note that the sum in the right-hand side contains only finitely many non-zero terms.

and for any other pair (E', p'), where

(1) $E' \in \mathrm{Ob}\,\mathcal{C}$;
(2) $p' \in \mathrm{Hom}_{\mathcal{C}}(E', X)$ such that $f \circ p' = g \circ p'$,

there exists a unique $u \in \mathrm{Hom}_{\mathcal{C}}(E', E)$ which makes the following diagram commute:

$$E \xrightarrow{\ p\ } X \underset{g}{\overset{f}{\rightrightarrows}} Y \qquad \text{i.e., } p \circ u = p'.$$

$$\underset{E'}{\overset{u\ \uparrow\quad \nearrow\ p'}{}}$$

We say that $\mathcal{C}$ is a *category with equalizers* or that $\mathcal{C}$ *has equalizers* if any pair of morphisms in $\mathcal{C}$ with the same domain and codomain has an equalizer.

Next we introduce coequalizers, the dual notion of equalizers.

Definition 2.1.7 A *coequalizer* of the morphisms $f, g \in \mathrm{Hom}_{\mathcal{C}}(X, Y)$ is a pair (Q, q), where

(1) $Q \in \mathrm{Ob}\,\mathcal{C}$;
(2) $q \in \mathrm{Hom}_{\mathcal{C}}(Y, Q)$ such that $q \circ f = q \circ g$,

and for any other pair (Q', q'), where

(1) $Q' \in \mathrm{Ob}\,\mathcal{C}$;
(2) $p' \in \mathrm{Hom}_{\mathcal{C}}(E', X)$ such that $f \circ p' = g \circ p'$,

there exists a unique $v \in \mathrm{Hom}_{\mathcal{C}}(Q, Q')$ which makes the following diagram commute:

$$X \underset{g}{\overset{f}{\rightrightarrows}} Y \xrightarrow{\ q\ } Q \qquad \text{i.e., } v \circ q = q'.$$

$$\underset{Q'}{\overset{\quad q'\ \searrow\quad \downarrow\ v}{}}$$

We say that $\mathcal{C}$ is a *category with coequalizers* or that $\mathcal{C}$ *has coequalizers* if any pair of morphisms in $\mathcal{C}$ with the same domain and codomain has a coequalizer.

At this point we only state the uniqueness up to isomorphism of (co)equalizers. The proof relies on the same idea used in proving Proposition 2.1.3. Furthermore, this uniqueness result will follow as a special case of Proposition 2.2.9.

Proposition 2.1.8 *When it exists, the (co)equalizer of two morphisms is unique up to isomorphism.*

Proposition 2.1.9 *If (E, p) is the equalizer (resp. coequalizer) of the pair of morphisms $f, g \in \mathrm{Hom}_{\mathcal{C}}(X, Y)$ in a category $\mathcal{C}$, then p is a monomorphism (resp. epimorphism).*

Proof Consider $h_1, h_2 \colon E' \to E$ such that $p \circ h_1 = p \circ h_2 := h$. First notice that $f \circ h = f \circ (p \circ h_1) = (f \circ p) \circ h_1 = (g \circ p) \circ h_1 = g \circ (p \circ h_1) = g \circ h$. Since p is the equalizer of f and g, there exists a unique morphism $u \colon E' \to E$ such that $p \circ u = h$.

$$
\begin{array}{ccc}
E & \xrightarrow{\ p\ } & X \begin{array}{c} \xrightarrow{\ f\ } \\ \xrightarrow[\ g\ ]{} \end{array} Y \\
{\scriptstyle u}\big\uparrow & \nearrow & \\
E' & {\scriptstyle h} &
\end{array}
$$

Now notice that both maps $h_1, h_2 \colon E' \to E$ fulfill the above equality. Due to the uniqueness of u we obtain $u = h_1 = h_2$, as desired.

For the second part we use duality. Indeed, (E, p) is the coequalizer of the pair of morphisms $f, g \colon X \to Y$ in category $\mathcal{C}$ if and only if (E, p^{op}) is the equalizer of the pair of morphisms $f^{op}, g^{op} \colon Y \to X$ in the category $\mathcal{C}^{op}$. The conclusion now follows from the first part of the proof and Proposition 1.4.3, (1). □

Examples 2.1.10

(1) In **Set** the equalizer of two functions $f, g \colon X \to Y$ is the pair (E, i), where $E = \{x \in X \mid f(x) = g(x)\}$ and $i \colon E \to X$ is the canonical inclusion. Indeed, consider $f, g \in \mathrm{Hom}_{\mathbf{Set}}(X, Y)$ and suppose $j \colon E' \to X$ is a morphism in **Set** such that $f \circ j = g \circ j$.

$$
\begin{array}{ccc}
E & \xrightarrow{\ i\ } & X \begin{array}{c} \xrightarrow{\ f\ } \\ \xrightarrow[\ g\ ]{} \end{array} Y \\
{\scriptstyle u}\big\uparrow & \nearrow & \\
E' & {\scriptstyle j} &
\end{array}
$$

Then $\mathrm{Im}(j) \subseteq E$ and the unique map $u \colon E' \to E$ such that $i \circ u = j$ is given by $u(e) = j(e)$ for all $e \in E'$.

(2) In **Grp**, **Ab**, $_R\mathcal{M}$ the underlying set of the equalizer is constructed as in the previous example. Consider for example the category **Grp** and let f, $g \in \mathrm{Hom}_{\mathbf{Grp}}(G, H)$. Then $E = \{x \in G \mid f(x) = g(x)\}$ is a subgroup of G with respect to the induced structure and the inclusion $i \colon E \to G$ is obviously a group homomorphism. Similar arguments can be used for the other two categories **Ab** and $_R\mathcal{M}$.

(3) Let $f, g \in \mathrm{Hom}_{\mathbf{Top}}(X, Y)$ and endow the set E constructed in the first example with the subspace topology.[2] This implies that the inclusion $i: E \to X$ is a continuous map and therefore (E, i) is the equalizer of the pair of morphisms (f, g) in **Top**.

(4) Equalizers in **Haus** and **KHaus** are constructed as in **Top**. Indeed, assume first that $f, g \in \mathrm{Hom}_{\mathbf{Haus}}(X, Y)$ and consider $E = \{x \in X \mid f(x) = g(x)\}$ endowed with the subspace topology. We only need to show that E is a Hausdorff space. To this end, let $x_1, x_2 \in E$ such that $x_1 \neq x_2$. As X is a Hausdorff space, we can find two disjoint neighborhoods of x_1 and x_2 in X, say U_1 and U_2 respectively. By definition of the subspace topology, the sets $U_1 \cap E$ and $U_2 \cap E$ are open in E and therefore neighborhoods in E of x_1 and x_2, respectively. Furthermore, we have $(U_1 \cap E) \cap (U_2 \cap E) = (U_1 \cap U_2) \cap E = \emptyset$, which proves that E is Hausdorff.

Consider now $u, v \in \mathrm{Hom}_{\mathbf{KHaus}}(K, H)$. We have already proved that E together with the subspace topology is Hausdorff; we are left to show that E is compact as well. It will suffice to prove that E is closed in X as any closed subspace of a compact space is compact as well [39, Theorem 26.2]. To this end, let $x_0 \in X - E$. As $f(x_0) \neq g(x_0)$ and Y is Hausdorff we can find two disjoint open sets U and V such that $f(x_0) \in U$ and $g(x_0) \in V$. Then $T = f^{-1}(U)$ and $W = g^{-1}(V)$ are open sets in X by continuity of f and g. Consequently $S = T \cap W$ is also an open set in X and $x_0 \in S$. Furthermore, S and E are disjoint; indeed, if there exists some $x \in S \cap E$ then we have $f(x) \in U$, $g(x) \in V$ and $f(x) = g(x)$, which is an obvious contradiction since this would imply $f(x) \in U \cap V = \emptyset$.

To summarize, we have proved that for each $x_0 \in X - E$ there exists an open set S such that $x_0 \in S \subset X - E$. This shows that $X - E$ is contained in the interior[3] of $X - E$, which allows us to conclude that $X - E = \mathrm{Int}(X - E)$. Hence $X - E$ is an open subset of X and therefore E is a closed set, as desired.

(5) Let $f, g \in \mathrm{Hom}_{\mathbf{Set}}(X, Y)$ be two functions. Consider $R = \{(f(x), g(x)) \mid x \in X\}$ and let $\sim_R$ be the equivalence relation on Y generated by R.[4] Then the pair $(Y/\sim_R, \pi)$ is the coequalizer of the maps f and g in **Set**, where $Y/\sim_R$ is the set of equivalence classes of Y with respect to $\sim_R$, while $\pi: Y \to Y/\sim_R$, $\pi(y) = \overline{y}$, for all $y \in Y$, is the canonical projection. To start with, for any $x \in X$, we have $\pi \circ f(x) = \overline{f(x)} = \overline{g(x)} = \pi \circ g(x)$. Assume now that $q: Y \to Q$ is another map such that $q \circ f = q \circ g$ and define $v: Y/\sim_R \to Q$ by $v(\overline{y}) = q(y)$ for all $\overline{y} \in Y/\sim_R$. First we will show that v is well-defined. To this end, let $y, y' \in Y$ such that $\overline{y} = \overline{y'}$. Then $y \sim_R y'$ and this implies that

---

[2] Let (X, τ) be a topological space and $Y \subseteq X$ a subset of X. Then $\tau_Y = \{Y \cap U \mid U \in \tau\}$ is a topology on Y called the *subspace topology* [39, p. 88].

[3] Recall that given a subset U of a topological space X, the *interior of* U, denoted by $\mathrm{Int}(U)$, is defined as the union of all open sets contained in U [39, p. 95]. Moreover, we have $\mathrm{Int}(U) \subset U$.

[4] The *equivalence relation generated by a binary relation* R on a set Y (regarded as a subset of $Y \times Y$) is defined as the intersection of all equivalence relations on Y which contain R.

there exists some positive integer n and $y_0, y_1, \ldots, y_n \in Y$ such that $y = y_0$, $y' = y_n$ and for any $i = 1, 2 \ldots, n$ we have either $y_i \sim y_{i+1}$ or $y_{i+1} \sim y_i$. Furthermore, note that if $y_i \sim y_{i+1}$ then $y_i = f(x_0)$, $y_{i+1} = g(x_0)$ for some $x_0 \in X$, which implies $q(y_i) = q \circ f(x_0) = q \circ g(x_0) = q(y_{i+1})$. Therefore, we obtain $q(y_0) = q(y_1) = \ldots = q(y_n)$, which leads to $q(y) = q(y')$. This shows that v is well-defined. Obviously we have $v \circ \pi(y) = q(y)$ for all $y \in Y$. If $v' : Y/\sim_R \to Q$ such that $v' \circ \pi = q$ then we easily obtain $v'(\overline{y}) = v' \circ \pi(y) = q(y) = v \circ \pi(y) = v(\overline{y})$ for all $y \in Y$. Thus $v = v'$, which completes the proof.

(6) Let $f, g \in \mathrm{Hom}_{\mathbf{Top}}(X, Y)$ and let $\pi : Y \to Y/\sim_R$ be the canonical projection constructed in the previous example. Now we endow $Y/\sim_R$ with the quotient topology[5] with respect to π and we will show that $(Y/\sim_R, \pi)$ is the coequalizer of the pair (f, g) in **Top**. In particular, $\pi : Y \to Y/\sim_R$ is a continuous map. Consider now $Q \in \mathrm{Ob}\ \mathbf{Top}$ and another continuous map $q : Y \to Q$ such that $q \circ f = q \circ g$. We only need to prove that the unique map $v : Y/\sim_R \to Q$ constructed in the previous example is continuous. To this end, let $U \subseteq Q$ be an open subset. Note that since $Y/\sim_R$ is endowed with the finest topology that makes π into a continuous map, in order to prove that $v^{-1}(U) \subseteq Y/\sim_R$ is an open subset it will suffice to show that $\pi^{-1}(v^{-1}(U)) \subseteq Y$ is an open set. To this end, we have $\pi^{-1}(v^{-1}(U)) = (\pi \circ v)^{-1}(U) = q^{-1}(U)$ and since q is continuous, the desired conclusion follows.

(7) Let $f, g \in \mathrm{Hom}_{\mathbf{Haus}}(X, Y)$ and consider $(Y/\sim_R, \pi)$ to be the coequalizer in **Top** constructed in the previous example. However, as the quotient topology does not behave well with respect to the Hausdorff property,[6] the coequalizer in **Haus** will be constructed by using the so-called *Hausdorff quotient* of a topological space (we refer to [40, 44] for further details). To this end, if we denote by (H, u) the Hausdorff quotient of $Y/\sim_R$, then (H, q) is the coequalizer in **Haus** of the pair (f, g), where $q = u \circ \pi$.

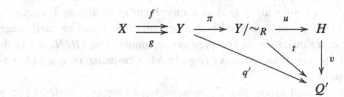

Indeed, to start with, we have $q \circ f = u \circ \pi \circ f = u \circ \pi \circ g = q \circ g$, where in the second equality we used the fact that $(Y/\sim_R, \pi)$ is the coequalizer in **Top** of (f, g). Consider now another Hausdorff space Q' and a morphism $q' \in \mathrm{Hom}_{\mathbf{Haus}}(Y, Q')$ such that $q' \circ f = q' \circ g$. As $(Y/\sim_R, \pi)$ is the coequalizer

---

[5] Let X be a topological space, Y a set and $f : X \to Y$ a surjective map. The *quotient topology* with respect to f is the finest topology on Y such that f is continuous. In other words, $U \subseteq Y$ is open if and only if $f^{-1}(U)$ is open in X [39, p. 138].

[6] See, for instance, [38, Examples 6.4.17, 6.4.19].

in **Top** of (f, g), there exists a unique morphism $t \in \mathrm{Hom}_{\mathbf{Top}}\big(Y/{\sim_R},\ Q'\big)$ such that $t \circ \pi = q'$. Now the universal property of the Hausdorff quotient (see [40]) yields a unique $v \in \mathrm{Hom}_{\mathbf{Haus}}(H,\ Q')$ such that $v \circ u = t$. If we put everything together we obtain $v \circ q = \underline{v \circ u} \circ \pi = \underline{t \circ \pi} = q'$. We are left to show that v is the unique morphism with this property. Indeed, assume there exists another $v' \in \mathrm{Hom}_{\mathbf{Haus}}(H,\ Q')$ such that $v' \circ q = q'$; then we have $(v' \circ u) \circ \pi = q'$ and since t is the unique morphism with this property it follows that $v' \circ u = t$ and therefore $v = v'$.

(8) Let $f, g \in \mathrm{Hom}_{\mathbf{Grp}}(G,\ H)$ and consider $H' = \{f(x)g(x)^{-1} \mid x \in X\} \subseteq H$. If we denote by $N = \bigcap_{H' \subseteq U \trianglelefteq H} U$ the normal subgroup generated by H' then $(H/N, \pi)$ is the coequalizer of the pair of morphisms (f, g) in the category **Grp** of groups, where $\pi \colon H \to H/N$ is the canonical projection. Indeed, since $f(x)g(x)^{-1} \in H' \subseteq H$ for all $x \in G$, we have $\widehat{f(x)} = \widehat{g(x)}$ in H/N, which comes down to $\pi \circ f = \pi \circ g$. Consider now $q' \in \mathrm{Hom}_{\mathbf{Grp}}(H,\ Q')$ such that $q' \circ f = q' \circ g$. This yields $q'\big(f(x)g(x)^{-1}\big) = 1$ for all $x \in G$. Therefore, we have $H' \subseteq \ker(q') \trianglelefteq H$ and thus $N \subseteq \ker(q')$. Now from the universal property of the quotient group H/N we obtain a unique morphism $v \in \mathrm{Hom}_{\mathbf{Grp}}\big(H/N,\ Q'\big)$ such that the following diagram is commutative:

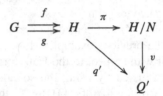

i.e., $(H/N, \pi)$ is the coequalizer of the pair of morphisms (f, g) in **Grp**.

(9) Let $f, g \in \mathrm{Hom}_{\mathbf{Ab}}(A,\ B)$. Then the map $(f - g) \colon A \to B$ defined by $(f - g)(a) = f(a) - g(a)$ for all $a \in A$ is a morphism of groups and, therefore, the set $N = \{f(a) - g(a) \mid a \in A\}$ is a subgroup of B. It can be easily seen, using the same arguments as in the previous example, that $\big(B/N, \pi\big)$ is the coequalizer of the pair of morphisms (f, g) in **Ab**. Coequalizers in ${}_R\mathcal{M}$ have a similar description.

(10) Let G be a non-trivial group and $\mathcal{G}$ the associated category as described in Example 1.2.2, (3). If $x, y \in G$ such that $x \neq y$ then the pair of morphisms (x, y) does not admit a (co)equalizer in $\mathcal{G}$. Indeed, this follows easily by noticing that there are no elements $z \in G$ such that $xz = yz$ or $zx = zy$.
□

In a pointed category we can introduce an important special case of (co)equalizers. First, recall from Remark 1.3.8 that in such categories there exists a morphism, called the zero-morphism, between any two given objects.

Definition 2.1.11 Let $\mathcal{C}$ be a pointed category and $f \in \text{Hom}_{\mathcal{C}}(A, B)$. The (co)equalizer of the pair of morphisms $(f, 0_{A, B})$ is called the *(co)kernel* of f.

Example 2.1.12 By applying the above definition to the pointed categories **Grp**, **Ab** and $_R\mathcal{M}$, we recover the familiar algebraic kernel of a morphism defined as the preimage of the neutral element. □

The last examples we provide before introducing (co)limits are pullbacks together with their duals, called pushouts.

Definition 2.1.13 Let $\mathcal{C}$ be a category and $f \in \text{Hom}_{\mathcal{C}}(B, A)$, $g \in \text{Hom}_{\mathcal{C}}(C, A)$. A *pullback* of (f, g) is a triple (P, f', g'), where

(1) $P \in \text{Ob}\,\mathcal{C}$;
(2) $f' \in \text{Hom}_{\mathcal{C}}(P, C)$, $g' \in \text{Hom}_{\mathcal{C}}(P, B)$ such that $f \circ g' = g \circ f'$,

and for any other triple (P', f'', g''), where

(1) $P' \in \text{Ob}\,\mathcal{C}$;
(2) $f'' \in \text{Hom}_{\mathcal{C}}(P', C)$, $g'' \in \text{Hom}_{\mathcal{C}}(P', B)$ such that $f \circ g'' = g \circ f''$,

there exists a unique $\Theta \in \text{Hom}_{\mathcal{C}}(P', P)$ such that $f'' = f' \circ \Theta$ and $g'' = g' \circ \Theta$.

The complete picture is captured by the diagram below:

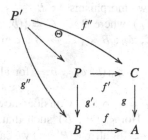

We say that $\mathcal{C}$ is a *category with pullbacks* or that $\mathcal{C}$ *has pullbacks* if any pair of morphisms in $\mathcal{C}$ with the same codomain has a pullback.

Definition 2.1.14 Let $\mathcal{C}$ be a category and $f \in \text{Hom}_{\mathcal{C}}(A, B)$, $g \in \text{Hom}_{\mathcal{C}}(A, C)$. A *pushout* of (f, g) is a triple (P, f', g'), where

(1) $P \in \text{Ob}\,\mathcal{C}$;
(2) $f' \in \text{Hom}_{\mathcal{C}}(C, P)$, $g' \in \text{Hom}_{\mathcal{C}}(B, P)$ such that $g' \circ f = f' \circ g$,

and for any other triple (P', f'', g''), where

(1) $P' \in \text{Ob}\,\mathcal{C}$;
(2) $f'' \in \text{Hom}_{\mathcal{C}}(C, P')$, $g'' \in \text{Hom}_{\mathcal{C}}(B, P')$ such that $g'' \circ f = f'' \circ g$,

there exists a unique $\Theta \in \text{Hom}_{\mathcal{C}}(P, P')$ such that $f'' = \Theta \circ f'$ and $g'' = \Theta \circ g'$.

The complete picture is captured by the diagram below:

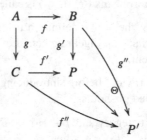

We say that $\mathcal{C}$ is a *category with pushouts* or that $\mathcal{C}$ *has pushouts* if any pair of morphisms in $\mathcal{C}$ with the same domain has a pushout.

As in the case of (co)products and (co)equalizers, both pullbacks and pushouts are unique up to isomorphism:

Proposition 2.1.15 *When it exists, the pullback (resp. pushout) of two morphisms is unique up to isomorphism.*

Examples 2.1.16

(1) In **Set** the pullback of two morphisms $f : B \to A$, $g : C \to A$ is given by the triple $(B \times_A C, \overline{\pi}_C, \overline{\pi}_B)$, where $B \times_A C = \{(b, c) \in B \times C \mid f(b) = g(c)\}$ and $\overline{\pi}_C : B \times_A C \to C, \overline{\pi}_B : B \times_A C \to B$ are given by

$$\overline{\pi}_C(b, c) = c, \qquad \overline{\pi}_B(b, c) = b, \quad \text{for all } (b, c) \in B \times_A C.$$

First note that $f \circ \overline{\pi}_B = g \circ \overline{\pi}_C$. Consider now $P' \in \mathrm{Ob\,Set}$ and $f'' \in \mathrm{Hom}_{\mathbf{Set}}(P', C)$, $g'' \in \mathrm{Hom}_{\mathbf{Set}}(P', B)$ such that $f \circ g'' = g \circ f''$. Then $(g''(x), f''(x)) \in B \times_A C$ for all $x \in P'$ and we can define $\Theta : P' \to B \times_A C$ by $\Theta(x) = (g''(x), f''(x))$. Moreover, we have

$$\overline{\pi}_C \circ \Theta(x) = f''(x),$$
$$\overline{\pi}_B \circ \Theta(x) = g''(x).$$

As $\overline{\pi}_B$ and $\overline{\pi}_C$ are the restrictions to $B \times_A C$ of the projections on B and C, respectively, Θ is obviously the unique morphism which renders the diagram below commutative:

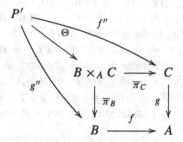

(2) Let $f \in \mathrm{Hom}_{\mathbf{Grp}}(B, A)$, $g \in \mathrm{Hom}_{\mathbf{Grp}}(C, A)$ and consider the set $B \times_A C$ constructed in the previous example endowed with the group structure given by the direct product. Then, the two projections $\overline{\pi}_C : B \times_A C \to C, \overline{\pi}_B : B \times_A C \to B$ defined in the previous example are group homomorphisms and $(B \times_A C, \overline{\pi}_C, \overline{\pi}_B)$ is the pullback of the pair of morphisms (f, g). Indeed, let $P' \in \mathrm{Ob}\,\mathbf{Grp}$, $f'' \in \mathrm{Hom}_{\mathbf{Grp}}(P', C)$, $g'' \in \mathrm{Hom}_{\mathbf{Grp}}(P', B)$ such that $f \circ g'' = g \circ f''$. The only thing left to check is that the map $\Theta : P' \to B \times_A C$ defined in the previous example is a group homomorphism. To this end, for all $x, y \in P'$ we have $\Theta(xy) = (g''(xy), f''(xy)) = (g''(x), f''(x))(g''(y), f''(y)) = \Theta(x)\Theta(y)$. Furthermore, $\Theta(1_P) = (g''(1_P), f''(1_P)) = (1_B, 1_C)$ and the desired conclusion follows.

(3) Let $f \in \mathrm{Hom}_{\mathbf{Top}}(B, A)$, $g \in \mathrm{Hom}_{\mathbf{Top}}(C, A)$ and consider the set $B \times_A C$ constructed in the first example. Furthermore, we see $B \times C$ as a topological space with respect to the product topology and endow $B \times_A C \subseteq B \times C$ with the subspace topology. Then $(B \times_A C, \overline{\pi}_C, \overline{\pi}_B)$ is the pullback in $\mathbf{Top}$ of the pair of morphisms (f, g). Indeed, note first that both $\overline{\pi}_B = \pi_B \circ i$ and $\overline{\pi}_C = \pi_C \circ i$ are compositions of continuous maps and are therefore continuous, where $i : B \times_A C \to B \times C$ is the inclusion map and π_B, π_C denote the projection maps. Consider now $P' \in \mathrm{Ob}\,\mathbf{Top}$ and $f'' \in \mathrm{Hom}_{\mathbf{Top}}(P', C)$, $g'' \in \mathrm{Hom}_{\mathbf{Top}}(P', B)$ such that $f \circ g'' = g \circ f''$. We are left to show that the map $\Theta : P' \to B \times_A C$ defined in the first example is continuous. To this end, recall that the topology on $B \times_A C$ is generated by the sets $\overline{\pi}_B^{-1}(U), \overline{\pi}_C^{-1}(V)$ for all open sets $U \subseteq B$ and $V \subseteq C$. In other words, the sets $\overline{\pi}_B^{-1}(U), \overline{\pi}_C^{-1}(V)$, where $U \subseteq B$ and $V \subseteq C$ are open sets, form a sub-basis for the topology on $B \times_A C$. Therefore, in order to show that Θ is a continuous map it will suffice to show that the preimage of any of these sets through Θ is an open set. Indeed, for all open sets $U \subseteq B$, $V \subseteq C$ we have

$$\Theta^{-1} \circ \overline{\pi}_B^{-1}(U) = (\overline{\pi}_B \circ \Theta)^{-1}(U) = g''^{-1}(U),$$

$$\Theta^{-1} \circ \overline{\pi}_C^{-1}(V) = (\overline{\pi}_C \circ \Theta)^{-1}(V) = f''^{-1}(V).$$

The conclusion now follows easily as both f'' and g'' are continuous maps.

(4) Assume R is a commutative ring and $_R\mathcal{M}$ is the category of left modules over R. Let $f \in \mathrm{Hom}_{_R\mathcal{M}}(B, A)$, $g \in \mathrm{Hom}_{_R\mathcal{M}}(C, A)$ and consider the submodule $B \times_A C = \{(b, c) \in B \times C \mid f(b) = g(c)\}$ of the direct product $B \times C$. Then $\left(B \times_A C, \overline{\pi}_C, \overline{\pi}_B\right)$ is the pullback in $_R\mathcal{M}$ of the pair of morphisms (f, g), where $\overline{\pi}_C \colon B \times_A C \to C$ and $\overline{\pi}_B \colon B \times_A C \to B$ are the restrictions of the canonical projections. It can be easily checked, as in the previous examples, that $f \circ \overline{\pi}_B = g \circ \overline{\pi}_C$. Consider now $P' \in \mathrm{Ob}\,_R\mathcal{M}$ and $f'' \in \mathrm{Hom}_{_R\mathcal{M}}(P', C)$, $g'' \in \mathrm{Hom}_{_R\mathcal{M}}(P', B)$ such that $f \circ g'' = g \circ f''$. We are left to show that the map $\Theta \colon P' \to B \times_A C$ defined as follows is R-linear:

$$\Theta(x) = (g''(x), f''(x)), \quad x \in P'.$$

Indeed, since both maps f'' and g'' are R-linear, for all $r, s \in R$ and $x, y \in P'$ we have

$$\begin{aligned}
\Theta(rx + sy) &= \left(g''(rx + sy), \ f''(rx + sy)\right) \\
&= \left(rg''(x) + sg''(y), \ rf''(x) + sf''(y)\right) \\
&= \left(rg''(x), \ rf''(x)\right) + \left(sg''(y), \ sf''(y)\right) \\
&= r\left(g''(x), \ f''(x)\right) + s\left(g''(y), \ f''(y)\right) \\
&= r\Theta(x) + s\Theta(y),
\end{aligned}$$

as desired.

(5) Let $f \in \mathrm{Hom}_{\mathbf{Set}}(A, B)$, $g \in \mathrm{Hom}_{\mathbf{Set}}(A, C)$ and consider the disjoint union of the sets B and C, denoted by $B \bigsqcup C = \{(b, 0) \mid b \in B\} \cup \{(c, 1) \mid c \in C\}$, together with the corresponding inclusion maps $j_0 \colon B \to B \bigsqcup C$, $j_1 \colon C \to B \bigsqcup C$ defined by $j_0(b) = (b, 0)$, $j_1(c) = (c, 1)$ for all $b \in B$, $c \in C$. Define $R = \{\left((f(a), 0), (g(a), 1)\right) \mid a \in A\} \subseteq (B \bigsqcup C) \times (B \bigsqcup C)$ and let $\sim_R$ be the equivalence relation on $B \bigsqcup C$ generated by R. Then the pushout of the pair (f, g) is given by the quotient set $(B \bigsqcup C)/\!\sim_R$ together with the maps $f' \colon C \to (B \bigsqcup C)/\!\sim_R$, $g' \colon B \to (B \bigsqcup C)/\!\sim_R$ defined as follows:

$$f' = \pi \circ j_1, \quad g' = \pi \circ j_0,$$

where $\pi \colon B \bigsqcup C \to (B \bigsqcup C)/\!\sim_R$ is the canonical projection. Indeed, first note that for all $a \in A$ we have $g' \circ f(a) = \pi(f(a), 0) = \pi(g(a), 1) = f' \circ g(a)$. Consider now $P' \in \mathrm{Ob}\,\mathbf{Set}$ and $f'' \in \mathrm{Hom}_{\mathbf{Set}}(C, P')$, $g'' \in \mathrm{Hom}_{\mathbf{Set}}(B, P')$ such that $g'' \circ f = f'' \circ g$ and define $\psi \colon B \bigsqcup C \to P'$ as follows:

$$\psi(x) = \begin{cases} f''(c) & \text{if } x = (c, 1) \text{ for some } c \in C, \\ g''(b) & \text{if } x = (b, 0) \text{ for some } b \in B. \end{cases} \tag{2.6}$$

Then for all $a \in A$ we have $\psi(f(a), 0) = g''(f(a)) = f''(g(a)) = \psi(g(a), 1)$. The universal property of the quotient set yields a unique map $\Theta : (B \coprod C)/{\sim}_R \to P'$ such that $\Theta \circ \pi = \psi$. Furthermore, for all $b \in B$ and $c \in C$ we have

$$\Theta \circ f'(c) = \Theta \circ \pi \circ j_1(c) = \psi(c, 1) = f''(c),$$
$$\Theta \circ g'(b) = \Theta \circ \pi \circ j_0(b) = \psi(b, 0) = g''(b).$$

Assume there exists another map $\overline{\Theta} : (B \coprod C)/{\sim}_R \to P'$ such that $\overline{\Theta} \circ f' = f''$ and $\overline{\Theta} \circ g' = g''$. Then, for all $b \in B$ we obtain $\overline{\Theta}(\widehat{(b, 0)}) = \overline{\Theta} \circ \pi(b, 0) = \overline{\Theta} \circ \pi \circ j_0(b) = \overline{\Theta} \circ g'(b) = g''(b) = \Theta \circ \pi \circ j_0(b) = \Theta(\widehat{(b, 0)})$, where $\widehat{(b, 0)}$ denotes the class of $(b, 0)$ in $(B \coprod C)/{\sim}_R$. Similarly, one can easily see that we also have $\overline{\Theta}(\widehat{(c, 1)}) = \Theta(\widehat{(c, 1)})$ for all $c \in C$ and therefore $\overline{\Theta} = \Theta$.

(6) Let $f \in \mathrm{Hom}_{\mathbf{Top}}(A, B)$, $g \in \mathrm{Hom}_{\mathbf{Top}}(A, C)$ and let $(B \coprod C)/{\sim}_R$ be the set constructed in the previous example together with the projection map $\pi : B \coprod C \to (B \coprod C)/{\sim}_R$. We endow $B \coprod C$ with the finest topology for which the inclusion maps $j_0 : B \to B \coprod C$ and $j_1 : C \to B \coprod C$ are continuous maps and we consider on $(B \coprod C)/{\sim}_R$ the quotient topology. Furthermore, note that both $f' = \pi \circ j_1$ and $g' = \pi \circ j_0$ are continuous as compositions of continuous maps. We will show that $\left((B \coprod C)/{\sim}_R, (f'\, g')\right)$ is the pushout of the pair (f, g). Indeed, consider $P' \in \mathrm{Ob}\,\mathbf{Top}$ and $f'' \in \mathrm{Hom}_{\mathbf{Top}}(C, P')$, $g'' \in \mathrm{Hom}_{\mathbf{Top}}(B, P')$ such that $g'' \circ f = f'' \circ g$. We are left to show that the unique map $\Theta : (B \coprod C)/{\sim}_R \to P'$ such that $\Theta \circ f' = f''$ and $\Theta \circ g' = g''$ defined in the previous example is continuous. To this end, let $U \subseteq P'$ be an open set. Note that since $(B \coprod C)/{\sim}_R$ is endowed with the finest topology that makes π into a continuous map, in order to prove that $\Theta^{-1}(U) \subseteq (B \coprod C)/{\sim}_R$ is an open subset it will suffice to show that $\pi^{-1}(\Theta^{-1}(U)) \subseteq B \coprod C$ is an open subset. We have $\pi^{-1}(\Theta^{-1}(U)) = (\Theta \circ \pi)^{-1}(U) = \psi^{-1}(U)$, where ψ is the map defined in (2.6). Our claim will be proved once we show that $\psi^{-1}(U) \subseteq B \coprod C$ is an open subset. Indeed, since $B \coprod C$ is endowed with the finest topology for which the canonical inclusions j_0 and j_1 are continuous maps, we only need to show that $j_0^{-1}(\psi^{-1}(U)) \subseteq B$ and $j_1^{-1}(\psi^{-1}(U)) \subseteq C$ are open subsets. We have $j_0^{-1}(\psi^{-1}(U)) = (\psi \circ j_0)^{-1}(U) = (g'')^{-1}(U)$ and $j_1^{-1}(\psi^{-1}(U)) = (\psi \circ j_1)^{-1}(U) = (f'')^{-1}(U)$ and the desired conclusion now follows as both f'' and g'' are continuous maps.

(7) Let $f \in \mathrm{Hom}_{\mathbf{Grp}}(A, B)$, $g \in \mathrm{Hom}_{\mathbf{Grp}}(A, C)$ and let $B \star C$ be the free product of the groups B and C together with the corresponding group homomorphisms $i_B : B \to B \star C$ and $i_C : C \to B \star C$. Furthermore, we let $N' = \{i_B(f(a))^{-1} i_C(g(a)) \mid a \in A\}$ and denote by N the normal subgroup of $B \star C$ generated by N'. Then the quotient group $(B \star C)/N$ together with the group homomorphisms $f' : C \to (B \star C)/N$, $g' : B \to (B \star C)/N$

is the pushout of the pair (f, g), where $f' = \pi \circ i_C$, $g' = \pi \circ i_B$ and $\pi : B \star C \to (B \star C)/N$ is the canonical projection. Indeed, having in mind that $N' \subseteq N$, we have

$$(f' \circ g)(a) = \pi\big(i_C(g(a))\big) = \pi\big(i_B(f(a))\big) = (g' \circ f)(a) \text{ for all } a \in A.$$

Consider now $P' \in \mathrm{Ob}\,\mathbf{Grp}$ and $f'' \in \mathrm{Hom}_{\mathbf{Grp}}(C, P')$, $g'' \in \mathrm{Hom}_{\mathbf{Grp}}(B, P')$ such that $g'' \circ f = f'' \circ g$. The definition of the free product of groups ([50, Chapter 11, p. 388]) yields a unique group homomorphism $\overline{\Theta} : B \star C \to P'$ such that the following diagram is commutative:

$$
\begin{array}{ccc}
C \xrightarrow{\ i_C\ } B \star C \xleftarrow{\ i_B\ } B & & \text{i.e., } \overline{\Theta} \circ i_B = g'' \text{ and } \overline{\Theta} \circ i_C = f''. \\
\quad {}_{f''}\searrow \quad \downarrow{\scriptstyle\overline{\Theta}} \quad \swarrow{}_{g''} & & \\
P' & &
\end{array}
$$

(2.7)

We show first that $N' \subseteq \ker \overline{\Theta}$. Indeed, for all $a \in A$ we have

$$\overline{\Theta}\big(i_B(f(a))^{-1} i_C(g(a))\big) = \overline{\Theta}\big(i_B(f(a))^{-1}\big)\overline{\Theta}\big(i_C(g(a))\big)$$
$$= g''(f(a))^{-1} f''(g(a)) = 1.$$

Therefore, $N \subseteq \ker \overline{\Theta}$ and the universal property of the quotient groups yields a unique group homomorphism $\Theta : (B \star C)/N \to P'$ such that $\Theta \circ \pi = \overline{\Theta}$. We obtain

$$\Theta \circ f' = \Theta \circ \pi \circ i_C = \overline{\Theta} \circ i_C = f'',$$
$$\Theta \circ g' = \Theta \circ \pi \circ i_B = \overline{\Theta} \circ i_B = g''.$$

We are left to show that Θ is the unique group homomorphism with this property. Assume $\Theta' : (B \star C)/N \to P'$ is another group homomorphism such that $\Theta' \circ f' = f''$ and $\Theta' \circ g' = g''$. This leads to $\Theta' \circ \pi \circ i_C = f''$ and $\Theta' \circ \pi \circ i_B = g''$ and since $\overline{\Theta}$ is the unique group homomorphism which makes diagram (2.7) commute, we obtain $\Theta' \circ \pi = \overline{\Theta}$. Now by the universal property of the quotient groups, Θ is the unique group homomorphism with this property and therefore $\Theta' = \Theta$.

(8) Assume R is a commutative ring and ${}_R\mathcal{M}$ is the category of left modules over R. Let $f \in \mathrm{Hom}_{{}_R\mathcal{M}}(A, B)$, $g \in \mathrm{Hom}_{{}_R\mathcal{M}}(A, C)$ and consider the submodule $S = \{(f(a), -g(a)) \mid a \in A\}$ of $B \times C$. Then the triple $\big((B \times C)/S, f', g'\big)$ is the pushout in ${}_R\mathcal{M}$ of the morphisms above, where $(B \times C)/S$ denotes the quotient module corresponding to S and $f' : C \to (B \times C)/S$, $g' : B \to$

$(B \times C)/S$ are given, for any $b \in B$ and $c \in C$, as follows:

$$f'(c) = \overline{(0, \, c)}, \qquad g'(b) = \overline{(b, \, 0)}.$$

Indeed, since $\bigl(f(a), \, 0\bigr) - \bigl(0, \, g(a)\bigr) = \bigl(f(a), \, -g(a)\bigr) \in S$, we get $\overline{\bigl(f(a), \, 0\bigr)} = \overline{\bigl(0, \, g(a)\bigr)}$ and thus $g' \circ f = f' \circ g$. Consider now $P' \in \mathrm{Ob} \, _R\mathcal{M}$ and $f'' \in \mathrm{Hom}_{R\mathcal{M}}(C, \, P')$, $g'' \in \mathrm{Hom}_{R\mathcal{M}}(B, \, P')$ such that $g'' \circ f = f'' \circ g$. The map defined for all $(b, \, c) \in B \times C$ as follows:

$$\chi : B \times C \to P', \qquad \chi(b, \, c) = g''(b) + f''(c),$$

is a morphism in $_R\mathcal{M}$ and moreover, $\chi(S) = 0$ since we have

$$\chi\bigl(f(a), \, -g(a)\bigr) = g''(f(a)) - f''(g(a)) = 0$$

for all $a \in A$. Now the universal property of the quotient module yields a unique morphism $\Theta \in \mathrm{Hom}_{R\mathcal{M}}\bigl((B \times C)/S, \, P'\bigr)$ such that $\Theta\bigl(\overline{(b, \, c)}\bigr) = g''(b) + f''(c)$ for all $(b, \, c) \in B \times C$. Moreover, we have

$$(\Theta \circ g')(b) = \Theta\bigl(\overline{(b, \, 0)}\bigr) = g''(b),$$

$$(\Theta \circ f')(c) = \Theta\bigl(\overline{(0, \, c)}\bigr) = f''(c),$$

i.e., the following diagram is commutative:

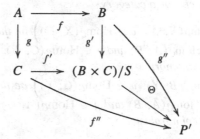

We are left to prove the uniqueness of Θ. Let $\Upsilon : (B \times C)/S \to P'$ such that $\Upsilon \circ g' = g''$ and $\Upsilon \circ f' = f''$. To this end, for all $(b, \, c) \in B \times C$, we have

$$\Upsilon\bigl(\overline{(b, \, c)}\bigr) = \Upsilon\bigl(\overline{(b, \, 0)}\bigr) + \Upsilon\bigl(\overline{(0, \, c)}\bigr)$$

$$= \Upsilon \circ g'(b) + \Upsilon \circ f'(c)$$

$$= g''(b) + f''(c)$$

$$= \Theta\bigl(\overline{(b, \, c)}\bigr).$$

(9) Let $\mathrm{PO}(X, \, \leqslant)$ be the category corresponding to the pre-ordered set $(X, \, \leqslant)$ and $a, \, b, \, c \in X$ such that $a \leq b$ and $a \leq c$. If it exists, the pushout of the above maps is some element $p \in X$ satisfying:

- $b \leq p$ and $c \leq p$;
- for any $x \in X$ such that $b \leq x$ and $c \leq x$ we have $p \leq x$.

In other words, if it exists, the pushout of the maps $b \leq a$ and $c \leq a$ is given by the supremum of b and c. Similarly, if it exists, the pullback of two maps above is given by the infimum of b and c.

(10) Let $M = \mathrm{End}(X)$ denote the endomorphisms on the set $X = \{x, y\}$. More precisely, $M = \{1_X, \tau, \psi_x, \psi_y\}$ is a monoid with respect to the composition of maps, where $\tau, \psi_x, \psi_y \colon M \to M$ are defined as follows:

$$\tau(x) = y, \ \tau(y) = x,$$

$$\psi_x(x) = \psi_x(y) = x,$$

$$\psi_y(x) = \psi_y(y) = y.$$

If $\mathcal{M}$ denotes the category associated to the monoid M as in Example 1.2.2, (3) then the pair of morphisms (ψ_x, ψ_y) does not have an equalizer in $\mathcal{M}$. Consequently, in $\mathcal{M}^{op}$ the pair of morphisms $(\psi_x^{op}, \psi_y^{op})$ does not have a coequalizer.   □

The next two results highlight the different connections existing between monomorphisms (resp. epimorphisms) and pullbacks (resp. pushouts) and will be useful in the sequel.

Proposition 2.1.17 *Let C be a category.*

(1) Let $\Big(X, u \in \mathrm{Hom}_C(X, A), v \in \mathrm{Hom}_C(X, C)\Big)$ be the pullback of the pair of morphisms $f \in \mathrm{Hom}_C(A, B)$ and $g \in \mathrm{Hom}_C(C, B)$. If g is a monomorphism then u is also a monomorphism.

(2) Let $\Big(Y, u \in \mathrm{Hom}_C(B, Y), v \in \mathrm{Hom}_C(C, Y)\Big)$ be the pushout of the pair of morphisms $f \in \mathrm{Hom}_C(A, B)$ and $g \in \mathrm{Hom}_C(A, C)$. If g is an epimorphism then u is also an epimorphism.

Proof (1) Let $Y \in \mathrm{Ob}\,C$ and assume there exists $\alpha, \beta \in \mathrm{Hom}_C(Y, X)$ such that $u \circ \alpha = u \circ \beta$. In particular, this leads to $f \circ u \circ \alpha = f \circ u \circ \beta$ and since $f \circ u = g \circ v$ we also have $g \circ v \circ \alpha = g \circ v \circ \beta$.

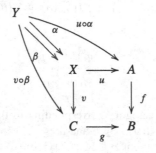

As g is a monomorphism, we obtain $v \circ \alpha = v \circ \beta$. Furthermore, this implies $f \circ u \circ \alpha = g \circ v \circ \beta$. Since (X, u, v) is the pullback of the pair (f, g), there exists a unique morphism $\gamma \in \mathrm{Hom}_{\mathcal{C}}(Y, X)$ such that $u \circ \gamma = u \circ \alpha$ and $v \circ \beta = v \circ \gamma$. As both morphisms α and β satisfy this property we obtain $\alpha = \beta$, which shows that u is indeed a monomorphism.

(2) By duality $\left(Y, u^{op} \in \mathrm{Hom}_{\mathcal{C}^{op}}(Y, B), v^{op} \in \mathrm{Hom}_{\mathcal{C}^{op}}(Y, C)\right)$ is the pullback in $\mathcal{C}^{op}$ of the pair of morphisms (f^{op}, g^{op}). If g is an epimorphism in $\mathcal{C}$ then g^{op} is a monomorphism in $\mathcal{C}^{op}$ and by 1) we obtain that u^{op} is also a monomorphism in $\mathcal{C}^{op}$. Therefore, u is an epimorphism in $\mathcal{C}$, as desired. □

Proposition 2.1.18 *Let $\mathcal{C}$ be a category and $f \in \mathrm{Hom}_{\mathcal{C}}(A, B)$.*

(1) f is a monomorphism if and only if $(A, 1_A, 1_A)$ is the pullback of the pair of morphisms (f, f).

(2) f is an epimorphism if and only if $(B, 1_B, 1_B)$ is the pushout of the pair of morphisms (f, f).

Proof (1) Assume first that f is a monomorphism and let $u, v \in \mathrm{Hom}_{\mathcal{C}}(P, A)$ such that $f \circ u = f \circ v$. As f is a monomorphism it follows that $u = v$ and therefore we have a unique morphism in $\mathrm{Hom}_{\mathcal{C}}(P, A)$, namely u, which makes the following diagram commutative:

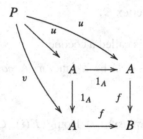

This shows that $(A, 1_A, 1_A)$ is the pullback of the pair of morphisms (f, f).

Conversely, assume that $(A, 1_A, 1_A)$ is the pullback of the pair of morphisms (f, f) and let $u, v \in \mathrm{Hom}_{\mathcal{C}}(P, A)$ such that $f \circ u = f \circ v$. Hence, we have a unique morphism $w \in \mathrm{Hom}_{\mathcal{C}}(P, A)$ such that $1_A \circ w = u$ and $1_A \circ w = v$. This shows that $u = v = w$ and therefore f is a monomorphism.

(2) Follows easily by duality; indeed, applying 1) to the morphism f^{op} gives the desired conclusion. □

2.2 (Co)limit of a Functor. (Co)complete Categories

Following the general pattern induced by the previous constructions (i.e., (co)products, (co)equalizers and pullbacks/pushouts) we can now introduce the concepts of limit and colimit, which unify all the above. We start by introducing the following:

Definition 2.2.1 Let $F : I \to C$ be a functor.[7] A *cone* on F consists of the following:

(1) $C \in \mathrm{Ob}\, C$;
(2) for every $i \in \mathrm{Ob}\, I$, a morphism $s_i \in \mathrm{Hom}_C(C, F(i))$,

such that for any morphism $d \in \mathrm{Hom}_D(i, j)$, the following diagram is commutative:

$$
\begin{array}{ccc}
 & C & \\
s_i \swarrow & & \searrow s_j \\
F(i) \xrightarrow[\;\;F(d)\;\;]{} & & F(j)
\end{array}
\qquad\qquad \text{i.e., } F(d) \circ s_i = s_j.
$$

The object C is called the *vertex of the cone*. If I is a small category and $C \in \mathrm{Ob}\, C$, we denote by

$$
\mathrm{Cone}(F, C) = \{ (s_i \in \mathrm{Hom}_C(C, F(i)))_{i \in \mathrm{Ob}\, I} \mid (C, (s_i)_{i \in \mathrm{Ob}\, I}) \text{ is a cone on } F \}
$$

the set[8] of cones on F with vertex C.

The dual notion to a cone is called a cocone:

Definition 2.2.2 Let $F : I \to C$ be a functor. A *cocone* on F consists of the following:

(1) $C \in \mathrm{Ob}\, C$,
(2) for every $i \in \mathrm{Ob}\, I$, a morphism $t_i \in \mathrm{Hom}_C(F(i), C)$,

such that for any morphism $d \in \mathrm{Hom}_D(i, j)$, the following diagram is commutative:

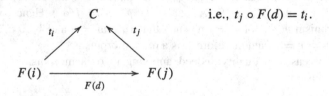

$$
\text{i.e., } t_j \circ F(d) = t_i.
$$

The object C is called the *vertex of the cocone*. If I is a small category and $C \in \mathrm{Ob}\, C$, we denote by

---

[7] The category I will almost always be considered small.
[8] Note that $\mathrm{Cone}(F, C)$ is indeed a set since $\mathrm{Ob}\, I$ is a set and $\mathrm{Cone}(F, C) \subseteq \prod_{i \in \mathrm{Ob}\, I} \mathrm{Hom}_C(C, F(i))$.

$\mathrm{Cocone}(F, C) = \{(t_i \in \mathrm{Hom}_{\mathcal{C}}(F(i), C))_{i \in \mathrm{Ob}\, I} \mid (C, (t_i)_{i \in \mathrm{Ob}\, I}) \text{ is a cocone on } F\}$

the set of cocones on F with vertex C.

The two notions are dual to each other in the following precise sense:

Lemma 2.2.3 *Let $F: I \to \mathcal{C}$ be a functor. Then $(L, (p_i)_{i \in \mathrm{Ob}\, I})$ is a cone on F if and only if $(L, (p_i^{op})_{i \in \mathrm{Ob}\, I})$ is a cocone on the dual functor $F^{op}: I^{op} \to \mathcal{C}^{op}$.*

Proof Let $(L, (p_i)_{i \in \mathrm{Ob}\, I})$ be a cone on F. Then for all $d \in \mathrm{Hom}_I(i, j)$ we have $F(d) \circ p_i = p_j$. Equivalently, this means that $p_i^{op} \circ^{op} F^{op}(d^{op}) = p_j^{op}$ for all $d^{op} \in \mathrm{Hom}_{I^{op}}(j, i)$, which shows that $(L, (p_i^{op})_{i \in \mathrm{Ob}\, I})$ is a cocone on F^{op}. The converse follows by replacing F by F^{op} in the argument above. $\square$

A much more elegant description of (co)cones can be obtained by using natural transformations:

Proposition 2.2.4 *Let $F: I \to \mathcal{C}$ be a functor.*

(1) The pair $(C, (s_i)_{i \in \mathrm{Ob}\, I})$ is a cone on F if and only if $\eta^C: \Delta_C \to F$ defined by $\eta_i^C = s_i$, for all $i \in \mathrm{Ob}\, I$, is a natural transformation.

(2) The pair $(C, (t_i)_{i \in \mathrm{Ob}\, I})$ is a cocone on F if and only if $^C\eta: F \to \Delta_C$ defined by $^C\eta_i = t_i$, for all $i \in \mathrm{Ob}\, I$, is a natural transformation.

Proof (1) $\eta^C: \Delta_C \to F$ is a natural transformation if and only if the following diagram is commutative for all $f \in \mathrm{Hom}_{\mathcal{D}}(i, j)$:

$$
\begin{array}{ccc}
\Delta_C(i) = C & \xrightarrow{\;\;\eta_i^C = s_i\;\;} & F(i) \\
{\scriptstyle \Delta_C(f) = 1_C} \downarrow & & \downarrow {\scriptstyle F(f)} \\
\Delta_C(j) = C & \xrightarrow{\;\;\eta_j^C = s_j\;\;} & F(j)
\end{array}
\tag{2.8}
$$

The commutativity of the above diagram comes down to $F(f) \circ s_i = s_j$, which means precisely that $(C, (s_i)_{i \in \mathrm{Ob}\, I})$ is a cone on F.

(2) $^C\eta: F \to \Delta_C$ is a natural transformation if and only if the following diagram is commutative for all $f \in \mathrm{Hom}_{\mathcal{D}}(i, j)$:

$$
\begin{array}{ccc}
F(i) & \xrightarrow{\;\;^C\eta_i = t_i\;\;} & \Delta_C(i) = C \\
{\scriptstyle F(f)} \downarrow & & \downarrow {\scriptstyle \Delta_C(f) = 1_C} \\
F(j) & \xrightarrow{\;\;^C\eta_j = t_j\;\;} & \Delta_C(i) = C
\end{array}
\tag{2.9}
$$

The commutativity of the above diagram comes down to $t_j \circ F(f) = t_i$, which means precisely that $\big(C, (t_i)_{i \in \mathrm{Ob}\, I}\big)$ is a cocone on F.   □

Definition 2.2.5 Let $F \colon I \to C$ be a functor.

(1) A *morphism between two cones* $\big(C, (s_i)_{i \in \mathrm{Ob}\, I}\big)$ and $\big(\overline{C}, (r_i)_{i \in \mathrm{Ob}\, I}\big)$ on F is a morphism $f \in \mathrm{Hom}_C(C, \overline{C})$ such that the following diagram is commutative for all $i \in \mathrm{Ob}\, I$:

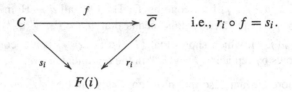

i.e., $r_i \circ f = s_i$.

The cones on F together with morphisms between them as defined above form a category denoted by $\mathcal{C}(F)$.

(2) A *morphism between two cocones* $\big(C, (t_i)_{i \in \mathrm{Ob}\, I}\big)$ and $\big(\overline{C}, (u_i)_{i \in \mathrm{Ob}\, I}\big)$ on F is a morphism $f \in \mathrm{Hom}_C(C, \overline{C})$ such that the following diagram is commutative for any $i \in \mathrm{Ob}\, I$:

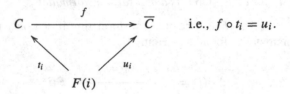

i.e., $f \circ t_i = u_i$.

The cocones on F together with morphisms between them as defined above form a category denoted by $\mathcal{CO}(F)$.

Under some assumptions, the category of (co)cones is in fact isomorphic to a certain comma category.

Theorem 2.2.6 *Let I be a small category and $F \colon I \to C$ a functor. If $T_F \colon \mathbf{1} \to \mathrm{Fun}\,(I, C)$ denotes the constant functor at F and $\Delta \colon C \to \mathrm{Fun}(I, C)$ is the diagonal functor, then:*

(1) the category $\mathcal{C}(F)$ of cones on F is isomorphic to the comma category $(\Delta \downarrow T_F)$;

(2) the category $\mathcal{CO}(F)$ of cocones on F is isomorphic to the comma category $(T_F \downarrow \Delta)$.

Proof Throughout, we denote by $*$ the unique object of the discrete category $\mathbf{1}$.

(1) Recall from Theorem 1.8.4 that the objects of the comma-category $(\Delta \downarrow T_F)$ are triples $(C, \alpha, *)$, where $C \in \mathrm{Ob}\, C$ and $\alpha \colon \Delta_C \to F$ is a natural transformation. Proposition 2.2.4, (1) implies that the pair $\big(C, (\alpha_i)_{i \in \mathrm{Ob}\, I}\big)$ is a cone on F. Furthermore, a morphism between two objects $(C, \alpha, *)$ and $(C', \alpha', *)$ is a

pair $(f, 1_*)$ consisting of a morphism $f \in \mathrm{Hom}_{\mathcal{C}}(C, C')$ and the identity morphism 1_* on $*$ such that the following diagram commutes:

$$
\begin{array}{ccc}
\Delta_C & \xrightarrow{\ \Delta(f)\ } & \Delta_{C'} \\
\alpha \downarrow & & \downarrow \alpha' \\
T_F(*) = F & \xrightarrow[T_F(1_*)=1_F]{} & T_F(*) = F
\end{array}
$$

where 1_F denotes the identity natural transformation on the functor F. This leads to the following equality between natural transformations: $\alpha' \circ \Delta(f) = \alpha$. Hence, for all $i \in \mathrm{Ob}\,I$ we have $\alpha'_i \circ f = \alpha_i$ i.e., f is a morphism in $\mathcal{C}(F)$ between the cones $\big(C, (\alpha_i)_{i \in \mathrm{Ob}\,I}\big)$ and $\big(C', (\alpha'_i)_{i \in \mathrm{Ob}\,I}\big)$. We can now define two functors $U : \mathcal{C}(F) \to (\Delta \downarrow T_F)$ and $V : (\Delta \downarrow T_F) \to \mathcal{C}(F)$ as follows:

$$U\big(C, (\alpha_i)_{i \in \mathrm{Ob}\,I}\big) = (C, \alpha, *), \qquad U(f) = (f, 1_*),$$

$$V(C, \alpha, *) = \big(C, (\alpha_i)_{i \in \mathrm{Ob}\,i}\big), \qquad V(f, 1_*) = f.$$

It is now straightforward to see that $U \circ V = 1_{(\Delta \downarrow T_F)}$ and $V \circ U = 1_{\mathcal{C}(F)}$. Hence the categories $(\Delta \downarrow T_F)$ and $\mathcal{C}(F)$ are isomorphic.

(2) Similarly, the objects of the comma-category $(T_F \downarrow \Delta)$ are triples $(*, \alpha, C)$, where $C \in \mathrm{Ob}\,\mathcal{C}$ and $\alpha : F \to \Delta_C$ is a natural transformation. By Proposition 2.2.4, (2) the pair $\big(C, (\alpha_i)_{i \in \mathrm{Ob}\,I}\big)$ is a cocone on F. Furthermore, a morphism between two objects $(*, \alpha, C)$ and $(*, \alpha', C')$ is a pair $(1_*, f)$ consisting of a morphism $f \in \mathrm{Hom}_{\mathcal{C}}(C, C')$ and the identity morphism 1_* on $*$ such that the following diagram commutes:

$$
\begin{array}{ccc}
T_F(*) = F & \xrightarrow{\ T_F(1_*)=1_F\ } & T_F(*) = F \\
\alpha \downarrow & & \downarrow \alpha' \\
\Delta_C & \xrightarrow[\Delta(f)]{} & \Delta_{C'}
\end{array}
$$

where 1_F denotes the identity natural transformation on the functor F. This leads to the following equality between natural transformations: $\Delta(f) \circ \alpha = \alpha'$. Hence, for all $i \in \mathrm{Ob}\,I$ we have $f \circ \alpha_i = \alpha'_i$, i.e., f is a morphism in $\mathcal{CO}(F)$ between the cocones $\big(C, (\alpha_i)_{i \in \mathrm{Ob}\,I}\big)$ and $\big(C', (\alpha'_i)_{i \in \mathrm{Ob}\,I}\big)$. We can now define two functors $U : \mathcal{CO}(F) \to (T_F \downarrow \Delta)$ and $V : (T_F \downarrow \Delta) \to \mathcal{CO}(F)$ as follows:

$$U\big(C, (\alpha_i)_{i \in \mathrm{Ob}\,I}\big) = (*, \alpha, C), \qquad U(f) = (1_*, f),$$

$$V(*, \alpha, C) = \big(C, (\alpha_i)_{i \in \mathrm{Ob}\,i}\big), \qquad V(1_*, f) = f.$$

It is straightforward to see that $U \circ V = 1_{(T_F \downarrow \Delta)}$ and $V \circ U = 1_{\mathcal{CO}(F)}$, which shows that the categories $(\Delta \downarrow T_F)$ and $\mathcal{CO}(F)$ are isomorphic.   □

We can now introduce (co)limits:

Definition 2.2.7 Let $F \colon I \to \mathcal{C}$ be a functor.

(1) A *limit* for the functor F is a final object in the category of cones on F, i.e., a cone on F denoted by $\left(\lim F, (p_i)_{i \in \mathrm{Ob}\, I}\right)$ such that for any other cone $\left(C, (s_i)_{i \in \mathrm{Ob}\, I}\right)$ on F there exists a unique morphism $f \in \mathrm{Hom}_{\mathcal{C}}(C, \lim F)$ such that the following diagram is commutative for any $i \in \mathrm{Ob}\, I$:

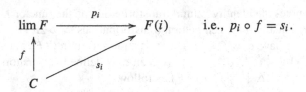

i.e., $p_i \circ f = s_i$.

A category $\mathcal{C}$ *has (small/finite) limits* if any functor $F \colon I \to \mathcal{C}$ has a limit for any (small/finite) category I. We say that a category $\mathcal{C}$ is *complete* if it has small limits.

(2) A *colimit* for the functor F is an initial object in the category of cocones on F, i.e., a cocone on F denoted by $\left(\mathrm{colim}\, F, (q_i)_{i \in \mathrm{Ob}\, I}\right)$ such that for every other cocone $\left(C, (t_i)_{i \in \mathrm{Ob}\, I}\right)$ on F there exists a unique morphism $f \in \mathrm{Hom}_{\mathcal{C}}(\mathrm{colim}\, F, C)$ such that the following diagram is commutative for any $i \in \mathrm{Ob}\, I$:

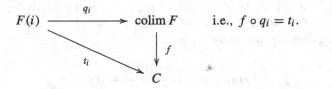

i.e., $f \circ q_i = t_i$.

A category $\mathcal{C}$ *has (small/finite) colimits* if any functor $F \colon I \to \mathcal{C}$ has a colimit for any (small/finite) category I. We say that a category $\mathcal{C}$ is *cocomplete* if it has small colimits.

In the sequel we will consider solely small (co)limits. Dealing with limits of functors $F \colon I \to \mathcal{C}$ for arbitrary categories I leads to set-theoretical issues which are far beyond our scope. The interested reader is referred to [51].

We can now recover all previously introduced special cases of (co)limits: initial and final objects, (co)products, (co)equalizers, pullbacks and pushouts.

Examples 2.2.8

(1) Consider the empty functor ϕ from the empty category to $\mathcal{C}$. Then the limit of ϕ, if it exists, is just the final object in $\mathcal{C}$. Indeed, note that a cone on the empty functor is just an object of $\mathcal{C}$. Furthermore, any morphism between two such

cones are just morphisms in $\mathcal{C}$ between the corresponding objects. Therefore, the category of cones on the empty functor is just the category $\mathcal{C}$ and the limit is a *final object* of $\mathcal{C}$. Analogously, the colimit of ϕ, if it exists, is just the *initial object* in $\mathcal{C}$.

(2) Take I to be a small discrete category. Then a functor $F: I \rightarrow \mathcal{C}$ is essentially nothing but a family of objects $(C_i)_{i \in \mathrm{Ob}\,I}$ in $\mathcal{C}$. A cone on F is a pair $\big(C, (u_i: C \rightarrow C_i)_{i \in \mathrm{Ob}\,I}\big)$ and since I is a discrete category, no further constrains are imposed on the family of morphisms $(u_i)_{i \in \mathrm{Ob}\,I}$. Now the limit of F, if it exists, is just a *product* in $\mathcal{C}$ of the family $(C_i)_{i \in \mathrm{Ob}\,I}$. Similarly, the colimit of F is just a *coproduct* in $\mathcal{C}$ of the family $(C_i)_{i \in I}$.

(3) Consider a category I with two objects A_1 and A_2 and four morphisms 1_{A_1}, 1_{A_2}, u, v, where $u, v \in \mathrm{Hom}_I(A_1, A_2)$ and define the functor $F: I \rightarrow \mathcal{C}$ as follows:

$$F(A_1) = X, \quad F(A_2) = Y, \quad F(u) = f, \quad F(v) = g.$$

A cone on F consists of an object $C \in \mathrm{Ob}\,\mathcal{C}$ and morphisms $\big(s \in \mathrm{Hom}_{\mathcal{C}}(C, X), t \in \mathrm{Hom}_{\mathcal{C}}(C, Y)\big)$ such that the following diagrams are commutative:

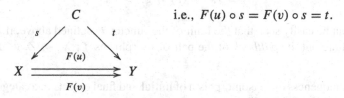

i.e., $F(u) \circ s = F(v) \circ s = t.$

As the morphism t is uniquely determined by s, a cone on F comes down to a pair $\big(C, s \in \mathrm{Hom}_{\mathcal{C}}(C, X)\big)$ such that $f \circ s = g \circ s$.

Therefore, the (co)limit of the functor F defined above, if it exists, is nothing but the *(co)equalizer* of the pair of morphisms $f, g: X \rightarrow Y$ in $\mathcal{C}$.

(4) Consider a category I with three objects A_1, A_2, A_3 and five morphisms 1_{A_1}, 1_{A_2}, 1_{A_3}, u, v, where $u \in \mathrm{Hom}_I(A_3, A_1)$, $v \in \mathrm{Hom}_I(A_3, A_2)$ and define the functor $F: I \rightarrow \mathcal{C}$ as follows:

$$F(A_1) = X, \quad F(A_2) = Y, \quad F(A_3) = Z,$$

$$F(u) = f \in \mathrm{Hom}_{\mathcal{C}}(Z, X), \quad F(v) = g \in \mathrm{Hom}_{\mathcal{C}}(Z, Y).$$

A cocone on F consists of an object C together with morphisms $\big(s \in \mathrm{Hom}_{\mathcal{C}}(X, C), t \in \mathrm{Hom}_{\mathcal{C}}(Y, C), l \in \mathrm{Hom}_{\mathcal{C}}(Z, C)\big)$ such that the following diagrams are commutative:

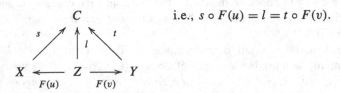

i.e., $s \circ F(u) = l = t \circ F(v)$.

As the morphism l is uniquely determined by s and t, a cocone on F comes down to a triple $\big(C,\, s \in \mathrm{Hom}_{\mathcal{C}}(X,\, C),\, t \in \mathrm{Hom}_{\mathcal{C}}(Y,\, C)\big)$ such that $s \circ f = t \circ g$. The colimit of the functor F defined above, if it exists, is nothing but the *pushout* of the pair of morphisms $f: Z \to X,\, g: Z \to Y$ in $\mathcal{C}$.

In a similar manner *pullbacks* can be obtained as a special case of limits. Consider a category J with three objects B_1, B_2, B_3 and five morphisms 1_{B_1}, 1_{B_2}, 1_{B_3}, u, v, where $u \in \mathrm{Hom}_{\mathrm{I}}(B_1, B_3)$, $v \in \mathrm{Hom}_{\mathrm{I}}(B_2, B_3)$ and define the functor $F: J \to \mathcal{C}$ as follows:

$$F(B_1) = X, \quad F(B_2) = Y, \quad F(B_3) = Z,$$

$$F(u) = f, \quad F(v) = g.$$

It can be easily seen that the limit of the functor F defined above, if it exists, is nothing but the *pullback* of the pair of morphisms $f: X \to Z,\, g: Y \to Z$ in $\mathcal{C}$. $\qquad\qquad\qquad\qquad\qquad\qquad\qquad\qquad\qquad\qquad\qquad\qquad$ □

The uniqueness up to isomorphism of initial and final objects in a category proved in Proposition 1.3.9 implies:

Proposition 2.2.9 *When it exists, the (co)limit of a functor is unique up to isomorphism.*

Remark 2.2.10 In light of Example 2.2.8, the uniqueness results stated in Propositions 2.1.3, 2.1.8 and 2.1.15 can now be easily derived from Proposition 2.2.9. □

Example 2.2.11 The category **Set**($\subseteq$) defined in Example 1.2.2, (2) is obviously not complete, as we have already established in Example 1.3.10, (6) that it has no final objects (i.e., the empty functor from the empty category to **Set**($\subseteq$) does not have a limit). However, **Set**($\subseteq$) is cocomplete. To this end, consider a functor $H: I \to$ **Set**($\subseteq$) where I is a small category. Then $\big(Z,\, (u_{H(i),\, Z})_{i \in \mathrm{Ob}\, I}\big)$ is the colimit of H, where $Z = \bigcup_{i \in \mathrm{Ob}\, I} H(i)$. Recall that for all sets A, B such that $A \subseteq B$, we denote by $u_{A,\, B}$ the unique morphism in **Set**($\subseteq$) from A to B. With this in mind, it is straightforward to see that $\big(Z,\, (u_{H(i),\, Z})_{i \in \mathrm{Ob}\, I}\big)$ is a cocone on H.

Assume now that $\big(W,\, (u_{H(i),\, W})_{i \in \mathrm{Ob}\, I}\big)$ is another cocone on H. In particular, as we have morphisms $u_{H(i),\, W}: H(i) \to W$ in **Set**($\subseteq$), it follows that $H(i) \subseteq W$ for all $i \in \mathrm{Ob}\, I$. This implies that $Z = \bigcup_{i \in \mathrm{Ob}\, I} H(i) \subseteq W$ and we have a unique morphism in **Set**($\subseteq$) between Z and W, namely $u_{Z,\, W}$. Furthermore, for all $i \in \mathrm{Ob}\, I$

we have $u_{Z,W} \circ u_{H(i),Z} = u_{H(i),W}$. This shows that $u_{Z,W}$ is the unique morphism in $\mathbf{Set}(\subseteq)$ which makes the following diagram commutative for all $i \in \mathrm{Ob}\, I$:

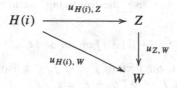

Hence, $\big(Z,\, (u_{H(i),Z})_{i \in \mathrm{Ob}\, I}\big)$ is the colimit of H.   □

Lemma 2.2.12 *Let $G\colon I \to \mathcal{C}$ be a functor and let $\big(L,\, (p_i)_{i \in \mathrm{Ob}\, I}\big)$ be a cone on G. Then $\big(L,\, (p_i)_{i \in \mathrm{Ob}\, I}\big)$ is the limit of G if and only if $\big(L,\, (p_i^{op})_{i \in \mathrm{Ob}\, I}\big)$ is the colimit of the dual functor $G^{op}\colon I^{op} \to \mathcal{C}^{op}$.*

Proof Assume that $\big(L,\, (p_i)_{i \in \mathrm{Ob}\, I}\big)$ is the limit of G and, furthermore, consider a cocone $\Big(M,\, \big(q_i^{op} \in \mathrm{Hom}_{\mathcal{C}^{op}}(G^{op}(i),\, M)\big)_{i \in \mathrm{Ob}\, I}\Big)$ on G^{op}. Lemma 2.2.3 implies that $\Big(M,\, \big(q_i \in \mathrm{Hom}_{\mathcal{C}}(M,\, G(i))\big)_{i \in \mathrm{Ob}\, I}\Big)$ is a cone on G and since $\big(L,\, (p_i)_{i \in \mathrm{Ob}\, I}\big)$ is its limit, there exists a unique $f \in \mathrm{Hom}_{\mathcal{C}}(M,\, L)$ such that the following diagram is commutative for all $i \in I$:

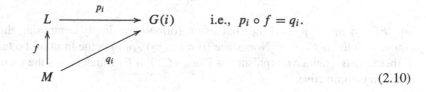

$$\text{i.e., } p_i \circ f = q_i.$$

$$(2.10)$$

In particular, this implies that we also have $f^{op} \circ^{op} p_i^{op} = q_i^{op}$ for all $i \in I$, i.e., the following diagram is commutative:

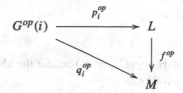

Moreover, the uniqueness of the morphism f^{op} which makes the above diagram commutative follows from the uniqueness of the morphism which makes diagram (2.10) commutative. Therefore, we can conclude that $\big(L,\, (p_i^{op})_{i \in \mathrm{Ob}\, I}\big)$ is the colimit of G^{op}.   □

As an easy consequence of Theorem 2.2.6 we obtain the following:

Corollary 2.2.13 *Let I be a small category and $F\colon I \to \mathcal{C}$ a functor. Then:*

(1) F has a limit of and only if the comma category $(\Delta \downarrow T_F)$ has a final object;

(2) F has a colimit if and only if the comma category $(T_F \downarrow \Delta)$ has an initial object.

We record here, for further use, the following useful results which generalize Proposition 2.1.9:

Proposition 2.2.14 *Let $F: I \to C$ be a functor and $C \in \mathrm{Ob}\, C$.*

(1) If $\bigl(\lim F, (p_i)_{i \in \mathrm{Ob}\, I}\bigr)$ is the limit of F and $f, g \in \mathrm{Hom}_C(C, \lim F)$ such that $p_i \circ f = p_i \circ g$ for all $i \in \mathrm{Ob}\, I$ then $f = g$.

(2) If $\bigl(\mathrm{colim}\, F, (q_i)_{i \in \mathrm{Ob}\, I}\bigr)$ is the colimit of $F: I \to C$ and $f, g \in \mathrm{Hom}_C(\mathrm{colim}\, F, C)$ such that $f \circ q_i = g \circ q_i$ for all $i \in \mathrm{Ob}\, I$ then $f = g$.

Proof (1) To start with, note that $\bigl(C, (p_i \circ f)_{i \in \mathrm{Ob}\, I}\bigr)$ is a cone on the functor F, i.e., $F(d) \circ p_i \circ f = p_j \circ f$ for any $d \in \mathrm{Hom}_I(i, j)$. Indeed, since $\bigl(\lim F, (p_i)_{i \in \mathrm{Ob}\, I}\bigr)$ is the limit of the functor F and in particular a cone on F, the following diagram is commutative:

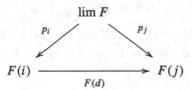

i.e., $F(d) \circ p_i = p_j$ and the conclusion follows by simply composing the last equality on the right by f. Now since $\bigl(\lim F, (p_i)_{i \in \mathrm{Ob}\, I}\bigr)$ is the limit of the functor F there exists a unique morphism $h \in \mathrm{Hom}_C(C, \lim F)$ which makes the following diagram commutative:

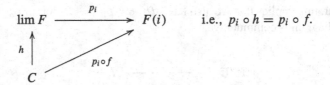

i.e., $p_i \circ h = p_i \circ f$.

As both morphisms $f, g \in \mathrm{Hom}_C(C, \lim F)$ render the above diagram commutative, the desired conclusion follows.

(2) Lemma 2.2.12 implies that $\bigl(\mathrm{colim}\, F, (q_i^{op})_{i \in \mathrm{Ob}\, I}\bigr)$ is the limit of the functor $F^{op}: I^{op} \to C^{op}$ and moreover, we have $q_i^{op} \circ^{op} f^{op} = q_i^{op} \circ^{op} g^{op}$ for all $i \in \mathrm{Ob}\, I$, where $\circ^{op}$ denotes the composition in C^{op}. The first part of the proof implies $f^{op} = g^{op}$ and therefore $f = g$, as desired.   □

Lemma 2.2.15 *Let $F, G: J \to C$ be two functors, where J is a small category, and denote by $\bigl(L, (p_j: L \to F(j))_{j \in \mathrm{Ob}\, J}\bigr)$ and $\bigl(C, (q_j: F(j) \to C)_{j \in \mathrm{Ob}\, J}\bigr)$ the limit and colimit, respectively, of F. If $\alpha: F \to G$ is a natural transformation, then:*

(1) $\bigl(L,\ (\alpha_j \circ p_j \colon L \to G(j))_{j \in \mathrm{Ob}\,J}\bigr)$ *is a cone on* G*. Furthermore, if* α *is a natural isomorphism then* $\bigl(L,\ (\alpha_j \circ p_j \colon L \to G(j))_{j \in \mathrm{Ob}\,J}\bigr)$ *is the limit of* G*.*

(2) *If* α *is a natural isomorphism then* $\bigl(C,\ (q_j \circ \alpha_j^{-1} \colon G(j) \to C)_{j \in \mathrm{Ob}\,J}\bigr)$ *is the colimit of* G*.*

Proof Given that in most of the previous proofs we have worked mainly with limits, here we will prove the second assertion concerning colimits.

(2) First we show that $\bigl(C,\ (q_j \circ \alpha_j^{-1} \colon G(j) \to C)_{j \in \mathrm{Ob}\,J}\bigr)$ is a cocone on G. To this end, consider $d \in \mathrm{Hom}_J(i,\, l)$; we need to prove the commutativity of the following diagram:

$$\begin{array}{ccc}
 & C & \\
{\scriptstyle q_i \circ \alpha_i^{-1}} \nearrow & & \nwarrow {\scriptstyle q_l \circ \alpha_l^{-1}} \\
G(i) & \xrightarrow{\quad G(d) \quad} & G(l)
\end{array}$$

i.e., $q_l \circ \alpha_l^{-1} \circ G(d) = q_i \circ \alpha_i^{-1}.$

$$(2.11)$$

Since $\bigl(C,\ (q_j \colon F(j) \to C)_{j \in \mathrm{Ob}\,J}\bigr)$ is in particular a cocone on F, the following diagram is commutative:

$$\begin{array}{ccc}
 & C & \\
{\scriptstyle q_i} \nearrow & & \nwarrow {\scriptstyle q_l} \\
F(i) & \xrightarrow{\quad F(d) \quad} & F(l)
\end{array}$$

i.e., $q_l \circ F(d) = q_i.$

$$(2.12)$$

Furthermore, as α is a natural transformation, the following diagram is commutative as well:

$$\begin{array}{ccc}
F(i) & \xrightarrow{\ \alpha_i\ } & G(i) \\
{\scriptstyle F(d)} \downarrow & & \downarrow {\scriptstyle G(d)} \\
F(l) & \xrightarrow[\ \alpha_l\ ]{} & G(l)
\end{array}$$

i.e., $G(d) \circ \alpha_i = \alpha_l \circ F(d).$

$$(2.13)$$

Putting all the above together yields

$$q_l \circ \alpha_l^{-1} \circ G(d) \overset{(2.13)}{=} q_l \circ F(d) \circ \alpha_i^{-1} \overset{(2.12)}{=} q_i \circ \alpha_i^{-1},$$

which proves that (2.11) holds and therefore $\left(C, (q_j \circ \alpha_j^{-1}: G(j) \to C)_{j\in \mathrm{Ob}\, J}\right)$ is indeed a cocone on G. Consider now another cocone $\left(C', (t_j: G(j) \to C')_{j\in\mathrm{Ob}\, J}\right)$ on G. Then $\left(C', (t_j \circ \alpha_j: F(j) \to C')_{j\in\mathrm{Ob}\, J}\right)$ is a cocone on F. Indeed, for all $d \in \mathrm{Hom}_J(i, l)$ we have

$$t_l \circ \alpha_l \circ F(d) \stackrel{(2.13)}{=} t_l \circ G(d) \circ \alpha_i = t_i \circ \alpha_i,$$

where in the last equality we used the fact that $\left(C', (t_j: G(j) \to C')_{j\in\mathrm{Ob}\, J}\right)$ is a cocone on G.

Now since $\left(C, (q_j: F(j) \to C)_{j\in\mathrm{Ob}\, J}\right)$ is the colimit of F, there exists a unique $f \in \mathrm{Hom}_C(C, C')$ such that the following diagram is commutative for all $j \in \mathrm{Ob}\, J$:

$$
\begin{array}{ccc}
F(j) & \xrightarrow{\;q_j\;} & C \\
 & \searrow^{t_j\circ\alpha_j} & \downarrow f \\
 & & C'
\end{array}
\qquad \text{i.e., } f \circ q_j = t_j \circ \alpha_j.
$$

Thus $f \in \mathrm{Hom}_C(C, C')$ is the unique morphism such that $f \circ (q_j \circ \alpha_j^{-1}) = t_j$, i.e., the unique morphism which makes the following diagram commutative for all $j \in \mathrm{Ob}\, J$:

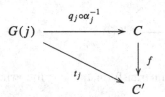

This proves that $\left(C, (q_j \circ \alpha_j^{-1}: G(j) \to C)_{j\in\mathrm{Ob}\, J}\right)$ is the initial object in the category of cocones on G, as desired.

(1) To start with, note that $\alpha^{op}: G^{op} \to F^{op}$ is also a natural isomorphism by Proposition 1.8.3. Furthermore, $(\alpha^{op})^{-1}: F^{op} \to G^{op}$ is again a natural isomorphism (see Example 1.7.2, (6)). Lemma 2.2.12 implies that $\left(L, (p_j^{op})_{j\in\mathrm{Ob}\, J}\right)$ is the colimit of F^{op} and using part (2) proved above we obtain that $\left(L, (p_j^{op} \circ^{op} \alpha_j^{op})_{j\in\mathrm{Ob}\, J}\right)$ is the colimit of G^{op}. Now since we have $p_j^{op} \circ^{op} \alpha_j^{op} = (\alpha_j \circ p_j)^{op}$, the conclusion follows by Lemma 2.2.12. $\square$

(Co)limit as a Functor

It turns out that taking (co)limits[9] yields a functor:

Theorem 2.2.16 *Let I be a small category and C a complete category. Then* $\lim: \text{Fun}(I, C) \to C$ *defined below is a functor:*

$$\lim(F) = \lim F, \qquad \lim(\alpha) = \overline{\alpha}$$

for all functors $F, G: I \to C$ and all natural transformations $\alpha: F \to G$,

where $\left(\lim F, (p_i)_{i \in \text{Ob } I}\right)$ and $\left(\lim G, (s_i)_{i \in \text{Ob } I}\right)$ are the limits of F and G respectively and $\overline{\alpha} \in \text{Hom}_C(\lim F, \lim G)$ is the unique morphism which makes the following diagram commute for all $i \in \text{Ob } I$:

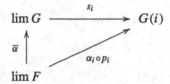

Proof Before going into the proof, we point out that in light of Lemma 2.2.15, (1) the pair $\left(\lim F, (\alpha_i \circ p_i)_{i \in \text{Ob } I}\right)$ is a cone on G. Therefore, the unique morphism $\overline{\alpha} \in \text{Hom}_C(\lim F, \lim G)$ which makes the diagram above commutative for all $i \in \text{Ob } I$ exists by virtue of Definition 2.2.7.

Now let $F: I \to C$ be a functor and 1_F the identity transformation. Then, $\lim(1_F)$ is the unique morphism in $\text{Hom}_C(\lim F, \lim F)$ such that the following diagram is commutative for all $i \in \text{Ob } I$:

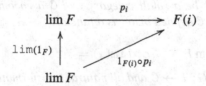

As $1_{\lim F}$ makes the same diagram commutative we obtain $\lim(1_F) = 1_{\lim F}$. This leads to $\lim(1_F) = 1_{\lim(F)}$, as desired.

Consider now functors $F, G, H: I \to C$ and natural transformations $\alpha: F \to G$, $\beta: G \to H$. The proof will be finished once we show that $\lim(\beta \circ \alpha) = \lim(\beta) \circ \lim(\alpha)$. To this end, let $\left(\lim F, (p_i)_{i \in \text{Ob } I}\right), \left(\lim G, (s_i)_{i \in \text{Ob } I}\right), \left(\lim H, (t_i)_{i \in \text{Ob } I}\right)$ be the limits of F, G and H, respectively. Recall that $\overline{\beta \circ \alpha}, \overline{\beta}, \overline{\alpha}$ are the unique

---

[9] Note that defining the (co)limit functor requires an arbitrary choice of a limit for each functor. This is always possible as we assumed the axiom of choice to hold.

morphisms which make the following diagrams commutative for all $i \in \mathrm{Ob}\, I$:

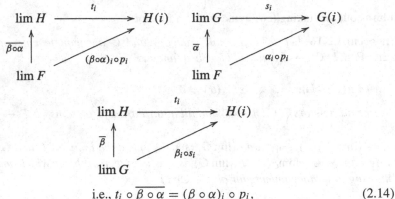

i.e., $t_i \circ \overline{\beta \circ \alpha} = (\beta \circ \alpha)_i \circ p_i,$ (2.14)

$$s_i \circ \overline{\alpha} = \alpha_i \circ p_i, \qquad\qquad\qquad (2.15)$$

$$t_i \circ \overline{\beta} = \beta_i \circ s_i. \qquad\qquad\qquad (2.16)$$

Putting everything together yields

$$t_i \circ \overline{\beta} \circ \overline{\alpha} \overset{(2.16)}{=} \beta_i \circ s_i \circ \overline{\alpha} \overset{(2.15)}{=} \beta_i \circ \alpha_i \circ p_i = (\beta \circ \alpha)_i \circ p_i \text{ for all } i \in \mathrm{Ob}\, I.$$

Now since $\overline{\beta \circ \alpha}$ is the unique morphism for which (2.14) holds, we obtain $\overline{\beta \circ \alpha} = \overline{\beta} \circ \overline{\alpha}$, and the proof is now complete.   □

For the sake of completeness we record below the dual result concerning colimits and leave the proof to the reader:

Theorem 2.2.17 *Let I be a small category and C a cocomplete category. Then* $colim\colon \mathrm{Fun}(I, C) \to C$ *defined below is a functor:*

$$colim(F) = \mathrm{colim}\, F, \qquad colim(\alpha) = \overline{\alpha}$$

for all functors $F, G\colon I \to C$ and all natural transformations $\alpha\colon F \to G$,

where $\left(\mathrm{colim}\, F, (q_i)_{i\in\mathrm{Ob}\, I}\right)$ and $\left(\mathrm{colim}\, G, (t_i)_{i\in\mathrm{Ob}\, I}\right)$ are the colimits of F and G respectively and $\overline{\alpha} \in \mathrm{Hom}_C(\mathrm{colim}\, F, \mathrm{colim}\, G)$ is the unique morphism which makes the following diagram commute for all $i \in \mathrm{Ob}\, I$:

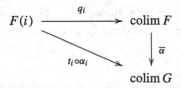

2.3   (Co)limit as a Representing Pair

In this section we show that the (co)limit of a functor arises as the representing pair of a certain functor. Throughout, I is a small category and $F\colon I \to C$ is a functor. We can define two functors, $\mathrm{Cocone}(F, -)\colon C \to \mathbf{Set}$ and $\mathrm{Cone}(F, -)\colon C \to \mathbf{Set}$, as follows:

$$\mathrm{Cocone}(F, C) = \mathrm{Cocone}(F, C),$$

$$\mathrm{Cocone}(F, f)\colon \mathrm{Cocone}(F, C) \to \mathrm{Cocone}(F, D),$$

$$\mathrm{Cocone}(F, f)\big((t_i)_{i \in \mathrm{Ob}\, I}\big) = \big(f \circ t_i\big)_{i \in \mathrm{Ob}\, I},$$

$$\mathrm{Cone}(F, C) = \mathrm{Cone}(F, C),$$

$$\mathrm{Cone}(F, f)\colon \mathrm{Cone}(F, D) \to \mathrm{Cone}(F, C),$$

$$\mathrm{Cone}(F, f)\big((s_i)_{i \in \mathrm{Ob}\, I}\big) = \big(s_i \circ f\big)_{i \in \mathrm{Ob}\, I}$$

for all $C, D \in \mathrm{Ob}\,C$, $f \in \mathrm{Hom}_C(C, D)$, $(t_i)_{i \in \mathrm{Ob}\, I} \in \mathrm{Cocone}(F, C)$ and $(s_i)_{i \in \mathrm{Ob}\, I} \in \mathrm{Cone}(F, D)$.

Note that $\mathrm{Cocone}(F, -)\colon C \to \mathbf{Set}$ is a covariant functor while $\mathrm{Cone}(F, -)\colon C \to \mathbf{Set}$ is contravariant.

Lemma 2.3.1 *Let* $F\colon I \to C$ *be a functor and* I *a small category. Then* $\Big(C, \big(t_i \in \mathrm{Hom}_C(C, F(i))\big)_{i \in \mathrm{Ob}\, I}\Big)$ *is the representing pair of the contravariant functor* $\mathrm{Cone}(F, -)$ *if and only if* $\Big(C, \big(t_i^{op} \in \mathrm{Hom}_{C^{op}}(F(i), C)\big)_{i \in \mathrm{Ob}\, I}\Big)$ *is the representing pair of the functor* $\mathrm{Cocone}(F^{op}, -)$.

Proof Assume that $\Big(C, \big(t_i^{op} \in \mathrm{Hom}_{C^{op}}(F(i), C)\big)_{i \in \mathrm{Ob}\, I}\Big)$ is the representing pair of the functor $\mathrm{Cocone}(F^{op}, -)\colon C^{op} \to \mathbf{Set}$ and $\big(u_i \in \mathrm{Hom}_C(C', F(i))_{i \in \mathrm{Ob}\, I}\big) \in \mathrm{Cone}(F, C')$, where $C' \in \mathrm{Ob}\,C$. Then Lemma 2.2.3 implies that $\big(u_i^{op}\big)_{i \in \mathrm{Ob}\, I} \in \mathrm{Cocone}(F^{op}, C')$ and using Proposition 1.7.7 we obtain a unique $f^{op} \in \mathrm{Hom}_{C^{op}}(C, C')$ such that

$$\mathrm{Cocone}(F^{op}, f^{op})\big((t_i^{op})_{i \in \mathrm{Ob}\, I}\big) = \big((u_i^{op})_{i \in \mathrm{Ob}\, I}\big).$$

This comes down to $(t_i \circ f)_{i \in \mathrm{Ob}\, I} = (u_i)_{i \in \mathrm{Ob}\, I}$. In other words, there exists a unique $f \in \mathrm{Hom}_C(C', C)$ such that $\mathrm{Cone}(F, f)\big((t_i)_{i \in \mathrm{Ob}\, I}\big) = \big((u_i)_{i \in \mathrm{Ob}\, I}\big)$ and it follows by Corollary 1.8.2 that $\Big(C, \big(t_i \in \mathrm{Hom}_C(C, F(i))\big)_{i \in \mathrm{Ob}\, I}\Big)$ is the representing pair of the contravariant functor $\mathrm{Cone}(F, -)\colon C \to \mathbf{Set}$.

The converse follows by similar arguments and is left to the reader.   □

The colimit of F, if it exists, is the representing pair of the functor $\mathrm{Cocone}(F, -)$; similarly, the limit of F, if it exists, is the representing pair of the functor $\mathrm{Cone}(F, -)$.

Theorem 2.3.2 *Let I be a small category and $F: I \to C$ a functor. Then:*

*(1) a cocone on F is the colimit of F if and only if is the representing pair of the functor $\mathrm{Cocone}(F, -): C \to$ **Set**;*

*(2) a cone on F is the limit of F if and only if is the representing pair of the contravariant functor $\mathrm{Cone}(F, -): C \to$ **Set**.*

Proof (1) Assume first that $\left(C, \left(t_i \in \mathrm{Hom}_C(F(i), C)\right)_{i \in \mathrm{Ob}\,I}\right)$ is the colimit of F. We define a natural transformation $\alpha: \mathrm{Hom}_C(C, -) \to \mathrm{Cocone}(F, -)$ as follows for all $D \in \mathrm{Ob}\,C$ and $f \in \mathrm{Hom}_C(C, D)$:

$$\alpha_D(f) = \left(f \circ t_i\right)_{i \in \mathrm{Ob}\,I} \in \mathrm{Cocone}(F, D).$$

In order to prove that α is indeed a natural transformation, consider $u \in \mathrm{Hom}_C(D, D')$; we need to show the commutativity of the following diagram:

$$
\begin{array}{ccc}
\mathrm{Hom}_C(C, D) & \xrightarrow{\ \ \alpha_D\ \ } & \mathrm{Cocone}(F, D) \\
\Big\downarrow {\scriptstyle \mathrm{Hom}_C(F, u)} & & \Big\downarrow {\scriptstyle \mathrm{Cocone}(F, u)} \\
\mathrm{Hom}_C(C, D') & \xrightarrow[\ \ \alpha_{D'}\ \ ]{} & \mathrm{Cocone}(F, D')
\end{array}
\tag{2.17}
$$

Indeed, for all $f \in \mathrm{Hom}_C(C, D)$ we have

$$\mathrm{Cocone}(F, u) \circ \alpha_D(f) = \mathrm{Cocone}(F, u)\left((f \circ t_i)_{i \in \mathrm{Ob}\,I}\right) = \left(u \circ f \circ t_i\right)_{i \in \mathrm{Ob}\,I}$$

$$= \alpha_{D'}(u \circ f) = \alpha_{D'} \circ \mathrm{Hom}_C(F, u)(f),$$

which shows that diagram (2.17) is indeed commutative. We are left to show that each α_D is an isomorphism. To this end, recall that since $\left(C, \left(t_i \in \mathrm{Hom}_C(F(i), C)\right)_{i \in \mathrm{Ob}\,I}\right)$ is the colimit of F, for any $\left(v_i\right)_{i \in \mathrm{Ob}\,I} \in \mathrm{Cocone}(F, D)$ there exists a unique $v \in \mathrm{Hom}_C(C, D)$ such that the following diagram is commutative for all $i \in \mathrm{Ob}\,I$:

$$
\begin{array}{ccc}
F(i) & \xrightarrow{\ \ t_i\ \ } & C \\
 & {\scriptstyle v_i}\searrow & \Big\downarrow {\scriptstyle v} \\
 & & D
\end{array}
\qquad \text{i.e., } v \circ t_i = v_i
\tag{2.18}
$$

We can now define a map $\beta_D\colon \mathrm{Cocone}(F, D) \to \mathrm{Hom}_{\mathcal{C}}(F, D)$ by $\beta_D\big((v_i)_{i\in\mathrm{Ob}\,I}\big) = v$, where $v \in \mathrm{Hom}_{\mathcal{C}}(C, D)$ is the unique morphism which makes diagram (2.18) commutative. Then, for all $(v_i)_{i\in\mathrm{Ob}\,I} \in \mathrm{Cocone}(F, D)$ and $f \in \mathrm{Hom}_{\mathcal{C}}(C, D)$ we have

$$\alpha_D \circ \beta_D\big((v_i)_{i\in\mathrm{Ob}\,I}\big) = \alpha_D(v) = \big(v \circ t_i\big)_{i\in\mathrm{Ob}\,I} = (v_i)_{i\in\mathrm{Ob}\,I},$$

$$\beta_D \circ \alpha_D(f) = \beta_D\big((f \circ t_i)_{i\in\mathrm{Ob}\,I}\big) = f,$$

which shows that β_D is the inverse of α_D. Therefore, α is a natural isomorphism. This shows that the functor $\mathrm{Cocone}(F, -)\colon \mathcal{C} \to \mathbf{Set}$ is representable, as desired.

Suppose now that $\Big(C,\ \big(t_i \in \mathrm{Hom}_{\mathcal{C}}(F(i), C)\big)_{i\in\mathrm{Ob}\,I}\Big)$ is the representing pair of the functor $\mathrm{Cocone}(F, -)\colon \mathcal{C} \to \mathbf{Set}$ and consider $D \in \mathrm{Ob}\,\mathcal{C}$ and $(v_i)_{i\in\mathrm{Ob}\,I} \in \mathrm{Cocone}(F, D)$. Using Proposition 1.7.7, there exists a unique $f \in \mathrm{Hom}_{\mathcal{C}}(C, D)$ such that

$$\mathrm{Cocone}(F, f)\big((t_i)_{i\in\mathrm{Ob}\,I}\big) = (v_i)_{i\in\mathrm{Ob}\,I}.$$

This comes down to $\big(f \circ t_i\big)_{i\in\mathrm{Ob}\,I} = (v_i)_{i\in\mathrm{Ob}\,I}$. In other words, there exists a unique $f \in \mathrm{Hom}_{\mathcal{C}}(C, D)$ such that the following diagram commutes for all $i \in \mathrm{Ob}\,I$:

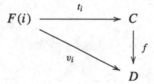

Therefore, $\big(C, (t_i)_{i\in\mathrm{Ob}\,I}\big)$ is the colimit of F.

(2) By Lemma 2.2.12, a cone $\Big(C,\ \big(s_i \in \mathrm{Hom}_{\mathcal{C}}(C, F(i))\big)_{i\in\mathrm{Ob}\,I}\Big)$ on F is the limit of F if and only if $\Big(C,\ \big(s_i^{op} \in \mathrm{Hom}_{\mathcal{C}^{op}}(F(i), C)\big)_{i\in\mathrm{Ob}\,I}\Big)$ is the colimit of the dual functor F^{op}. Using 1) it follows that $\Big(C,\ \big(s_i^{op} \in \mathrm{Hom}_{\mathcal{C}^{op}}(F(i), C)\big)_{i\in\mathrm{Ob}\,I}\Big)$ is the colimit of F^{op} if and only if $\Big(C,\ \big(s_i^{op} \in \mathrm{Hom}_{\mathcal{C}^{op}}(F(i), C)\big)_{i\in\mathrm{Ob}\,I}\Big)$ is the representing pair of the functor $\mathrm{Cocone}(F^{op}, -)\colon \mathcal{C}^{op} \to \mathbf{Set}$ and by Lemma 2.3.1 this is equivalent to $\Big(C,\ \big(s_i \in \mathrm{Hom}_{\mathcal{C}}(C, F(i))\big)_{i\in\mathrm{Ob}\,I}\Big)$ being the representing pair of the functor $\mathrm{Cone}(F, -)\colon \mathcal{C} \to \mathbf{Set}$.   □

2.4   (Co)limits by (Co)equalizers and (Co)products

(Co)products and (co)equalizers are perhaps the most important among the special cases of (co)limits. This is due to the fact that all small (co)limits can be constructed out of (co)products and (co)equalizers. We start by presenting an example which hints at this construction.

Example 2.4.1 Let $F: I \to$ **Set** be a functor, where I is a small category. Then: $\left(\lim F, \ (p_k)_{k \in \mathrm{Ob}I}\right)$ is the limit of F, where

$$\lim F = \{(x_k)_{k \in \mathrm{Ob}I} \in \prod_{k \in \mathrm{Ob}I} F(k) \mid F(f)(x_i) = x_j \text{ for all } f \in \mathrm{Hom}_I(i, \ j)\},$$

$$p_j: \lim F \to F(j), \ p_j((x_k)_{k \in \mathrm{Ob}I}) = x_j, \text{ for all } j \in \mathrm{Ob}\,I.$$

We start by proving that $\left(\lim F, \ (p_k)_{k \in \mathrm{Ob}I}\right)$ is a cone on F. Indeed, for any $f \in \mathrm{Hom}_I(i, \ j)$ we have $F(f)(x_i) = x_j$ for all $(x_k)_{k \in \mathrm{Ob}\,I} \in \lim F$, which can be written equivalently as $F(f)\big(p_i((x_k)_{k \in \mathrm{Ob}I})\big) = p_j((x_k)_{k \in \mathrm{Ob}I})$. Thus we obtain $F(f) \circ p_i = p_j$, as desired.

Assume now that $\left(C, \ (s_k)_{k \in \mathrm{Ob}\,I}\right)$ is another cone on F, i.e., $C \in \mathrm{Ob}\,\mathbf{Set}$ and $s_k \in \mathrm{Hom}_{\mathbf{Set}}(C, \ F(k))$ such that $F(f) \circ s_i = s_j$ for all $f \in \mathrm{Hom}_I(i, \ j)$. Now recall that $\left(\prod_{k \in \mathrm{Ob}I} F(k), \ (p_k)_{k \in \mathrm{Ob}I}\right)$ is the product in **Set** of the family $\left((F(k))_{k \in \mathrm{Ob}I}\right)$; hence, there exists a unique morphism $g: C \to \prod_{k \in \mathrm{Ob}\,I} F(k)$ in **Set** such that the following diagram is commutative for all $j \in \mathrm{Ob}I$:

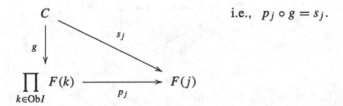

i.e., $p_j \circ g = s_j$.

We are left to prove that $\mathrm{Im}g \subseteq \lim F$. To this end, notice that $g(c) = \big(s_k(c)\big)_{k \in \mathrm{Ob}\,I}$ for all $c \in C$. Moreover, since $F(f) \circ s_i = s_j$ for any $f \in \mathrm{Hom}_I(i, \ j)$ we get $F(f)\big(s_i(c)\big) = s_j(c)$ for all $c \in C$. Thus $g(c) = \big(s_k(c)\big)_{k \in \mathrm{Ob}\,I} \in \lim F$ for all $c \in C$. $\square$

The previous example shows that the limit of a functor $F: I \to$ **Set**, where I is a small category, can be constructed as a subset of the product $\prod_{i \in \mathrm{Ob}I} F(i)$. This suggests that equalizers are being used in order to construct the limit. The next theorem shows that this method can be generalized to arbitrary categories allowing for the construction of small limits out of products and equalizers.

Theorem 2.4.2 *A category $\mathcal{C}$ is (co)complete if and only if it has (co)products and (co)equalizers.*

Proof We will only prove the assertion regarding completeness and leave the (dual) one about cocompleteness to the reader. Obviously, if a category is complete then it has products and equalizers, as shown in Example 2.2.8, (2) and (3). Conversely, assume $\mathcal{C}$ is a category with products and equalizers and let $F : I \to \mathcal{C}$ be a functor, where I is a small category. For any morphism f in I we will denote by $d(f)$ the domain of f and by $c(f)$ the codomain of f; in other words we have $f \in \mathrm{Hom}_I(d(f), c(f))$. We start by constructing the products in $\mathcal{C}$ of the families of objects $(F(i))_{i \in \mathrm{Ob} I}$ and $\big(F(c(f))\big)_{f \in \mathrm{Hom}_I(d(f), c(f))}$, respectively:

$$\left(\prod_i F(i), (u_i)_{i \in \mathrm{Ob}\, I}\right), \qquad \left(\prod_f F(c(f)), (v_{c(f)})_{f \in \mathrm{Hom}_I(d(f), c(f))}\right).$$

As $\big(\prod_f F(c(f)), (v_{c(f)})_{f \in \mathrm{Hom}_I(d(f), c(f))}\big)$ is the product in $\mathcal{C}$ of the family of objects $\big(F(c(f))\big)_{f \in \mathrm{Hom}_I(d(f), c(f))}$, there exists a unique morphism $\alpha : \prod_i F(i) \to \prod_f F(c(f))$ in $\mathcal{C}$ such that the following diagram is commutative for all $g \in \mathrm{Hom}_I(d(g), c(g))$:

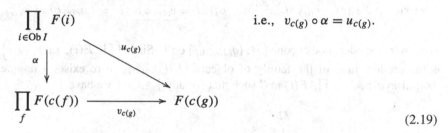

i.e., $v_{c(g)} \circ \alpha = u_{c(g)}.$

$$\tag{2.19}$$

Similarly, there exists a unique morphism $\beta : \prod_i F(i) \to \prod_f F(c(f))$ in $\mathcal{C}$ such that the following diagram is commutative for all $g \in \mathrm{Hom}_I(d(g), c(g))$:

$$\prod_{i \in \mathrm{Ob}\, I} F(i)$$

i.e., $v_{c(g)} \circ \beta = F(g) \circ u_{d(g)}.$

$\beta \downarrow \qquad \xrightarrow{F(g) \circ u_{d(g)}}$

$$\prod_f F(c(f)) \xrightarrow{\;\;v_{c(g)}\;\;} F(c(g))$$

$$\tag{2.20}$$

Now consider (L, l) to be the equalizer in $\mathcal{C}$ of the pair of morphisms (α, β). The complete picture is captured by the following diagram:

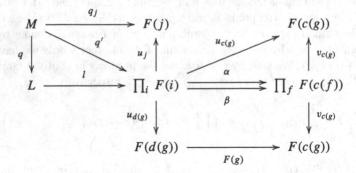

We will prove that $\big(L, (p_i = u_i \circ l)_{i \in \mathrm{Ob}\, I}\big)$ is the limit of the functor F. First we prove that $\big(L, (p_i)_{i \in \mathrm{Ob}\, I}\big)$ is a cone on F. Indeed, if $r \in \mathrm{Hom}_I(d(r), c(r))$ we have

$$F(r) \circ p_{d(r)} = \underline{F(r) \circ u_{d(r)}} \circ l \overset{(2.20)}{=} v_{c(r)} \circ \beta \circ l = \underline{v_{c(r)} \circ \alpha \circ l} \overset{(2.19)}{=} u_{c(r)} \circ l = p_{c(r)}.$$

Moreover, consider another cone $\big(M, (q_i)_{i \in \mathrm{Ob}I}\big)$ on F. Since $\big(\prod_i F(i), (u_i)_{i \in \mathrm{Ob}I}\big)$ is the product in $\mathcal{C}$ of the family of objects $(F(i))_{i \in \mathrm{Ob}I}$, there exists a unique morphism $q' : M \to \prod_i F(i)$ in $\mathcal{C}$ such that for any $j \in \mathrm{Ob}\, I$ we have

$$
\begin{array}{ccc}
 & M & \\
 & \downarrow q' \quad \searrow^{q_j} & \\
\prod_{i \in \mathrm{Ob}I} F(i) & \xrightarrow{\quad u_j \quad} & F(j)
\end{array}
\tag{2.21}
$$

Now for any $r \in \mathrm{Hom}_I(d(r), c(r))$ we have

$$\underline{v_{c(r)} \circ \alpha \circ q'} \overset{(2.19)}{=} \underline{u_{c(r)} \circ q'} \overset{(2.21)}{=} q_{c(r)} = F(r) \circ q_{d(r)}$$

$$\overset{(2.21)}{=} \underline{F(r) \circ u_{d(r)}} \circ q' \overset{(2.20)}{=} v_{c(r)} \circ \beta \circ q',$$

where in the third equality we used the fact that $\big(M, (q_i)_{i \in \mathrm{Ob}I}\big)$ is a cone on F. Therefore we have $v_{c(r)} \circ \alpha \circ q' = v_{c(r)} \circ \beta \circ q'$ for any $r \in \mathrm{Hom}_I(d(r), c(r))$ and according to Proposition 2.2.14, (1) we obtain $\alpha \circ q' = \beta \circ q'$. Since (L, l) is the equalizer of the pair of morphisms (α, β) in $\mathcal{C}$ we obtain a unique morphism $q : M \to L$ such that

$$l \circ q = q'.\tag{2.22}$$

It turns out that q is the unique morphism in C which makes the following diagram commute for all $j \in \mathrm{Ob}\,I$:

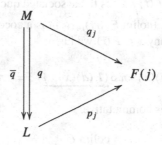

Indeed, for any $j \in \mathrm{Ob}\,I$ we have $p_j \circ q = u_j \circ l \circ q \overset{(2.22)}{=} u_j \circ q' \overset{(2.21)}{=} q_j$. Finally, we are left to prove the uniqueness of q. To this end, assume $\overline{q} \in \mathrm{Hom}_C(M, L)$ is another morphism such that $p_j \circ \overline{q} = q_j$ for all $j \in \mathrm{Ob}\,I$. Hence, we obtain $u_j \circ l \circ \overline{q} = q_j$ for all $j \in \mathrm{Ob}\,I$ and since q' is the unique morphism in C which makes diagram (2.21) commute, we obtain $l \circ \overline{q} = q'$. Now the uniqueness of the morphism in C for which (2.22) holds implies $\overline{q} = q$, as desired.   □

Examples 2.4.3

(1) Products and equalizers for the categories **Set**, **Grp**, **Ab**, $_R\mathcal{M}$, **Top** are constructed in Examples 2.1.5 and 2.1.10, respectively. Hence, Theorem 2.4.2 shows that all these categories are complete.
(2) Similarly, coproducts and coequalizers for the categories **Set**, **Grp**, **Ab**, $_R\mathcal{M}$, **Top** are constructed in Examples 2.1.5 and 2.1.10, respectively. In light of Theorem 2.4.2 we can conclude that all these categories are cocomplete.   □

The next example gives a precise description of colimits in **Set**.

Example 2.4.4 Let I be a small category and $F: I \to$ **Set** a functor. Let $\left(\coprod_{i \in I} F(i),\ (q_i)_{i \in \mathrm{Ob}\,I}\right)$ be the coproduct in **Set** of the family $\left(F(i)\right)_{i \in \mathrm{Ob}\,\mathbf{Set}}$ as described in Example 2.1.5, (9) and let R be the relation on $\coprod_{i \in I} F(i)$ defined as follows:

$$(x, i)\ \mathrm{R}\ (y, j) \text{ if and only if } F(f)(x) = y, \text{ for some } f \in \mathrm{Hom}_I(i, j). \tag{2.23}$$

R is obviously reflexive and transitive but not necessarily symmetric. We denote by $\sim_R$ the equivalence relation on $\coprod_{i \in I} F(i)$ generated by R. Then, the pair $(\mathrm{colim}\, F,\ (r_i)_{i \in \mathrm{Ob}\,I})$ defined below is the colimit of F:

$$\mathrm{colim}\, F = \coprod_{i \in I} F(i) \Big/ {\sim_R},$$

$$r_j: F(j) \to \mathrm{colim},\ r_j = \pi \circ q_j, \text{ for all } j \in \mathrm{Ob}\,I,$$

where $\coprod_{i \in I} F(i) \big/_{\sim_R}$ denotes the quotient set by the equivalence relation $\sim_R$ and $\pi: \coprod_{i \in \mathrm{Ob}\, I} F(i) \to \coprod_{i \in I} F(i) \big/_{\sim_R}$ is the associated quotient function.

We start by proving that $(\mathrm{colim}\, F, (r_i)_{i \in \mathrm{Ob}\, I})$ is a cocone on F. To this end, let $d \in \mathrm{Hom}_I(i,\, j)$; then, for any $x \in F(i)$ we have

$$r_j \circ F(d)(x) = \pi \circ q_j \circ F(d)(x) = \pi \circ \underline{(F(d)(x),\, j)} \stackrel{(2.23)}{=} \pi(x,\, i) = \pi \circ q_i(x) = r_i(x),$$

i.e., the following diagram is commutative:

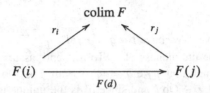

Hence $(\mathrm{colim}\, F, (r_i)_{i \in \mathrm{Ob}\, I})$ is indeed a cocone on F. Consider now another cocone $(C, (t_i)_{i \in \mathrm{Ob}\, I})$ on F. Definition 2.1.2 yields a unique $\psi \in \mathrm{Hom}_{\mathrm{Set}}(\coprod_{i \in I} F(i),\, C)$ which renders the following diagram commutative for all $j \in \mathrm{Ob}\, I$:

$$F(j) \xrightarrow{\quad q_j \quad} \coprod_{i \in \mathrm{Ob}\, I} F(i) \qquad \text{i.e., } \psi \circ q_j = t_j.$$

$$\downarrow \psi$$

$$t_j \searrow \quad C \tag{2.24}$$

Let $(x,\, i),\, (y,\, j) \in \coprod_{i \in \mathrm{Ob}\, I} F(i)$ such that $(x,\, i) \sim_R (y,\, j)$. Thus we have either $(x,\, i)\, R\, (y,\, j)$ or $(y,\, j)\, R\, (x,\, i)$. In the first case, there exists some $f \in \mathrm{Hom}_I(i,\, j)$ such that $F(f)(x) = y$. Then, we obtain

$$\psi(x,\, i) = \underline{\psi \circ q_i}(x) \stackrel{(2.24)}{=} t_i(x) = t_j \circ F(f)(x) = t_j(y) \stackrel{(2.24)}{=} \psi \circ q_j(y) = \psi(y,\, j)$$

where in the third equality we used the fact that $(C, (t_i)_{i \in \mathrm{Ob}\, I})$ is a cocone on F. On the other hand, if $(y,\, j)\, R\, (x,\, i)$ there exists some $g \in \mathrm{Hom}_I(j,\, i)$ such that $F(g)(y) = x$. This leads to

$$\psi(y,\, j) = \underline{\psi \circ q_j}(y) \stackrel{(2.24)}{=} t_j(y) = t_i \circ \underline{F(g)(y)} = t_i(x) \stackrel{(2.24)}{=} \psi \circ q_i(x) = \psi(x,\, i),$$

where the third equality follows from the fact that $(C, (t_i)_{i \in \mathrm{Ob}\, I})$ is a cocone on F.

Putting all together, we proved that $(x,\, i) \sim_R (y,\, j)$ implies $\psi(x,\, i) = \psi(y,\, j)$. Therefore, by the universal property of the quotient set $\coprod_{i \in I} F(i) \big/_{\sim_R}$ there exists a unique map $\varphi: \coprod_{i \in I} F(i) \big/ _{\sim_R} \to C$ such that the following diagram commutes:

$$\coprod_{i \in \mathrm{Ob}\, I} F(i) \xrightarrow{\;\;\pi\;\;} \coprod_{i \in I} F(i) \big/ _{\sim_R} \qquad \text{i.e., } \varphi \circ \pi = \psi.$$

$$\psi \searrow \quad \downarrow \varphi$$

$$C \tag{2.25}$$

Now it can be easily seen that $\varphi: \coprod_{i \in I} F(i) \big/ _{\sim_R} \to C$ is the unique morphism which makes the following diagram commutative for all $i \in \mathrm{Ob}\, I$:

$$F(i) \xrightarrow{\;\;r_i\;\;} \coprod_{i \in I} F(i) \big/ _{\sim_R}$$

$$t_i \searrow \quad \downarrow \varphi$$

$$C$$

Indeed, for all $i \in \mathrm{Ob}\, I$ we have

$$\varphi \circ r_i = \varphi \circ \pi \circ q_i \overset{(2.25)}{=} \psi \circ q_i \overset{(2.24)}{=} t_i.$$

$\square$

2.5   (Co)limit Preserving Functors

Definition 2.5.1 A functor $F: C \to D$ *preserves (small) limits/colimits* when for every functor $G: I \to C$, where I is a (small) category, if $(L, (p_i)_{i \in \mathrm{Ob}\, I})$ is the limit/colimit of G then $(F(L), (F(p_i))_{i \in \mathrm{Ob}\, I})$ is the limit/colimit of FG.

Lemma 2.5.2 *Let $F: C \to D$ be a functor. Then F preserves limits if and only if the dual functor $F^{op}: C^{op} \to D^{op}$ preserves colimits.*

Proof Assume F preserves limits and let I be a small category. If $G: I \to C^{op}$ is a functor whose colimit we denote by $\left(C, (q_i^{op} \in \mathrm{Hom}_{C^{op}}(G(i), C))_{i \in \mathrm{Ob}\, I}\right)$ then using Lemma 2.2.12 we obtain that $\left(C, (q_i \in \mathrm{Hom}_C(C, G(i)))_{i \in \mathrm{Ob}\, I}\right)$ is the limit of $G^{op}: I^{op} \to C$. As F preserves limits, it follows that $\left(F(C), (F(q_i))_{i \in \mathrm{Ob}\, I}\right)$ is the limit of FG^{op}. Using again Lemma 2.2.12 we obtain that $\left(F(C), (F^{op}(q_i^{op}))_{i \in \mathrm{Ob}\, I}\right)$ is the colimit of $(FG^{op})^{op} \overset{(1.25)}{=} F^{op} G^{op\, op} = F^{op} G$. This shows that F^{op} preserves colimits, as desired.

Assume now that F^{op} preserves colimits and let $H: I \to C$ be a functor whose limit we denote by $\left(L, (p_i \in \text{Hom}_C(L, H(i)))_{i \in \text{Ob} I}\right)$. Then Lemma 2.2.12 implies that $\left(L, (p_i^{op} \in \text{Hom}_{C^{op}}(H(i), L))_{i \in \text{Ob} I}\right)$ is the colimit of H^{op}. As F^{op} is colimit preserving we obtain that $\left(F(L), (F^{op}(p_i^{op}))_{i \in \text{Ob} I}\right)$ is the colimit of $F^{op} H^{op} \overset{(1.25)}{=} (FH)^{op}$. Again by Lemma 2.2.12 it follows that $\left(F(L), (F(p_i))_{i \in \text{Ob} I}\right)$ is the limit of FH. Therefore, F is limit preserving and the proof is now finished. $\square$

As a consequence of Theorem 2.4.2 we obtain the following:

Proposition 2.5.3 *Let C and D be two categories such that C is (co)complete. A functor $F: C \to D$ preserves small (co)limits if and only if it preserves (co)products and (co)equalizers.*

Proof Theorem 2.4.2 proves that the limit of a functor $H: I \to C$, for a small category I, can be constructed as the equalizer of certain morphisms between two products of some families of objects in C. As F preserves both products and equalizers we can conclude that it preserves limits. $\square$

Example 2.5.4 Let $\text{Bil}_{M,N}: \textbf{Ab} \to \textbf{Ab}$ be the functor defined in Example 1.5.3, (28). We will show that $\text{Bil}_{M,N}$ preserves limits. In light of Proposition 2.5.3 it will suffice to show that it preserves products and equalizers. To this end, let $(A_i)_{i \in I}$ be a family of abelian groups, where I is a set, and consider its product $\left(\prod_{i \in I} A_i, (p_j: \prod_{i \in I} A_i \to A_j)_{j \in I}\right)$ in $\textbf{Ab}$. Recall from Example 2.1.5, (2) that the underlying set of $\prod_{i \in I} A_i$ is just the cartesian product of the $A_i's$ while p_j is the j-th projection. We aim to prove that

$$\left(\text{Bil}_{M,N}(\prod_{i \in I} A_i), (\text{Bil}_{M,N}(p_j): \text{Bil}_{M,N}(\prod_{i \in I} A_i) \to \text{Bil}_{M,N}(A_j))_{j \in I}\right)$$

is the product in $\textbf{Ab}$ of the family $\left(\text{Bil}_{M,N}(A_i)\right)_{i \in I}$. Denote by

$$\left(\prod_{i \in I} \text{Bil}_{M,N}(A_i), (\pi_j: \prod_{i \in I} \text{Bil}_{M,N}(A_i) \to \text{Bil}_{M,N}(A_j))_{j \in I}\right)$$

the product in $\textbf{Ab}$ of the family $\left(\text{Bil}_{M,N}(A_i)\right)_{i \in I}$. Again by Example 2.1.5, (2) we know that the underlying set of $\prod_{i \in I} \text{Bil}_{M,N}(A_i)$ is the cartesian product of the $\text{Bil}_{M,N}(A_j)'s$ and π_j is the j-th projection. Now define $\psi: \text{Bil}_{M,N}(\prod_{i \in I} A_i) \to \prod_{i \in I} \text{Bil}_{M,N}(A_i)$ by $\psi(\alpha) = (p_i \circ \alpha)_{i \in I}$ for all $\alpha \in \text{Bil}_{M,N}(\prod_{i \in I} A_i)$. It can be easily seen that for all $i \in I$ we have $p_i \circ \alpha \in \text{Bil}_{M,N}(A_i)$, which shows that ψ is well-defined. Furthermore, ψ is bijective. Indeed, if $\alpha, \beta \in \text{Bil}_{M,N}(\prod_{i \in I} A_i)$ such that $\psi(\alpha) = \psi(\beta)$, we obtain $p_i \circ \alpha = p_i \circ \beta$ for all $i \in I$. Now Proposition 2.2.14, (1) implies $\alpha = \beta$, which shows that ψ is injective. Consider now $(u_i)_{i \in I} \in \prod_{i \in I} \text{Bil}_{M,N}(A_i)$, where $u_j \in \text{Bil}_{M,N}(A_j)$ for all $j \in I$. Then $\psi(u) = (u_i)_{i \in I}$, where $u \in \text{Bil}_{M,N}(\prod_{i \in I} A_i)$ is defined by $u(m, n) = (u_i(m, n))_{i \in I}$ for all $m \in M$, $n \in N$, which shows that ψ is also surjective. To conclude, we have an isomorphism of groups $\psi \colon \mathrm{Bil}_{M,N}(\prod_{i \in I} A_i) \to \prod_{i \in I} \mathrm{Bil}_{M,N}(A_i)$ which makes the following diagram commutative for all $j \in I$:

$$
\begin{array}{ccc}
\mathrm{Bil}_{M,N}(\prod_{i \in I} A_i) & & \\
\psi \downarrow & \searrow^{\mathrm{Bil}_{M,N}(p_j)} & \\
\prod_{i \in I} \mathrm{Bil}_{M,N}(A_i) & \xrightarrow[\pi_j]{} & \mathrm{Bil}_{M,N}(A_j)
\end{array}
$$

Indeed, for all $\alpha \in \mathrm{Bil}_{M,N}(\prod_{i \in I} A_i)$ and $j \in I$, we have

$$
(\pi_j \circ \psi)(\alpha) = \pi_j\big((p_i \circ \alpha)_{i \in I}\big) = p_j \circ \alpha = \mathrm{Bil}_{M,N}(p_j)(\alpha).
$$

Now Proposition 2.1.3 allows us to conclude that

$$
\Big(\mathrm{Bil}_{M,N}(\prod_{i \in I} A_i),\ (\mathrm{Bil}_{M,N}(p_j))_{j \in I}\Big)
$$

is the product in **Ab** of the family $\big(\mathrm{Bil}_{M,N}(A_i)\big)_{i \in I}$, as desired.

Next we show that $\mathrm{Bil}_{M,N}$ preserves equalizers. To this end, let f, $g \in \mathrm{Hom}_{\mathbf{Ab}}(A, B)$ and consider the equalizer (E, i) of the pair (f, g) in **Ab**, i.e., $E = \{a \in A \mid f(a) = g(a)\}$ and $i \colon E \to A$ is the inclusion morphism (see Example 2.1.10, (2)). Moreover, denote by (Q, j) the equalizer in **Ab** of the pair of morphisms $\mathrm{Bil}_{M,N}(f)$, $\mathrm{Bil}_{M,N}(g) \in \mathrm{Hom}_{\mathbf{Ab}}(\mathrm{Bil}_{M,N}(A), \mathrm{Bil}_{M,N}(B))$. We have

$$
Q = \{u \in \mathrm{Bil}_{M,N}(A) \mid \mathrm{Bil}_{M,N}(f)(u) = \mathrm{Bil}_{M,N}(g)(u)\},
$$

while $j \colon Q \to \mathrm{Bil}_{M,N}(A)$ denotes the inclusion. Now define $\varphi \colon \mathrm{Bil}_{M,N}(E) \to Q$ by $\varphi(v) = i \circ v$ for all $v \in \mathrm{Bil}_{M,N}(E)$. Consider $w \in Q$, i.e., $w \colon M \times N \to A$ is a bilinear map such that $f \circ w = g \circ w$. Then, for all $m \in M$, $n \in N$ we have $f(w(m, n)) = g(w(m, n))$, which implies $w(m, n) \in E$. If we denote by $\overline{w}$ the map obtained from w by restricting its codomain to E, we have $\varphi(\overline{w}) = i \circ \overline{w} = w$. Hence φ is surjective and is also trivially injective as i is a monomorphism by Proposition 2.1.9. Moreover, φ is the unique group morphism which makes the following diagram commutative:

$$
\begin{array}{ccc}
Q & \xrightarrow{j} & \mathrm{Bil}_{M,N}(A) \xrightarrow[\mathrm{Bil}_{M,N}(g)]{\mathrm{Bil}_{M,N}(f)} \mathrm{Bil}_{M,N}(B) \\
\varphi \uparrow & \nearrow_{\mathrm{Bil}_{M,N}(i)} & \\
\mathrm{Bil}_{M,N}(E) & &
\end{array}
$$

Indeed, for all $u \in \mathrm{Bil}_{M,N}(E)$ we have $\mathrm{Bil}_{M,N}(i)(u) = i \circ u = j \circ i \circ u = j \circ \varphi(u)$. This shows that $\big(\mathrm{Bil}_{M,N}(E),\ \mathrm{Bil}_{M,N}(i)\big)$ is the equalizer of the pair of morphisms $\big(\mathrm{Bil}_{M,N}(f),\ \mathrm{Bil}_{M,N}(g)\big)$ and therefore the functor $\mathrm{Bil}_{M,N}$ preserves equalizers as well.   □

The following lemma will be useful in the sequel:

Lemma 2.5.5 *Let* $F\colon \mathcal{C} \to \mathcal{D}$ *and* $G\colon I \to \mathcal{C}$ *be two functors. If* $\big(X,\ (s_i)_{i \in \mathrm{Ob}\,I}\big)$ *is a (co)cone on* G*, then* $\big(F(X),\ (F(s_i))_{i \in \mathrm{Ob}\,I}\big)$ *is a (co)cone on* $FG\colon I \to \mathcal{D}$*.*

Proof We will only prove the statement regarding cones. To this end consider $f \in \mathrm{Hom}_I(i,\ j)$. The proof will be finished once we show that the following diagram is commutative:

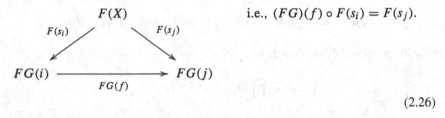

$$\text{i.e., } (FG)(f) \circ F(s_i) = F(s_j).$$

(2.26)

Indeed, as $\big(X,\ (s_i)_{i \in \mathrm{Ob}\,I}\big)$ is a cone on G, the following diagram is commutative:

$$\text{i.e., } G(f) \circ s_i = s_j.$$

(2.27)

Now it is straightforward to see that (2.26) holds true just by applying F to the identity (2.27).   □

One of the most important examples of functors which preserve limits are the hom functors.

Theorem 2.5.6 *Let* $\mathcal{C}$ *be a category and* $C \in \mathrm{Ob}\,\mathcal{C}$*.*

(1) The hom functor $\mathrm{Hom}_{\mathcal{C}}(C,\ -)\colon \mathcal{C} \to \mathbf{Set}$ *preserves all existing small limits.*
(2) The contravariant hom functor $\mathrm{Hom}_{\mathcal{C}}(-,\ C)\colon \mathcal{C} \to \mathbf{Set}$ *maps existing small colimits to small limits.*

Proof (1) Consider a functor $G\colon I \to \mathcal{C}$, where I is a small category, and $\big(L,\ (p_i)_{i \in \mathrm{Ob}\,I}\big)$ its limit. The proof will be finished once we show that

$$\big(\mathrm{Hom}_{\mathcal{C}}(C,\ L),\ (\mathrm{Hom}_{\mathcal{C}}(C,\ p_i))_{i \in \mathrm{Ob}\,I}\big)$$

is the limit of the functor $\mathrm{Hom}_\mathcal{C}(C, G(-))\colon I \to \mathbf{Set}$. To start with, by Lemma 2.5.5 we obtain that $\big(\mathrm{Hom}_\mathcal{C}(C, L), (\mathrm{Hom}_\mathcal{C}(C, p_i))_{i \in \mathrm{Ob}\,I}\big)$ is a cone on $\mathrm{Hom}_\mathcal{C}(C, G(-))$. Consider now another cone $\big(M, (q_i)_{i \in \mathrm{Ob}\,I}\big)$ on $\mathrm{Hom}_\mathcal{C}(C, G(-))$, where $M \in \mathrm{Ob}\,\mathbf{Set}$ and $q_i \in \mathrm{Hom}_{\mathbf{Set}}\big(M, \mathrm{Hom}_\mathcal{C}(C, G(i))\big)$ for all $i \in \mathrm{Ob}\,I$. Hence, the following diagram is commutative for any $f \in \mathrm{Hom}_I(i, j)$:

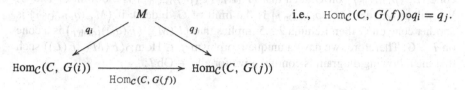

i.e., $\mathrm{Hom}_\mathcal{C}(C, G(f)) \circ q_i = q_j$.

Therefore, for all $m \in M$ we have $\mathrm{Hom}_\mathcal{C}(C, G(f))\big(q_i(m)\big) = q_j(m)$, which leads to $G(f) \circ q_i(m) = q_j(m)$. This implies that for each $m \in M$, $\big(C, (q_i(m))_{i \in \mathrm{Ob}\,I}\big)$ is a cone on G, where $q_i(m) \in \mathrm{Hom}_\mathcal{C}(C, G(i))$ for all $i \in \mathrm{Ob}\,I$. As $\big(L, (p_i)_{i \in \mathrm{Ob}\,I}\big)$ is the limit of G, it yields a unique morphism $q(m) \in \mathrm{Hom}_\mathcal{C}(C, L)$ such that the following diagram is commutative for all $i \in \mathrm{Ob}\,I$:

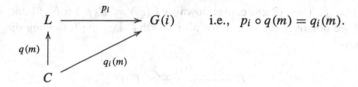

i.e., $p_i \circ q(m) = q_i(m)$.

Putting all this together we have defined a function $q\colon M \to \mathrm{Hom}_\mathcal{C}(C, L)$ (i.e., a morphism in $\mathbf{Set}$) satisfying $\mathrm{Hom}_\mathcal{C}(C, p_i) \circ q = q_i$ for any $i \in \mathrm{Ob}\,I$, i.e., the following diagram commutes:

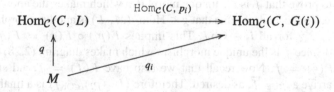

Furthermore, the uniqueness of q with this property follows from that of the $q(m)$'s.

(2) Showing that the contravariant hom functor maps existing small colimits to small limits follows the strategy used in the proof above and is left to the reader. □

Definition 2.5.7 A functor $F\colon \mathcal{C} \to \mathcal{D}$ *reflects (small) limits/colimits* when for every functor $G\colon I \to \mathcal{C}$, where I is a (small) category, and every cone/cocone $\big(L, (p_i)_{i \in \mathrm{Ob}\,I}\big)$ on G, if $\big(F(L), (F(p_i))_{i \in \mathrm{Ob}\,I}\big)$ is the limit/colimit of FG, then $\big(L, (p_i)_{i \in \mathrm{Ob}\,I}\big)$ is the limit/colimit of G.

An important class of (co)limit reflecting functors is the class of fully faithful functors:

Theorem 2.5.8 *A fully faithful functor $F: C \to D$ reflects small (co)limits.*

Proof Let $G: I \to C$ be a functor where I is a small category and consider a cone $\left(L, (p_i)_{i \in \mathrm{Ob}\, I}\right)$ on G such that $\left(F(L), (F(p_i))_{i \in \mathrm{Ob}\, I}\right)$ is the limit of $F \circ G$. We will prove that $\left(L, (p_i)_{i \in \mathrm{Ob}\, I}\right)$ is the limit of G. Indeed, if $\left(M, (q_i)_{i \in \mathrm{Ob}\, I}\right)$ is another cone on G then Lemma 2.5.5 implies that $\left(F(M), (F(q_i))_{i \in \mathrm{Ob}\, I}\right)$ is a cone on $F \circ G$. Therefore, we have a unique morphism $f \in \mathrm{Hom}_D(F(M), F(L))$ such that the following diagram is commutative for all $i \in \mathrm{Ob}\, I$:

$$
\begin{array}{ccc}
F(L) & \xrightarrow{\;\;F(p_i)\;\;} & (F \circ G)(i) \\
\end{array}
\qquad \text{i.e., } F(p_i) \circ f = F(q_i).
$$

(with $f: F(M) \to F(L)$ and $F(q_i): F(M) \to (F \circ G)(i)$)

$$ \tag{2.28} $$

Since F is fully faithful there exists a unique morphism $\overline{f} \in \mathrm{Hom}_C(M, L)$ such that $F(\overline{f}) = f$. Then (2.28) comes down to $F(q_i) = F(p_i) \circ F(\overline{f})$ and since F is faithful we obtain $q_i = p_i \circ \overline{f}$ for all $i \in \mathrm{Ob}\, I$, i.e., the following diagram is commutative:

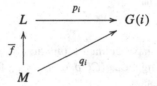

We are left to prove that $\overline{f}$ is the unique morphism which makes the above diagram commutative. To this end, assume that $g \in \mathrm{Hom}_C(M, L)$ is another morphism such that $q_i = p_i \circ g$ for all $i \in \mathrm{Ob}\, I$. This implies $F(p_i) \circ F(g) = F(q_i)$ for all $i \in \mathrm{Ob}\, I$ and since f is the unique morphism which makes diagram (2.28) commute we obtain $F(g) = f$. Now recall that we also have $F(\overline{f}) = f$ and since F is faithful we arrive at $g = \overline{f}$, as desired. Therefore $\left(L, (p_i)_{i \in \mathrm{Ob}\, I}\right)$ is a final object in the category of cones on G, as desired.

The dual statement will be settled as usual by the duality principle. Indeed, let $H: I \to C$ be a functor, where I is a small category, and consider a cocone $\left(Q, (q_i)_{i \in \mathrm{Ob}\, I}\right)$ on H such that $\left(F(Q), (F(q_i))_{i \in \mathrm{Ob}\, I}\right)$ is the colimit of $F \circ H$. In particular, by Lemma 2.2.3, $\left(Q, (q_i^{op})_{i \in \mathrm{Ob}\, I}\right)$ is a cone on H^{op}. Lemma 2.2.12 implies that $\left(F(Q), (F^{op}(q_i^{op}))_{i \in \mathrm{Ob}\, I}\right)$ is the limit of $(F \circ H)^{op} \overset{(1.25)}{=} F^{op} \circ H^{op}$. Since F^{op} is obviously also fully faithful and by the first part of the proof any fully faithful functor reflects limits, we obtain that $\left(Q, (q_i^{op})_{i \in \mathrm{Ob}\, I}\right)$ is the limit of H^{op}. By applying Lemma 2.2.12 once more it follows that $\left(Q, (q_i)_{i \in \mathrm{Ob}\, I}\right)$ is the colimit of H. This shows that F is colimit reflecting, as desired. $\qquad \square$

Proposition 2.5.9 *Let $F : C \to D$ be a (co)limit preserving functor. If C is (co)complete and F reflects isomorphisms then F also reflects small (co)limits.*

Proof Assume first that F is limit preserving, isomorphisms reflecting and C is complete. Consider a functor $G : I \to C$, where I is a small category, and let $(M, (q_i)_{i \in \mathrm{Ob}\, I})$ be a cone on G such that $(F(M), (F(q_i))_{i \in \mathrm{Ob}\, I})$ is the limit of $F \circ G$. The proof will be finished once we show that $(M, (q_i)_{i \in \mathrm{Ob}\, I})$ is the limit of G. According to the completeness assumption on C the functor G has a limit, say $(L, (p_i)_{i \in \mathrm{Ob}\, I})$. Thus, there exists a unique morphism $f \in \mathrm{Hom}_C(M, L)$ such that the following diagram is commutative for all $i \in \mathrm{Ob}\, I$:

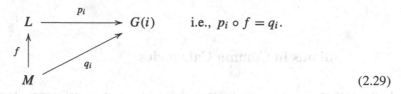

$$L \xrightarrow{\;\;p_i\;\;} G(i) \qquad \text{i.e., } p_i \circ f = q_i.$$

(2.29)

In particular, this implies the following:

$$F(p_i) \circ F(f) = F(q_i), \quad \text{for all } i \in \mathrm{Ob}\, I.$$

Since F is a limit preserving functor then $(F(L), (F(p_i))_{i \in \mathrm{Ob}\, I})$ is also a limit of $F \circ G$. Exactly as in the proof of Proposition 2.1.3, one can show that there exists a unique isomorphism $g \in \mathrm{Hom}_D(F(M), F(L))$ such that the following diagram is commutative for all $i \in \mathrm{Ob}\, I$:

$$F(L) \xrightarrow{\;\;F(p_i)\;\;} F \circ G(i) \qquad \text{i.e., } F(p_i) \circ g = F(q_i).$$

Hence $F(f) = g$ is an isomorphism in D. Our assumption implies that f is an isomorphism in C and thus $(M, (q_i)_{i \in \mathrm{Ob}\, I})$ is also a limit of G, as desired.

For the dual statement, let $H : I \to C$ be a functor, where I is a small category, and consider a cocone $(Q, (q_i)_{i \in \mathrm{Ob}\, I})$ on H such that $(F(Q), (F(q_i))_{i \in \mathrm{Ob}\, I})$ is the colimit of $F \circ H$. Lemma 2.2.12 implies that $(F(Q), (F^{op}(q_i^{op}))_{i \in \mathrm{Ob}\, I})$ is the limit of $(F \circ H)^{op} \overset{(1.25)}{=} F^{op} \circ H^{op}$. Furthermore, by Lemma 2.2.3, $(Q, (q_i^{op})_{i \in \mathrm{Ob}\, I})$ is a cone on H^{op} while Lemma 2.5.2 implies that F^{op} is limit preserving. As F^{op} is obviously also isomorphism reflecting, the first part of the proof implies that $(Q, (q_i^{op})_{i \in \mathrm{Ob}\, I})$ is the limit of H^{op}. Now Lemma 2.2.12 shows that $(Q, (q_i)_{i \in \mathrm{Ob}\, I})$ is the colimit of H. $\qquad\square$

Examples 2.5.10

(1) The forgetful functor $U:$ **Top** $\rightarrow$ **Set** preserves products and equalizers (see Examples 2.1.5 and 2.1.10). Therefore U preserves small limits by Proposition 2.5.3. Similar arguments show that the forgetful functor $U:$ **Grp** $\rightarrow$ **Set** is also limit preserving.

(2) The category **Ab** is complete and the inclusion functor $I:$ **Ab** $\rightarrow$ **Grp** preserves preserves products and equalizers (see Examples 2.1.5 and 2.1.10). Therefore I preserves small limits by Proposition 2.5.3. Furthermore, I reflects isomorphisms and, according to Proposition 2.5.9, I also reflects small limits.

$\square$

2.6   (Co)limits in Comma Categories

In this section we discuss the (co)completeness of comma categories. Throughout, $F: \mathcal{A} \rightarrow \mathcal{C}, G: \mathcal{B} \rightarrow \mathcal{C}$ are two functors and $(F \downarrow G)$ denotes the corresponding comma category (see Theorem 1.8.4). We consider the forgetful functors $U: (F \downarrow G) \rightarrow \mathcal{A}$ and $V: (F \downarrow G) \rightarrow \mathcal{B}$ defined as follows for all $(A, f, B) \in \mathrm{Ob}(F \downarrow G)$ and all morphisms (a, b) in $(F \downarrow G)$:

$$U(A, f, B) = A, \qquad U(a, b) = a,$$
$$V(A, f, B) = B, \qquad V(a, b) = b.$$

Lemma 2.6.1 *There exists a natural transformation* $\alpha: FU \rightarrow GV$ *between the functors* $FU, GV: (F \downarrow G) \rightarrow \mathcal{C}$ *given by* $\alpha_{(A, f, B)} = f$, *for all* $(A, f, B) \in \mathrm{Ob}(F \downarrow G)$.

Proof Let $(a, b): (A, f, B) \rightarrow (A', f', B')$ be a morphism in $(F \downarrow G)$. We need to show that the following diagram is commutative:

$$
\begin{array}{ccc}
FU(A, f, B) & \xrightarrow{\ \alpha_{(A, f, B)}\ } & GV(A, f, B) \\
{\scriptstyle FU(a,b)}\big\downarrow & & \big\downarrow{\scriptstyle GV(a,b)} \\
FU(A', f', B') & \xrightarrow[\ \alpha_{(A', f', B')}\ ]{} & GV(A', f', B')
\end{array}
$$

Indeed, note that the above diagram simplifies to the following:

$$
\begin{array}{ccc}
F(A) & \xrightarrow{\ \ f\ \ } & G(B) \\
{\scriptstyle F(a)}\big\downarrow & & \big\downarrow{\scriptstyle G(b)} \\
F(A') & \xrightarrow[\ \ f'\ \ ]{} & G(B')
\end{array}
$$

and the latter diagram is obviously commutative as a consequence of $(a,\ b)$ being a morphism in $(F \downarrow G)$.   □

Let I be a category and $H\colon I \to (F \downarrow G)$ a functor. We will use the following notation for all $i \in \mathrm{Ob}\, I$:

$$
H(i) = \big(UH(i), \alpha_{H(i)}, VH(i)\big), \tag{2.30}
$$

where α is the natural transformation defined in Lemma 2.6.1.

We record a new useful result in the following:

Lemma 2.6.2 *Let $H\colon I \to (F \downarrow G)$ be a functor, where I is a small category. If $\mathcal{A}$ is a complete category and $\big(L, (p_i)_{i\in\mathrm{Ob}\,I}\big)$ is the limit of $UH\colon I \to \mathcal{A}$, then the pair $\big(F(L), (\alpha_{H(i)} \circ F(p_i))_{i\in\mathrm{Ob}\,I}\big)$ is a cone on $GVH\colon I \to \mathcal{C}$.*

Proof Let $t \in \mathrm{Hom}_I(i,\ j)$ and let $H(t) = (a_{ij},\ b_{ij})$, where

$$
(a_{ij},\ b_{ij})\colon \big(UH(i), \alpha_{H(i)}, VH(i)\big) \to \big(UH(j), \alpha_{H(j)}, VH(j)\big)
$$

is a morphism in $(F \downarrow G)$. This implies that $a_{ij} \in \mathrm{Hom}_{\mathcal{A}}\big(UH(i),\ UH(j)\big)$ and $b_{ij} \in \mathrm{Hom}_{\mathcal{B}}\big(VH(i),\ VH(j)\big)$ such that the following diagram is commutative:

$$
\begin{array}{ccc}
FUH(i) & \xrightarrow{\ \ F(a_{ij})\ \ } & FUH(j) \\
{\scriptstyle \alpha_{H(i)}}\big\downarrow & & \big\downarrow{\scriptstyle \alpha_{H(j)}} \\
GVH(i) & \xrightarrow[\ \ G(b_{ij})\ \ ]{} & GVH(j)
\end{array}
\tag{2.31}
$$

Furthermore, as $\big(L, (p_i)_{i\in\mathrm{Ob}\,I}\big)$ is, in particular, a cone on $UH\colon I \to \mathcal{A}$, the following diagram is commutative as well:

$$
\begin{array}{ccc}
 & L & \\
{\scriptstyle p_i}\swarrow & & \searrow{\scriptstyle p_j} \\
UH(i) & \xrightarrow[\ UH(t)=a_{ij}\ ]{} & UH(j)
\end{array}
\tag{2.32}
$$

The proof will be finished once we show the commutativity of the following diagram:

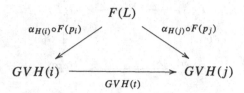

Indeed, we have

$$GVH(t) \circ \alpha_{H(i)} \circ F(p_i) = \underline{G(b_{ij}) \circ \alpha_{H(i)}} \circ F(p_i)$$

$$\overset{(2.31)}{=} \alpha_{H(j)} \circ \underline{F(a_{ij}) \circ F(p_i)}$$

$$\overset{(2.32)}{=} \alpha_{H(j)} \circ F(p_j),$$

as desired.   □

Lemma 2.6.3 *Let $H: I \to (F \downarrow G)$ be a functor, where I is a small category. If $\big((A, f, B), ((p_i, q_i))_{i \in \mathrm{Ob}\, I}\big)$ is a (co)cone on H then $\big(A, (p_i)_{i \in \mathrm{Ob}\, I}\big)$ and $\big(B, (q_i)_{i \in \mathrm{Ob}\, I}\big)$ are (co)cones on $UH: I \to \mathcal{A}$ and $VH: I \to \mathcal{B}$, respectively.*

Proof We only prove the statement concerning cones. To this end, consider $t \in \mathrm{Hom}_I(i, j)$ and let $H(t) = (a_{ij}, b_{ij})$, where $(a_{ij}, b_{ij}) \in \mathrm{Hom}_{(F \downarrow G)}(H(i), H(j))$. In particular, we have

$$a_{ij} \in \mathrm{Hom}_{\mathcal{A}}(UH(i), UH(j)) \text{ and } b_{ij} \in \mathrm{Hom}_{\mathcal{B}}(VH(i), VH(j)).$$

The proof will be finished once we show that the following diagrams commute:

$$\begin{array}{ccc} & A & \\ {\scriptstyle p_i}\swarrow & & \searrow{\scriptstyle p_j} \\ UH(i) & \xrightarrow[UH(t)=a_{ij}]{} & UH(j) \end{array} \qquad \text{i.e., } a_{ij} \circ p_i = p_j,$$

$$\tag{2.33}$$

$$\begin{array}{ccc} & B & \\ {\scriptstyle q_i}\swarrow & & \searrow{\scriptstyle q_j} \\ VH(i) & \xrightarrow[VH(t)=b_{ij}]{} & VH(j) \end{array} \qquad \text{i.e., } b_{ij} \circ q_i = q_j.$$

$$\tag{2.34}$$

Now since $\Big((A,\, f,\, B),\, \big((p_i,\, q_i)\big)_{i\in\mathrm{Ob}\, I}\Big)$ is a cone on H, we have $H(t)\circ(p_i,\, q_i) =$ $(p_j,\, q_j)$, which componentwise comes down to (2.33) and (2.34).   □

We can now state and prove the main result of this section:

Theorem 2.6.4 *Let $F\colon \mathcal{A} \to \mathcal{C}$ and $G\colon \mathcal{B} \to \mathcal{C}$ be two functors.*

(1) If $\mathcal{A}$ and $\mathcal{B}$ are complete categories and G preserves small limits then $(F \downarrow G)$ is also complete and both forgetful functors $U\colon (F \downarrow G) \to \mathcal{A}$ and $V\colon (F \downarrow G) \to \mathcal{B}$ preserve small limits.

(2) If $\mathcal{A}$ and $\mathcal{B}$ are cocomplete categories and F preserves small colimits then $(F \downarrow G)$ is also cocomplete and both forgetful functors $U\colon (F \downarrow G) \to \mathcal{A}$ and $V\colon (F \downarrow G) \to \mathcal{B}$ preserve small colimits.

Proof Throughout, we use the notation introduced in (2.30). Let I be a small category and $H\colon I \to (F \downarrow G)$ a functor.

(1) We need to show that H has a limit. Recall that both categories $\mathcal{A}$ and $\mathcal{B}$ are complete and therefore the functors $UH\colon I \to \mathcal{A}$ and $VH\colon I \to \mathcal{B}$ have limits, say $\big(L,\, (p_i)_{i\in\mathrm{Ob}\, I}\big)$ and $\big(M,\, (q_i)_{i\in\mathrm{Ob}\, I}\big)$ respectively, where:

$$L \in \mathrm{Ob}\,\mathcal{A}, \quad p_i \in \mathrm{Hom}_{\mathcal{A}}(L,\, UH(i)),$$
$$M \in \mathrm{Ob}\,\mathcal{B}, \quad q_i \in \mathrm{Hom}_{\mathcal{B}}(M,\, VH(i)).$$

Using Lemma 2.6.2, we have that $\big(F(L),\, (\alpha_{H(i)} \circ F(p_i))_{i\in\mathrm{Ob}\, I}\big)$ is a cone on $GVH\colon I \to \mathcal{C}$, where $\alpha_{H(i)} \circ F(p_i) \in \mathrm{Hom}_{\mathcal{C}}(F(L),\, GVH(i))$. Furthermore, $\big(G(M),\, (G(q_i))_{i\in\mathrm{Ob}\, I}\big)$ is the limit of the functor $GVH\colon I \to \mathcal{C}$ as G is limit preserving. Hence, there exists a unique morphism $h \in \mathrm{Hom}_{\mathcal{C}}(F(L),\, G(M))$ which makes the following diagram commutative for all $i \in \mathrm{Ob}\, I$:

$$
\begin{array}{ccc}
F(L) & \xrightarrow{\;\alpha_{H(i)}\circ F(p_i)\;} & GVH(i) \\
{\scriptstyle h}\Big\downarrow & {\scriptstyle G(q_i)}\nearrow & \\
G(M) & &
\end{array}
\qquad \text{i.e., } \alpha_{H(i)} \circ F(p_i) = G(q_i) \circ h.
$$

$$(2.35)$$

Note that, in particular, we have $(L,\, h,\, M) \in \mathrm{Ob}\,(F \downarrow G)$. Furthermore, (2.35) implies the commutativity of the following diagram:

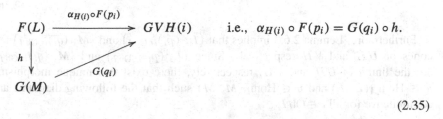

$$
\begin{array}{ccc}
F(L) & \xrightarrow{\;F(p_i)\;} & FUH(i) \\
{\scriptstyle h}\Big\downarrow & & \Big\downarrow{\scriptstyle \alpha_{H(i)}} \\
G(M) & \xrightarrow[\;G(q_i)\;]{} & GVH(i)
\end{array}
$$

which proves that $(p_i,\ q_i)\colon (L,\ h,\ M) \to H(i)$ is in fact a morphism in the comma category $(F \downarrow G)$, for all $i \in \mathrm{Ob}\, I$.

We will show that $\Big((L,\ h,\ M),\ \big((p_i,\ q_i)\big)_{i\in\mathrm{Ob}\, I}\Big)$ is the limit of the functor H. We start by proving that $\Big((L,\ h,\ M),\ \big((p_i,\ q_i)\big)_{i\in\mathrm{Ob}\, I}\Big)$ is a cone on H. To this end, let $t \in \mathrm{Hom}_I(i,\ j)$ and let $H(t) = (a_{ij},\ b_{ij})$, where $(a_{ij},\ b_{ij})\colon H(i) \to H(j)$ is a morphism in $(F \downarrow G)$. Now observe that the commutativity of the following diagram is trivially implied by the fact that $\big(L,\ (p_i)_{i\in\mathrm{Ob}\, I}\big)$ and $\big(M,\ (q_i)_{i\in\mathrm{Ob}\, I}\big)$ are in particular cones on UH and VH, respectively:

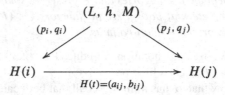

Thus, $\Big((L,\ h,\ M),\ \big((p_i,\ q_i)\big)_{i\in\mathrm{Ob}\, I}\Big)$ is indeed a cone on H. Consider now another cone $\Big((\overline{L},\ \overline{h},\ \overline{M}),\ \big((\overline{p_i},\ \overline{q_i})\big)_{i\in\mathrm{Ob}\, I}\Big)$ on H. Since $(\overline{p_i},\ \overline{q_i})\colon (\overline{L},\ \overline{h},\ \overline{M}) \to H(i)$ is a morphism in $(F \downarrow G)$ for all $i \in \mathrm{Ob}\, I$, the following diagram is commutative:

$$
\begin{array}{ccc}
F(\overline{L}) & \xrightarrow{\;\;F(\overline{p_i})\;\;} & FUH(i) \\[2pt]
{\scriptstyle \overline{h}}\big\downarrow & & \big\downarrow{\scriptstyle \alpha_{H(i)}} \\[2pt]
G(\overline{M}) & \xrightarrow[\;\;G(\overline{q_i})\;\;]{} & GVH(i)
\end{array}
\qquad\text{i.e.,}\quad \alpha_{H(i)} \circ F(\overline{p_i}) = G(\overline{q_i}) \circ \overline{h}.
$$

$$(2.36)$$

Furthermore, Lemma 2.6.3 implies that $\big(\overline{L},\ (\overline{p_i})_{i\in\mathrm{Ob}\, I}\big)$ and $\big(\overline{M},\ (\overline{q_i})_{i\in\mathrm{Ob}\, I}\big)$ are cones on UH and VH respectively. Since $\big(L,\ (p_i)_{i\in\mathrm{Ob}\, I}\big)$ and $\big(M,\ (q_i)_{i\in\mathrm{Ob}\, I}\big)$ are the limits of UH and VH, respectively, there exist two unique morphisms $u \in \mathrm{Hom}_{\mathcal{A}}(\overline{L},\ L)$ and $v \in \mathrm{Hom}_{\mathcal{B}}(\overline{M},\ M)$ such that the following diagrams are commutative for all $i \in \mathrm{Ob}\, I$:

$$
\begin{array}{ccc}
\overline{L} & \xrightarrow{\;\;\overline{p_i}\;\;} & UH(i) \\[2pt]
{\scriptstyle u}\big\downarrow & \nearrow & \\[2pt]
L & {\scriptstyle p_i} &
\end{array}
\qquad\text{i.e.,}\quad p_i \circ u = \overline{p_i},
$$

$$(2.37)$$

$$\overline{M} \xrightarrow{\overline{q}_i} VH(i) \qquad \text{i.e., } q_i \circ v = \overline{q}_i.$$

$$v \downarrow \qquad \nearrow q_i$$

$$M \tag{2.38}$$

We will show that (u, v) is a morphism in $(F \downarrow G)$ from $(\overline{L}, \overline{h}, \overline{M})$ to (L, h, M), i.e., the following diagram is commutative:

$$F(\overline{L}) \xrightarrow{F(u)} F(L) \qquad \text{i.e., } h \circ F(u) = G(v) \circ \overline{h}.$$

$$\overline{h} \downarrow \qquad \downarrow h$$

$$G(\overline{M}) \xrightarrow{G(v)} G(M) \tag{2.39}$$

As $\big(G(M), (G(q_i))_{i \in \mathrm{Ob}\, I}\big)$ is the limit of the functor $GVH: I \to \mathcal{C}$, using Proposition 2.2.14, (1) it will suffice to show that the following holds for all $i \in \mathrm{Ob}\, I$:

$$G(q_i) \circ h \circ F(u) = G(q_i) \circ G(v) \circ \overline{h}.$$

Indeed, we have

$$G(q_i) \circ h \circ F(u) \overset{(2.35)}{=} \alpha_{H(i)} \circ F(p_i) \circ F(u) = \alpha_{H(i)} \circ F(p_i \circ u)$$

$$\overset{(2.37)}{=} \alpha_{H(i)} \circ F(\overline{p}_i)$$

$$\overset{(2.36)}{=} G(\overline{q}_i) \circ \overline{h}$$

$$\overset{(2.38)}{=} G(q_i \circ v) \circ \overline{h} = G(q_i) \circ G(v) \circ \overline{h},$$

as desired. Moreover, in light of (2.37) and (2.38), (u, v) is obviously the unique morphism in $(F \downarrow G)$ which makes following diagram commute for all $i \in \mathrm{Ob}\, I$:

$$(\overline{L}, \overline{h}, \overline{M}) \xrightarrow{(\overline{p}_i, \overline{q}_i)} H(i)$$

$$(u, v) \downarrow \qquad \nearrow (p_i, q_i)$$

$$(L, h, M)$$

which shows that $\left((L, h, M), ((p_i, q_i))_{i \in \mathrm{Ob}\, I}\right)$ is indeed the limit of the functor H. Finally, note that both functors U and V are obviously limit preserving.

(2) We use the duality principle. As $\mathcal{A}^{op}$ and $\mathcal{B}^{op}$ are complete categories and $F^{op} \colon \mathcal{A}^{op} \to \mathcal{C}^{op}$ preserves limits (Lemma 2.5.2) we obtain, by applying 1), that the comma-category $(G^{op} \downarrow F^{op})$ is complete. Now by Proposition 1.8.10 we have an isomorphism of categories between $(G^{op} \downarrow F^{op})$ and $(F \downarrow G)^{op}$ and therefore $(F \downarrow G)^{op}$ is complete as well. This shows that $(F \downarrow G)$ is cocomplete, as desired.
                                                                                                        □

As an easy consequence of the previous result we have:

Corollary 2.6.5

(1) Let $\mathcal{B}$ be a complete category and $G \colon \mathcal{B} \to \mathcal{C}$ a functor which preserves small limits. Then, for all $C_0 \in \mathrm{Ob}\,\mathcal{C}$, the category $(C_0 \downarrow G)$ is complete.

(2) Let $\mathcal{A}$ be a cocomplete category and $F \colon \mathcal{A} \to \mathcal{C}$ a functor which preserves small colimits. Then, for all $C_0 \in \mathrm{Ob}\,\mathcal{C}$, the category $(F \downarrow C_0)$ is cocomplete.

Proof Recall from Corollary 1.8.6 that the category $(C_0 \downarrow G)$ is obtained as a special case of the comma-category $(F \downarrow G)$ by considering $\mathcal{A}$ to be the discrete category with one object while the functor F is the object C_0 of $\mathcal{C}$. Note that the discrete category with one object is obviously complete and the desired conclusion now follows from Theorem 2.6.4, (1). Similarly, (2) follows from Theorem 2.6.4, (2).
                                                                                                        □

As the identity functor on any category preserves all small (co)limits, we obtain:

Corollary 2.6.6 *Let $\mathcal{C}$ be a category and $C_0 \in \mathrm{Ob}\,\mathcal{C}$.*

(1) If $\mathcal{C}$ is complete then the coslice category $(C_0 \downarrow \mathcal{C})$ is also complete.

(2) If $\mathcal{C}$ is cocomplete then the slice category $(\mathcal{C} \downarrow C_0)$ is also cocomplete.

Proof (1) Follows by considering $\mathcal{B} = \mathcal{C}$ and $F = 1_{\mathcal{C}}$ in Corollary 2.6.5, (1). Similarly, (2) can be obtain by specializing $\mathcal{A} = \mathcal{C}$ and $F = 1_{\mathcal{C}}$ in Corollary 2.6.5, (2).
                                                                                                        □

2.7   (Co)limits in Functor Categories

Functor categories form another class of categories which behave well with respect to (co)limits. In what follows I and J are small categories and $\mathcal{C}$ is an arbitrary category. We start by introducing the following induced functors:

Lemma 2.7.1 *Let $F \colon I \to \mathrm{Fun}(J, \mathcal{C})$ be a functor. Then, we have*

(1) *each* $j \in \text{Ob } J$ *induces a functor* $F_j \colon I \to C$ *defined as follows for all* i,
$k \in \text{Ob } I$ *and* $t \in \text{Hom}_I(i, k)$:

$$F_j(i) = F(i)(j), \qquad F_j(t) = F(t)_j, \tag{2.40}$$

where $F(t)_j$ *denotes the j-component of the natural transformation*
$F(t) \colon F(i) \to F(k)$;
(2) *each* $f \in \text{Hom}_J(j, s)$ *induces a natural transformation* $\mathcal{F}_f \colon F_j \to F_s$ *defined
for all* $i \in \text{Ob } I$ *by* $(\mathcal{F}_f)_i = F(i)(f)$.

Proof (1) Given $j \in \text{Ob } J$, we show first that F_j preserves identity morphisms.
Since F itself is a functor and it preserves identities, for all $i \in \text{Ob } I$ we have

$$F_j(1_i) = F(1_i)_j = \left(1_{F(i)}\right)_j = 1_{F(i)(j)} = 1_{F_j(i)}.$$

Furthermore, for any $u \in \text{Hom}_I(i, k)$ and $v \in \text{Hom}_I(k, l)$ we have

$$F_j(v \circ u) = F(v \circ u)_j = \left(F(v) \circ F(u)\right)_j = F(v)_j \circ F(u)_j = F_j(v) \circ F_j(u),$$

where the second equality holds because F is a functor. This proves that F_j is a
functor.
  (2) Let $t \in \text{Hom}_I(i, r)$. The proof will be finished once we show the
commutativity of the following diagram:

$$
\begin{array}{ccc}
F_j(i) & \xrightarrow{\ (\mathcal{F}_f)_i\ } & F_s(i) \\
{\scriptstyle F_j(t)}\Big\downarrow & & \Big\downarrow{\scriptstyle F_s(t)} \\
F_j(r) & \xrightarrow[\ (\mathcal{F}_f)_r\ ]{} & F_s(r)
\end{array}
\qquad \text{i.e., } F(t)_s \circ F(i)(f) = F(r)(f) \circ F(t)_j.
$$

$$\tag{2.41}$$

Note that $F(t)$ is a natural transformation between the functors $F(i), F(r) \colon J \to C$.
Writing down the naturality of $F(t)$ for the morphism $f \in \text{Hom}_J(j, s)$ yields the
following commutative diagram:

$$
\begin{array}{ccc}
F(i)(j) & \xrightarrow{\ F(t)_j\ } & F(r)(j) \\
{\scriptstyle F(i)(f)}\Big\downarrow & & \Big\downarrow{\scriptstyle F(r)(f)} \\
F(i)(s) & \xrightarrow[\ F(t)_s\ ]{} & F(r)(s)
\end{array}
\qquad \text{i.e., } F(t)_s \circ F(i)(f) = F(r)(j) \circ F(t)_j.
$$

$$\tag{2.42}$$

Clearly, the commutativity of the diagram (2.42) implies the commutativity of (2.41), as desired.   □

We are now ready to prove the main result of this section, which states that, under some conditions, the (co)limit of a functor $F: I \to \text{Fun}(J, C)$ is constructed *pointwise*. Loosely speaking, this means that the (co)limit of F will be constructed by putting together the (co)limits of the functors F_j, for each $j \in \text{Ob } J$, as defined in (2.40).

Theorem 2.7.2 *Let* $F: I \to \text{Fun}(J, C)$ *be a functor and for each* $j \in \text{Ob } J$ *consider the functor* $F_j: I \to C$ *defined in (2.40). If for all* $j \in \text{Ob } J$ *the functor* F_j *has a (co)limit then* F *has a (co)limit as well and this (co)limit is computed pointwise.*

Proof We start by proving the claim concerning limits. For each $j \in \text{Ob } J$, let $\left(L_j, (p_i^j: L_j \to F_j(i))_{i \in \text{Ob } I}\right)$ be the limit of the functor $F_j: I \to C$, where $L_j \in \text{Ob } C$. We will construct a functor $L: J \to C$ together with a family of natural transformations $\left(L \to F(i)\right)_{i \in \text{Ob } I}$ which will form the limit of F. First we define the functor L on objects by $L(j) = L_j$, for all $j \in \text{Ob } J$. In order to define L on morphisms we show first that for any $f \in \text{Hom}_J(j, s)$, the pair

$$\left(L_j, \left((Ff)_i \circ p_i^j: L_j \to F_s(i)\right)_{i \in \text{Ob } I}\right)$$

is a cone on $F_s: I \to C$. Indeed, we aim to prove that for any $t \in \text{Hom}_I(i, k)$ the following diagram is commutative:

i.e., $F_s(t) \circ (\mathcal{F}_f)_i \circ p_i^j = (\mathcal{F}_f)_k \circ p_k^j.$

(2.43)

To this end, the naturality of $\mathcal{F}_f$ applied to t yields the following commutative diagram:

i.e., $F(t)_s \circ F(i)(f) = F(k)(f) \circ F(t)_j.$

(2.44)

Furthermore, as $\left(L_j, (p_i^j : L_j \to F_j(i))_{i \in \mathrm{Ob}\, I}\right)$ is in particular a cone on F_j, the following diagram is commutative:

$$
\begin{array}{ccc}
 & L_j & \\
\swarrow p_i^j & & \searrow p_k^j \\
F_j(i) \xrightarrow{\quad F_j(t) \quad} & & F_j(k)
\end{array}
\qquad \text{i.e.,} \quad F(t)_j \circ p_i^j = p_k^j.
$$

(2.45)

Putting all the above together yields

$$
F_s(t) \circ (\mathcal{F}_f)_i \circ p_i^j = \underline{F(t)_s \circ F(i)(f) \circ p_i^j} \overset{(2.44)}{=} \underline{F(k)(f) \circ F(t)_j \circ p_i^j}
$$

$$
\overset{(2.45)}{=} F(k)(f) \circ p_k^j = (\mathcal{F}_f)_k \circ p_k^j
$$

and we have proved that $\left(L_j, \left((\mathcal{F}_f)_i \circ p_i^j : L_j \to F_s(i)\right)_{i \in \mathrm{Ob}\, I}\right)$ is indeed a cone on F_s. As $\left(L_s, (p_i^s : L_s \to F_s(i))_{i \in \mathrm{Ob}\, I}\right)$ is the limit of F_s, there exists a unique morphism, denoted by $L(f)$, which makes the following diagram commutative for all $i \in \mathrm{Ob}\, I$:

$$
\begin{array}{ccc}
L_j & \xrightarrow{\ F(i)(f) \circ p_i^j\ } & F_s(i) \\
\downarrow{\scriptstyle L(f)} & \nearrow{\scriptstyle p_i^s} & \\
L_s & &
\end{array}
\qquad \text{i.e.,} \quad p_i^s \circ L(f) = F(i)(f) \circ p_i^j.
$$

(2.46)

We can now define a functor $L : J \to C$ by setting $L(j) = L_j$ while for all $f \in \mathrm{Hom}_J(j, s)$, $L(f)$ is the unique morphism which makes (2.46) commutative. However, we still need to show that $L : J \to C$ is indeed a functor. To start with, we prove that L preserves identities. As $F(i) : J \to C$ is a functor for any $i \in \mathrm{Ob}\, I$, we have $F(i)(1_j) = 1_{F(i)(j)}$ and $L(1_j)$ is the unique morphism such that $p_i^j \circ L(1_j) = 1_{F(i)(j)} \circ p_i^j$. Clearly this implies $L(1_j) = 1_{L(j)}$, as desired. Consider now $f \in \mathrm{Hom}_J(j, s)$ and $g \in \mathrm{Hom}_J(s, l)$. Then $L(f)$, $L(g)$ and $L(g \circ f)$, respectively, are the unique morphisms such that for all $i \in \mathrm{Ob}\, I$ we have

$$
p_i^s \circ L(f) = F(i)(f) \circ p_i^j, \tag{2.47}
$$

$$
p_i^l \circ L(g) = F(i)(g) \circ p_i^s, \tag{2.48}
$$

$$
p_i^l \circ L(g \circ f) = F(i)(g \circ f) \circ p_i^j. \tag{2.49}
$$

Hence, for all $i \in \mathrm{Ob}\,I$ the following holds:

$$\underline{p_i^l \circ L(g)} \circ L(f) \stackrel{(2.48)}{=} F(i)(g) \circ \underline{p_i^s \circ L(f)}$$

$$\stackrel{(2.47)}{=} F(i)(g) \circ F(i)(f) \circ p_i^j = F(i)(g \circ f) \circ p_i^j,$$

where in the last equality we used the fact that $F(i)$ is a functor. This shows that $L(g) \circ L(f)$ fulfills (2.49), which implies that $L(g) \circ L(f) = L(g \circ f)$. Hence $L \colon J \to C$ is a functor and therefore an object in $\mathrm{Fun}(J, C)$. Next, for each $i \in \mathrm{Ob}\,I$, we define a natural transformation $p_i \colon L \to F(i)$ by $(p_i)_j = p_i^j$ for all $j \in \mathrm{Ob}\,J$. In order to prove that the family of morphisms $\left(p_i^j \colon L(j) \to F(i)(j)\right)_{j \in \mathrm{Ob}\,J}$ indeed form a natural transformation we need to show the commutativity of the following diagram for all $f \in \mathrm{Hom}_J(j, s)$:

$$
\begin{array}{ccc}
L(j) & \xrightarrow{\;\;p_i^j\;\;} & F(i)(j) \\
{\scriptstyle L(f)}\Big\downarrow & & \Big\downarrow{\scriptstyle F(i)(f)} \\
L(s) & \xrightarrow[\;\;p_i^s\;\;]{} & F(i)(s)
\end{array}
\qquad \text{i.e.,} \quad F(i)(f) \circ p_i^j = p_i^s \circ L(f).
$$

The commutativity of the above diagram follows from the commutativity of (2.46). Hence, $p_i \colon L \to F(i)$ is a natural transformation for any $i \in \mathrm{Ob}\,I$ and therefore a morphism in the category $\mathrm{Fun}(J, C)$. We will prove that $\left(L, \left(p_i \colon L \to F(i)\right)_{i \in \mathrm{Ob}\,I}\right)$ is the limit of F. We start by showing that the above pair is a cone on F. Indeed, let $t \in \mathrm{Hom}_I(i, k)$; we need to prove that the following diagram is commutative:

$$
\begin{array}{ccc}
 & L & \\
{\scriptstyle p_i}\swarrow & & \searrow{\scriptstyle p_k} \\
F(i) & \xrightarrow[\;F(t)\;]{} & F(k)
\end{array}
\qquad \text{i.e.,} \quad F(t) \circ p_i = p_k.
$$

(2.50)

By the commutativity of diagram (2.45) we have $F(t)_j \circ p_i^j = p_k^j$ for all $j \in \mathrm{Ob}\,J$. This implies that the natural transformations $F(t) \circ p_i$ and p_k are equal and therefore diagram (2.50) is commutative, as desired.

Consider now another cone $\left(H, \left(q_i \colon H \to F(i)\right)_{i \in \mathrm{Ob}\,I}\right)$ on F, where $H \colon J \to$

$\mathcal{C}$ is a functor, $q_i : H \to F(i)$ is a natural transformation for each $i \in \mathrm{Ob}\, I$ and we denote $(q_i)_j$ by q_i^j for all $j \in \mathrm{Ob}\, J$. Note for further use that the naturality of q_i applied to a morphism $f \in \mathrm{Hom}_J(j, s)$ yields the following commutative diagram:

$$
\begin{array}{ccc}
H(j) & \xrightarrow{\;\;q_i^j\;\;} & F(i)(j) \\
{\scriptstyle H(f)}\big\downarrow & & \big\downarrow{\scriptstyle F(i)(f)} \\
H(s) & \xrightarrow[\;\;q_i^s\;\;]{} & F(i)(s)
\end{array}
\qquad \text{i.e.,} \quad F(i)(f) \circ q_i^j = q_i^s \circ H(f).
$$

(2.51)

Moreover, as $\left(H, \big(q_i : H \to F(i)\big)_{i \in \mathrm{Ob}\, I}\right)$ is a cone on F, the following diagram is commutative for any $t \in \mathrm{Hom}_I(i, k)$:

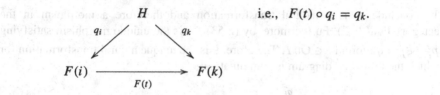

i.e., $F(t) \circ q_i = q_k$.

Therefore, for any $j \in \mathrm{Ob}\, J$, we have

$$F_j(t) \circ q_i^j = q_k^j, \tag{2.52}$$

which implies that $\left(H(j), \big(q_i^j : H(j) \to F_j(i)\big)_{i \in \mathrm{Ob}\, I}\right)$ is a cone on F_j. Recall now that $\left(L(j), \big(p_i^j : L(j) \to F_j(i)\big)_{i \in \mathrm{Ob}\, I}\right)$ is the limit of F_j, so there exists a unique morphism $\xi_j \in \mathrm{Hom}_{\mathcal{C}}\big(H(j), L(j)\big)$ such that the following diagram is commutative for all $i \in \mathrm{Ob}\, I$:

$$
\begin{array}{ccc}
H_j & \xrightarrow{\;\;q_i^j\;\;} & F_j(i) \\
{\scriptstyle \xi_j}\big\downarrow & \nearrow{\scriptstyle p_i^j} & \\
L_j & &
\end{array}
\qquad \text{i.e.,} \quad p_i^j \circ \xi_j = q_i^j.
$$

(2.53)

The proof will be finished once we show that the morphisms ξ_i, $i \in \mathrm{Ob}\, I$ form a natural transformation $\xi : H \to L$ or, equivalently, that ξ is a morphism in the category $\mathrm{Fun}(J, \mathcal{C})$. To this end, we are left to prove the commutativity of the following diagram for any $f \in \mathrm{Hom}_J(j, s)$:

$$
\begin{array}{ccc}
H(j) & \xrightarrow{\ \xi_j\ } & L(j) \\
{\scriptstyle H(f)}\downarrow & & \downarrow{\scriptstyle L(f)} \\
H(s) & \xrightarrow[\ \xi_s\ ]{} & L(s)
\end{array}
\qquad\text{i.e.,}\quad L(f)\circ\xi_j = \xi_s \circ H(f).
$$

$$(2.54)$$

By Proposition 2.2.14, (1) it is enough to show that for all $i \in \mathrm{Ob}\,I$ we have $p_i^s \circ L(f) \circ \xi_j = p_i^s \circ \xi_s \circ H(f)$. Indeed, we have

$$
p_i^s \circ L(f) \circ \xi_j \overset{(2.46)}{=} F(i)(f) \circ p_i^j \circ \xi_j \overset{(2.53)}{=} F(i)(f) \circ q_i^j
$$

$$
\overset{(2.51)}{=} q_i^s \circ H(f) \overset{(2.53)}{=} p_i^s \circ \xi_s \circ H(f).
$$

To conclude, ξ is a natural transformation and therefore a morphism in the category $\mathrm{Fun}(J, C)$. Furthermore, by (2.53), ξ_j is the unique morphism satisfying $p_i^j \circ \xi_j = q_i^j$ for all $i \in \mathrm{Ob}\,I$. Therefore, ξ is the unique natural transformation for which the following diagram is commutative:

$$
\begin{array}{ccc}
H & \xrightarrow{\ q_i\ } & F(i) \\
{\scriptstyle \xi}\downarrow & \nearrow{\scriptstyle p_i} & \\
L & &
\end{array}
\qquad\text{i.e.,}\quad p_i \circ \xi = q_i,
$$

which shows that $\left(L, \left(p_i : L \to F(i)\right)_{i\in\mathrm{Ob}\,I}\right)$ is indeed the limit of F. $\square$

Corollary 2.7.3 *If C is a (co)complete category then the functor category $\mathrm{Fun}(J, C)$ is also (co)complete, for all small categories J.*

Proof Let $F\colon I \to \mathrm{Fun}(J, C)$ be a functor. The (co)completeness assumption on C ensures the existence of a (co)limit for any functor $F_j\colon I \to C$. The desired conclusion now follows from Theorem 2.7.2. $\square$

Example 2.7.4 An important consequence that can be derived from Corollary 2.7.3 concerns the category of presheaves on an arbitrary small category C as defined in Example 1.9.2, (1). More precisely, as **Set** is both complete and cocomplete (see Examples 2.4.1 and 2.4.4) we can conclude that the category of presheaves $\mathrm{Fun}(C^{op}, \mathbf{Set})$ on any small category C is also complete and cocomplete.

Furthermore, this allows us to embed any arbitrary small category C into the (co)complete category of presheaves on C through the Yoneda embedding functor $Y\colon C \to \mathrm{Fun}(C^{op}, \mathbf{Set})$ (see Definition 1.10.7). As the category of presheaves on C inherits most properties of **Set**, the aforementioned embedding creates the appropriate setting for using set theoretical tools for studying the category C or, in certain situations, even to replace C by $\text{Fun}(C^{op}, \textbf{Set})$. For further important applications of this approach we refer the reader to [1, 32]. ☐

We end this section with a characterization of monomorphisms (resp. epimorphisms) in functor categories, where the fact that (co)limits in the aforementioned categories are computed pointwise is crucial.

Proposition 2.7.5 *Let I and C be two categories with I small, F, $G: I \to C$ functors and $\psi: F \to G$ a natural transformation.*

(1) If C is a category with pullbacks then ψ is a monomorphism in the functor category $\text{Fun}(I, C)$ if and only if $\psi_i: F(i) \to G(i)$ is a monomorphism in C, for all $i \in \text{Ob}\, I$.

(2) If C is a category with pushouts then ψ is an epimorphism in the functor category $\text{Fun}(I, C)$ if and only if $\psi_i: F(i) \to G(i)$ is an epimorphism in C, for all $i \in \text{Ob}\, I$.

Proof (1) Assume first that $\psi_i: F(i) \to G(i)$ is a monomorphism for all $i \in \text{Ob}\, I$ and consider natural transformations $\alpha, \beta: H \to F$ such that $\psi \circ \alpha = \psi \circ \beta$, where $H: I \to C$ is a functor. This implies $\psi_i \circ \alpha_i = \psi_i \circ \beta_i$ for all $i \in \text{Ob}\, I$. Since $\psi_i: F(i) \to G(i)$ is a monomorphism, we obtain $\alpha_i = \beta_i$ for all $i \in \text{Ob}\, I$.

Conversely, assume that ψ is a monomorphism in the functor category $\text{Fun}(I, C)$. Consider $i_0 \in \text{Ob}\, I$, $C \in \text{Ob}\, C$ and $f, g \in \text{Hom}_C(C, F(i_0))$ such that $\psi_{i_0} \circ f = \psi_{i_0} \circ g$. We know from Proposition 2.1.18, (1) that $(F, 1_F, 1_F)$ is the pullback of (ψ, ψ) in $\text{Fun}(I, C)$ and since limits in the functor category $\text{Fun}(I, C)$ are computed pointwise we obtain that $(F(i_0), 1_{F(i_0)}, 1_{F(i_0)})$ is the pullback of (ψ_{i_0}, ψ_{i_0}) in C. Therefore, we have a unique $u \in \text{Hom}_C(C, F(i_0))$ such that the following diagram is commutative:

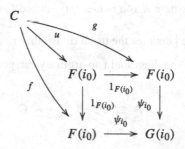

Hence we have $1_{F(i_0)} \circ u = g$ and $1_{F(i_0)} \circ u = f$, which shows that $f = g$, as desired.

(2) Follows easily by the duality principle and Lemma 1.9.6. Indeed, assume C is a category with pushouts and ψ is an epimorphism in the functor category $\text{Fun}(I, C)$. By Lemma 1.9.6, this is equivalent to C^{op} being a category with pullbacks and ψ^{op} being a monomorphism in the functor category $\text{Fun}(I^{op}, C^{op})$.

By the first part of the proof, the last statement is equivalent to $(\psi^{op})_i : F^{op}(i) \to G^{op}(i)$ being a monomorphism in $\mathcal{C}^{op}$ for all $i \in \text{Ob}\,I$. As $(\psi^{op})_i = (\psi_i)^{op}$, this is the same as $\psi_i : F(i) \to G(i)$ being an epimorphism in $\mathcal{C}$ for all $i \in \text{Ob}\,I$. $\square$

2.8   Exercises

2.1 Consider the forgetful functor $U : \mathbf{Ring}^c \to \mathbf{Set}$. Decide if U is:

     (a) faithful;   (b) representable;   (c) epimorphism preserving.

2.2 Let $\mathcal{C}$ be a category with a final object T and binary products. Prove that there is a natural isomorphism between the identity functor $1_{\mathcal{C}}$ on $\mathcal{C}$ and the product functor $- \times T : \mathcal{C} \to \mathcal{C}$.

2.3 Let $\mathcal{C}$ be a category, $f, g \in \text{Hom}_{\mathcal{C}}(A, B)$ and (E, e) the equalizer of the pair (f, g). Then the following are equivalent:

    a. $f = g$;
    b. e is an epimorphism;
    c. e is an isomorphism;
    d. $(A, 1_A)$ is the equalizer of (f, g).

2.4 Let $\mathcal{C}$ be a category and $F : \mathcal{C} \to \mathbf{Set}$ a functor. Prove that

    a. if F is representable then it preserves monomorphisms;
    b. if F is contravariant representable then it maps epimorphisms to monomorphisms.

2.5 Describe binary products in the following categories:

    a. $PO(\mathcal{P}(X), \subseteq)$, where X is a non-empty set and $\subseteq$ denotes the inclusion of sets;
    b. $PO(\mathbb{N}, \mid)$, where $\mid$ denotes the usual divisibility relation on $\mathbb{N}$.

2.6 Consider the following diagram in an arbitrary category $\mathcal{C}$:

$$A \underset{\underset{u}{\overset{g}{\longleftarrow}}}{\overset{f}{\rightrightarrows}} B \underset{v}{\overset{h}{\rightrightarrows}} C$$

such that the following hold:

$$h \circ f = h \circ g, \quad h \circ v = 1_C, \quad g \circ u = 1_B, \quad f \circ u = v \circ h.$$

In this case, the pair (C, h) is called the *split coequalizer* of (f, g). Prove that (C, h) is the coequalizer of the pair of morphisms (f, g). State and prove the dual statement.

2.7 Let $F: \mathcal{C} \to \mathcal{D}$ be a functor and suppose that the following diagram is a split coequalizer in $\mathcal{C}$:

$$A \underset{g}{\overset{f}{\rightrightarrows}} B \underset{v}{\overset{h}{\rightrightarrows}} C$$

Show that $\big(F(C), F(h)\big)$ is the coequalizer of the pair of morphisms $\big(F(f), F(g)\big)$.[10] State and prove the dual statement.

2.8 Let $\mathcal{C}$ be a complete category, $f, g \in \mathrm{Hom}_{\mathcal{C}}(X, Y)$ and let $(X \times Y, (p_X, p_Y))$ denote the product in $\mathcal{C}$ of X and Y. Show that if (E, p) is the equalizer of the pair (f, g) then (E, p, p) is the pullback of the pair $(\overline{f}, \overline{g})$, where $\overline{f}$, $\overline{g} \in \mathrm{Hom}_{\mathcal{C}}(X, X \times Y)$ are the unique morphisms such that the following hold:

$$p_X \circ \overline{f} = 1_X, \ p_Y \circ \overline{f} = f, \ p_X \circ \overline{g} = 1_X, \ p_Y \circ \overline{g} = g.$$

2.9 Let $\mathcal{C}$ be a category, I a set, $(A_i)_{i \in I}$ a family of objects in $\mathcal{C}$ and define the functor $F = \prod_{i \in I} \mathrm{Hom}_{\mathcal{C}}(A_i, -): \mathcal{C} \to \mathbf{Set}$ as follows:

$$F(C) = \prod_{i \in I} \mathrm{Hom}_{\mathcal{C}}(A_i, C),\text{[11]}$$
$$F(u)\big((\eta_i)_{i \in I}\big) = \big(u \circ \eta_i\big)_{i \in I}$$

for all $C, D \in \mathrm{Ob}\,\mathcal{C}$, $u \in \mathrm{Hom}_{\mathcal{C}}(C, D)$ and $\eta_i \in \mathrm{Hom}_{\mathcal{C}}(A_i, C)$, $i \in I$. Prove that F is representable if and only if the family $(A_i)_{i \in I}$ has a coproduct in the category $\mathcal{C}$.

2.10 Let $\mathcal{C}$ be a category. Prove that

   a. if $\mathcal{C}$ has an initial object and pushouts then $\mathcal{C}$ has binary coproducts and coequalizers;
   b. if $\mathcal{C}$ has a final object and pullbacks then $\mathcal{C}$ has binary products and equalizers.

2.11 Let $\mathcal{C}$ be a category. Prove that

   a. if $\mathcal{C}$ has binary coproducts and coequalizers then $\mathcal{C}$ has pushouts;
   b. if $\mathcal{C}$ has binary products and equalizers then $\mathcal{C}$ has pullbacks.

2.12 Let $\mathcal{C}$ be a category. Prove that

   a. if $\mathcal{C}$ has binary coproducts, coequalizers and an initial object then $\mathcal{C}$ has finite colimits;
   b. if $\mathcal{C}$ has binary products, equalizers and a final object then $\mathcal{C}$ has finite limits.

---

[10] In other words, split (co)equalizers are preserved by any functor.

[11] $\prod_{i \in I} \mathrm{Hom}_{\mathcal{C}}(A_i, C)$ denotes the product in the category $\mathbf{Set}$ of sets.

2.13 Let C be a category. Prove that

    a. if C has an initial object and pushouts then C has finite colimits;

    b. if C has a final object and pullbacks and then C has finite limits.

2.14 A morphism $e \in \mathrm{Hom}_C(X, Y)$ is called a *regular monomorphism* if there exist $Z \in \mathrm{Ob}\,C$ and $f, g \in \mathrm{Hom}_C(Y, Z)$ such that (X, e) is the equalizer of (f, g). Dually, a morphism $q \in \mathrm{Hom}_C(X, Y)$ is called a *regular epimorphism* if q is a regular monomorphism in C^{op}. Show that

    a. any split epimorphism (resp. monomorphism) is a regular epimorphism (resp. monomorphism);

    b. any regular epimorphism (resp. monomorphism) is a strong epimorphism (resp. monomorphism).

2.15 An epimorphism $f \in \mathrm{Hom}_C(A, B)$ is called an *extremal epimorphism* if it does not factor through any proper subobject of B (i.e., if $f = g \circ h$ and g is a monomorphism then g must be an isomorphism). Show that in a balanced category any epimorphism is extremal. Formulate and prove the dual statement.

2.16 Give an example to show that a category with pullbacks (resp. pushouts) does not necessarily have equalizers (resp. coequalizers).

2.17 Let C be a category and $f, g \in \mathrm{Hom}_C(A, B)$. Show that if the triple (C, u, u) is the pushout of (f, g) then the pair (C, u) is the coequalizer of (f, g).

2.18 Consider the following commutative diagram in an arbitrary category C and assume that the right-side square is a pullback:

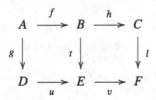

Show that the left-side square is a pullback if and only if the outer rectangle is a pullback.

2.19 Let G_1 and G_2 be two groups with a common subgroup H and $i : H \to G_1$, $j : H \to G_2$ the inclusion morphisms.

    a. Show that $\left(G_1 *_H G_2, f', g'\right)$ is the pushout of (i, j), where $G_1 *_H G_2$ denotes the free product with amalgamated subgroup and $f' : G_2 \to G_1 *_H G_2$ and $g' : G_1 \to G_1 *_H G_2$ are its corresponding group morphisms.

    b. Describe the pushout of (i, j) when H (resp. G_1) is the trivial group.

2.20 Let C be a category with equalizers and $F : C \to D$ a functor which preserves equalizers and reflects isomorphisms. Prove that F is faithful.

2.21 Let I and C be two arbitrary categories and assume that I is small and has an initial object. Show that any functor $F : I \to C$ has a limit.

2.22 Let $(X, \leqslant)$ be a pre-ordered set and PO$(X, \leqslant)$ the corresponding category. Describe (co)limits in PO$(X, \leqslant)$.

2.23 Let G be a group and $\mathcal{G}$ the corresponding category. Is $\mathcal{G}$ (co)complete?

2.24 Let PO$(\mathbb{Z}, \leqslant)$ be the category corresponding to the poset $(\mathbb{Z}, \leqslant)$, where $\leqslant$ is the usual ordering on the integers. Decide if the identity functor Id: PO$(\mathbb{Z}, \leqslant)$ $\rightarrow$ PO$(\mathbb{Z}, \leqslant)$ has a (co)limit.

2.25 Decide if the following categories are (co)complete: **Grp**, **Ab**, **Top**, **Ring**, $_R\mathcal{M}$, **Field**. Describe (co)limits whenever they exist.

Chapter 3
Adjoint Functors

Adjoint functors were first defined by Kan ([31]) in the 50s, motivated by homological algebra ([19, 28]). Nowadays they are present in most fields of mathematics, as will be shown in the forthcoming examples. The terminology was inspired by adjoint operators, whose definition is somewhat similar to the correspondence in Definition 3.1.1.

3.1 Definition and Generic Examples

We start by introducing the main characters of this chapter: adjoint functors.

Definition 3.1.1 An *adjunction* consists of a pair of functors $F: C \to D$, $G: D \to C$ and for any $X \in \mathrm{Ob}\, C$, $Y \in \mathrm{Ob}\, D$ a bijective map

$$\theta_{X, Y}: \mathrm{Hom}_D(F(X), Y) \to \mathrm{Hom}_C(X, G(Y))$$

which is natural in both variables. In this case, we say that F *is left adjoint to* G or equivalently that G *is right adjoint to* F and the notation $F \dashv G$ is used to designate such a pair of adjoint functors.

Unpacking the above naturality assumption in the two variables comes down to the following: for any $Y \in \mathrm{Ob}\, D$, $\theta_{-, Y}$ is a natural isomorphism between the (contravariant) functors $\mathrm{Hom}_D(F(-), Y)$ and $\mathrm{Hom}_C(-, G(Y))$ and for any $X \in \mathrm{Ob}\, C$, $\theta_{X, -}$ is a natural isomorphism between the (covariant) functors $\mathrm{Hom}_D(F(X), -)$ and $\mathrm{Hom}_C(X, G(-))$. In particular, this amounts to the commutativity of the following diagrams for all $f \in \mathrm{Hom}_C(X', X)$ and $g \in$

© The Author(s), under exclusive license to Springer Nature Switzerland AG 2023   153
A. Agore, *A First Course in Category Theory*, Universitext,
https://doi.org/10.1007/978-3-031-42899-9_3

$\text{Hom}_{\mathcal{D}}(Y, Y')$:

$$
\begin{array}{ccc}
\text{Hom}_{\mathcal{D}}(F(X), Y) & \xrightarrow{\theta_{X,Y}} & \text{Hom}_{C}(X, G(Y)) \\
\text{\scriptsize Hom}_{\mathcal{D}}(F(f), Y) \downarrow & & \downarrow \text{\scriptsize Hom}_{C}(f, G(Y)) \\
\text{Hom}_{\mathcal{D}}(F(X'), Y) & \xrightarrow{\theta_{X',Y}} & \text{Hom}_{C}(X', G(Y))
\end{array}
\tag{3.1}
$$

$$
\begin{array}{ccc}
\text{Hom}_{\mathcal{D}}(F(X), Y) & \xrightarrow{\theta_{X,Y}} & \text{Hom}_{C}(X, G(Y)) \\
\text{\scriptsize Hom}_{\mathcal{D}}(F(X), g) \downarrow & & \downarrow \text{\scriptsize Hom}_{C}(X, G(g)) \\
\text{Hom}_{\mathcal{D}}(F(X), Y') & \xrightarrow{\theta_{X,Y'}} & \text{Hom}_{C}(X, G(Y'))
\end{array}
\tag{3.2}
$$

Definition 3.1.1 can also be stated in an equivalent manner in terms of bifunctors. Indeed, recall from Example 1.5.6, (1) that the right and the left associated functors (as defined in Proposition 1.5.5) of the Hom bifunctor $\text{Hom}_{C}(-, -)\colon C^{op} \times C \to$ **Set** with respect to an object C in C are precisely $\text{Hom}_{C}(C, -)$ and $\text{Hom}_{C}(-, C)$, respectively. Furthermore, Proposition 1.7.4 shows that for bifunctors naturality is equivalent to the naturality of both the left and the right associated functors. Therefore, an adjunction between two functors $F\colon C \to \mathcal{D}$ and $G\colon \mathcal{D} \to C$ can be defined equivalently as a natural isomorphism θ between the bifunctors $\text{Hom}_{\mathcal{D}}(F^{op}(-), -)\colon C^{op} \times \mathcal{D} \to$ **Set** and $\text{Hom}_{C}(-, G(-))\colon C^{op} \times \mathcal{D} \to$ **Set**, where $F^{op}\colon C^{op} \to \mathcal{D}^{op}$ denotes the dual functor as defined in Proposition 1.8.1.

3.2 Adjoints Via Free Objects

Most of the categories we have considered so far are categories of sets endowed with some extra structure (e.g., groups, rings, vector spaces, algebras, topological spaces etc.) which allow for various forgetful functors: for example, from **Grp** to **Set**, from $_K\mathcal{M}$ to **Set**, from Alg_K (or Alg_K^c) to $_K\mathcal{M}$. It turns out that all the forgetful functors mentioned above do have left adjoints and this phenomenon can be explained by the existence of the so-called *free objects*; to be more precise, we have a free group (vector space) on any set, a free algebra (i.e., the tensor algebra) on any vector space and a free commutative algebra (i.e., the symmetric algebra) on any vector space. These free objects, together with their universal property, will be the main ingredients in the construction of the left adjoints for the aforementioned forgetful functors. We consider below the case of the forgetful functor from **Grp** to **Set**, but the same strategy works in general.

Example 3.2.1 Let $U : \mathbf{Grp} \to \mathbf{Set}$ be the forgetful functor. We will see that U has a left adjoint $F : \mathbf{Set} \to \mathbf{Grp}$ called the *free group functor*. More precisely, F is constructed as follows:

- for any $X \in \mathrm{Ob}\,\mathbf{Set}$, define $F(X) = FX$, the free group on the set X;[1]
- given $f \in \mathrm{Hom}_{\mathbf{Set}}(X, Y)$, define $F(f) : FX \to FY$ by $F(f) = \overline{f}$, where $\overline{f}$ is obtained from the universal property of the free group FX, i.e., $\overline{f} \in \mathrm{Hom}_{\mathbf{Grp}}(FX, FY)$ is the unique group homomorphism which makes the following diagram commute:

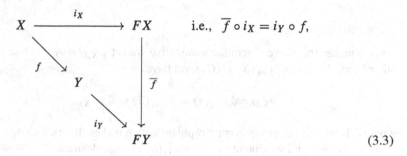

i.e., $\overline{f} \circ i_X = i_Y \circ f,$

$$(3.3)$$

where i_X and i_Y are the inclusion maps.

Let $X \in \mathrm{Ob}\,\mathbf{Set}$ and $G \in \mathrm{Ob}\,\mathbf{Grp}$. We will prove that there is a bijection

$$\theta_{X,G} : \mathrm{Hom}_{\mathbf{Grp}}(FX, G) \to \mathrm{Hom}_{\mathbf{Set}}(X, U(G)), \quad \text{given by } \theta_{X,G}(v) = v \circ i_X$$

for any $v \in \mathrm{Hom}_{\mathbf{Grp}}(FX, G)$. The inverse of $\theta_{X,G}$, denoted by $\psi_{X,G}$, is defined as follows:

$$\psi_{X,G} : \mathrm{Hom}_{\mathbf{Set}}(X, U(G)) \to \mathrm{Hom}_{\mathbf{Grp}}(FX, G),$$

$$\psi_{X,G}(u) = \overline{u}, \quad \text{for all } u \in \mathrm{Hom}_{\mathbf{Set}}(X, U(G)),$$

where $\overline{u} \in \mathrm{Hom}_{\mathbf{Grp}}(FX, G)$ is the unique group homomorphism which makes the following diagram commute:

$$
\begin{array}{ccc}
X & \xrightarrow{\;i_X\;} & FX \\
{\scriptstyle u}\downarrow & \swarrow{\scriptstyle \overline{u}} & \\
G & &
\end{array}
\qquad \text{i.e., } \overline{u} \circ i_X = u.
$$

$$(3.4)$$

---

[1] A group G containing X as a subset is called the *free group* on X if for every group G' and every function $f : X \to G'$, there exists a unique group homomorphism $\psi : G \to G'$ such that $f = \psi \circ i_X$, where $i_X : X \to G$ is the inclusion map ([49, Section 5.5]).

Indeed, for any $v \in \mathrm{Hom}_{\mathbf{Grp}}(FX,\ G)$ we have

$$\psi_{X,G} \circ \theta_{X,G}(v) = \psi_{X,G}(v \circ i_X) = \overline{v \circ i_X},$$

where $\overline{v \circ i_X}$ is the unique group homomorphism which makes the following diagram commute:

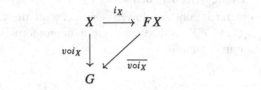

Since v makes the above diagram commutative we get $\psi_{X,G} \circ \theta_{X,G}(v) = v$. On the other hand, if $u \in \mathrm{Hom}_{\mathbf{Set}}(X,\ U(G))$, we have

$$\theta_{X,G} \circ \psi_{X,G}(u) = \theta_{X,G}(\overline{u}) = \overline{u} \circ i_X,$$

where $\overline{u}$ is the unique group homomorphism which makes diagram (3.4) commute. Thus $\overline{u} \circ i_X = u$ and we obtain $\theta_{X,G} \circ \psi_{X,G}(u) = u$, as desired.

Finally we check that the isomorphism θ is natural in both variables. First, fix $G \in \mathrm{Ob}\,\mathbf{Grp}$ and consider $f \in \mathrm{Hom}_{\mathbf{Set}}(X',\ X)$. We need to prove the commutativity of the following diagram:

$$
\begin{array}{ccc}
\mathrm{Hom}_{\mathbf{Grp}}(FX,\ G) & \xrightarrow{\ \theta_{X,G}\ } & \mathrm{Hom}_{\mathbf{Set}}(X,\ U(G)) \\
{\scriptstyle \mathrm{Hom}_{\mathbf{Grp}}(F(f),\,G)} \Big\downarrow & & \Big\downarrow {\scriptstyle \mathrm{Hom}_{\mathbf{Set}}(f,\,U(G))} \\
\mathrm{Hom}_{\mathbf{Grp}}(FX',\ G) & \xrightarrow[\ \theta_{X',G}\ ]{} & \mathrm{Hom}_{\mathbf{Set}}(X',\ U(G))
\end{array}
$$

i.e., $\mathrm{Hom}_{\mathbf{Set}}(f,\ U(G)) \circ \theta_{X,G} = \theta_{X',G} \circ \mathrm{Hom}_{\mathbf{Grp}}(F(f),\ G)$.

To this end, consider $r \in \mathrm{Hom}_{\mathbf{Grp}}(FX,\ G)$; we have

$$\mathrm{Hom}_{\mathbf{Set}}(f,\ U(G)) \circ \theta_{X,G}(r) = \mathrm{Hom}_{\mathbf{Set}}(f,\ U(G))(r \circ i_X)$$

$$= r \circ \underline{i_X \circ f}$$

$$\overset{(3.3)}{=} r \circ \overline{f} \circ i_{X'} = r \circ F(f) \circ i_{X'}$$

$$= \theta_{X',G}(r \circ F(f)) = \theta_{X',G} \circ \mathrm{Hom}_{\mathbf{Grp}}(F(f),\ G)(r).$$

Finally, fix $X \in \mathrm{Ob}\,\mathbf{Set}$ and consider $g \in \mathrm{Hom}_{\mathbf{Grp}}(G, G')$. We are left to prove that the following diagram is commutative:

$$
\begin{array}{ccc}
\mathrm{Hom}_{\mathbf{Grp}}(F(X),\, G) & \xrightarrow{\;\;\theta_{X,G}\;\;} & \mathrm{Hom}_{\mathbf{Set}}(X,\, U(G)) \\
{\scriptstyle \mathrm{Hom}_{\mathbf{Grp}}(F(X),\, g)}\Big\downarrow & & \Big\downarrow{\scriptstyle \mathrm{Hom}_{\mathbf{Set}}(X,\, U(g))} \\
\mathrm{Hom}_{\mathbf{Grp}}(F(X),\, G') & \xrightarrow[\;\;\theta_{X,G'}\;\;]{} & \mathrm{Hom}_{\mathbf{Set}}(X,\, U(G'))
\end{array}
$$

i.e., $\mathrm{Hom}_{\mathbf{Set}}(X, U(g)) \circ \theta_{X,G} = \theta_{X,G'} \circ \mathrm{Hom}_{\mathbf{Grp}}(F(X), g)$.

Let $t \in \mathrm{Hom}_{\mathbf{Grp}}(F(X), G)$; we have

$$
\begin{aligned}
\mathrm{Hom}_{\mathbf{Set}}(X, U(g)) \circ \theta_{X,G}(t) &= U(g) \circ t \circ i_X \\
&= g \circ t \circ i_X \\
&= \theta_{X,G'}(g \circ t) = \theta_{X,G'} \circ \mathrm{Hom}_{\mathbf{Grp}}(F(X), g)(t).
\end{aligned}
$$

$\square$

3.3 Galois Connections

Another class of generic examples of adjoint functors can be obtained from pre-ordered sets regarded as categories (see Example 1.2.2, (2)). The general context is the following: $(X, \leqslant)$ and $(Y, \ll)$ are two pre-ordered sets and we consider the corresponding induced categories $\mathbf{PO}(X, \leqslant)$ and $\mathbf{PO}(Y, \ll)$, respectively. Moreover, functors between such categories are nothing but order-preserving functions between the underlying pre-ordered sets, as we have seen in Example 1.5.3, (35).

If $F \colon \mathbf{PO}(X, \leqslant) \to \mathbf{PO}(Y, \ll)^{op}$ and $G \colon \mathbf{PO}(Y, \ll)^{op} \to \mathbf{PO}(X, \leqslant)$ are two functors then F is left adjoint to G if and only if for all $x \in X$ and $y \in Y$ we have

$$F(x) \gg y \text{ in } Y \text{ if and only if } x \leqslant G(y) \text{ in } X. \tag{3.5}$$

Indeed, recall that the hom sets in any category induced by a pre-ordered set have at most one element. Therefore, condition (3.5) can be equivalently expressed as a bijection between $\mathrm{Hom}_{\mathbf{PO}(Y, \ll)^{op}}(F(x), y)$ and $\mathrm{Hom}_{\mathbf{PO}(X, \leqslant)}(x, G(y))$. This bijection is trivially natural as we have at most one element in each hom set.

A pair of adjoint functors as above is called a *Galois connection* from $(X, \leqslant)$ to $(Y, \ll)$. Important examples of Galois connections can be found all across the

mathematical landscape. For instance, these include the classical Galois correspondence for field extensions as well as the correspondence between algebraic sets and
radical ideals in algebraic geometry.

An important example concerns the *Galois correspondence for field extensions*.

Example 3.3.1 Throughout this example, $K \subseteq L$ is a field extension and for any
field $S \subseteq L$ we define

$$\mathrm{Gal}(L/S) = \{\psi \in \mathrm{Aut}\,(L) \mid \psi(x) = x, \text{ for all } x \in S\} \text{ and } G = \mathrm{Gal}(L/K).$$

Furthermore, let

$$\mathcal{A} = \{S \subseteq L \text{ subfield} \mid K \subseteq S \subseteq L\} \text{ and}$$

$$\mathcal{B} = \{H \subseteq G \mid H \text{ subgroup in } G\}.$$

Then $(\mathcal{A}, \subseteq)$ and $(\mathcal{B}, \subseteq)$ are pre-ordered sets, where by a slight abuse of notation
we use "$\subseteq$" to denote both inclusions. We can now construct a Galois connection
from $(\mathcal{A}, \subseteq)$ to $(\mathcal{B}, \subseteq)$ as follows for all $S \in \mathcal{A}$ and $H \in \mathcal{B}$:

$$F: (\mathcal{A}, \subseteq) \to (\mathcal{B}, \subseteq)^{op}, \quad F(S) = \mathrm{Gal}(L/S),$$

$$G: (\mathcal{B}, \subseteq)^{op} \to (\mathcal{A}, \subseteq), \quad G(H) = \mathrm{Fix}(H) = \{l \in L \mid \tau(l) = l \text{ for all } \tau \in H\}.$$

First note that if $S \in \mathcal{A}$ then any ψ in $\mathrm{Gal}(L/S)$ fixes all elements of S and, in
particular, those of K, which shows that $\mathrm{Gal}(L/S)$ is a subgroup of $\mathrm{Gal}(L/K) = G$;
thus $F(S) \in \mathcal{B}$. Similarly, if H is a subgroup of G, and all elements of G fix all
elements of K, it follows that $K \subseteq \mathrm{Fix}(H)$; we can now conclude that $G(H) \in
\mathcal{A}$.

Moreover, if $S \subseteq S'$ and $\psi \in \mathrm{Gal}(L/S')$ then $\psi(x) = x$ for all $x \in S'$ and, in
particular, the same holds for all $x \in S$. This shows that $\mathrm{Gal}(L/S') \subseteq \mathrm{Gal}(L/S)$
and therefore F is order-preserving, i.e., a functor between the corresponding
categories. Now if $H \subseteq H'$ and $l \in \mathrm{Fix}(H')$ then $\tau(l) = l$ for all $\tau \in H'$
and, in particular, the same holds for all $\tau \in H$. Therefore, $\mathrm{Fix}(H') \subseteq \mathrm{Fix}(H)$
and G is also order-preserving. We are left to show that (3.5) also holds. Indeed
for all $S \in \mathcal{A}$ and $H \in \mathcal{B}$ we have $G(H) = \mathrm{Fix}(H) \supseteq S$ if and only if
$\tau(y) = y$ for all $\tau \in H$ and $y \in S$ if and only if $H \subseteq F(S)$, as desired.
□

In fact, the above bijective correspondence is not specific to the theory of
fields and can be generalized by replacing the automorphism group in the previous
example by an arbitrary group. To this end, let G be a group and $\triangleright \colon G \times X \to
X$ a group action on a set X. If we consider $\mathcal{A} = \mathcal{P}(X)$ and $\mathcal{B} = \{H \subseteq
G \mid H \text{ subgroup in } G\}$, then $(\mathcal{A}, \subseteq)$ and $(\mathcal{B}, \subseteq)$ are pre-ordered sets, where again
we use "$\subseteq$" to denote both inclusions. We can now define a Galois connection from

$(\mathcal{A}, \subseteq)$ to $(\mathcal{B}, \subseteq)$ as follows for all $Y \in \mathcal{A}$ and $H \in \mathcal{B}$:

$$F: (\mathcal{A}, \subseteq) \to (\mathcal{B}, \subseteq)^{op}, \quad F(Y) = \{h \in G \mid h \rhd y = y \text{ for all } y \in Y\},$$

$$G: (\mathcal{B}, \subseteq)^{op} \to (\mathcal{A}, \subseteq), \quad G(H) = \{x \in X \mid h \rhd x = x \text{ for all } h \in H\}.$$

First, note that the properties of a group action imply that the set

$$\{h \in G \mid h \rhd y = y \text{ for all } y \in Y\}$$

is in fact a subgroup of G and therefore $F(Y) \in \mathcal{B}$ for all $Y \in \mathcal{A}$. Furthermore, if $Y \subseteq Y'$ and $h \in F(Y')$ then $h \rhd y = y$ for all $y \in Y'$ and, in particular, for all $y \in Y$; this implies that $F(Y') \subseteq F(Y)$ and therefore, F is a well-defined functor from $(\mathcal{A}, \subseteq)$ to $(\mathcal{B}, \subseteq)^{op}$. Similarly, if H, $H' \in \mathcal{B}$ with $H \subseteq H'$ and $x \in G(H')$ then $h \rhd x = x$ for all $h \in H'$ and, in particular, for all $h \in H$. Hence $G(H') \subseteq G(H)$ and G is a well-defined functor from $(\mathcal{B}, \subseteq)^{op}$ to $(\mathcal{A}, \subseteq)$. Finally, the two functors fulfill condition (3.5); indeed, given $Y \in \mathcal{A}$ and $H \in \mathcal{B}$ we have $F(Y) \supseteq H$ if and only if $h \rhd y = y$ for all $y \in Y$ and $h \in H$ if and only if $Y \subseteq G(H)$. To conclude, we have proved that the functors F and G form a Galois connection.

3.4   More Examples and Properties of Adjoint Functors

We start this section with more examples of adjoint functors, spanning various fields.

Examples 3.4.1

(1) For any non-empty set X, the functor $- \times X \colon \mathbf{Set} \to \mathbf{Set}$ has a right adjoint given by $\mathrm{Hom}_{\mathbf{Set}}(X, -) \colon \mathbf{Set} \to \mathbf{Set}$. Indeed, for all Y, $Z \in \mathrm{Ob}\,\mathbf{Set}$, define

$$\theta_{Y, Z} \colon \mathrm{Hom}_{\mathbf{Set}}(Y \times X, Z) \to \mathrm{Hom}_{\mathbf{Set}}\big(Y, \mathrm{Hom}_{\mathbf{Set}}(X, Z)\big),$$

$$\theta_{Y, Z}(f)(y)(x) = f(y, x),$$

for all $f \in \mathrm{Hom}_{\mathbf{Set}}(Y \times X, Z)$, and all $y \in Y$, $x \in X$. Then $\theta_{Y, Z}$ is bijective with inverse given as follows:

$$\psi_{Y, Z} \colon \mathrm{Hom}_{\mathbf{Set}}\big(Y, \mathrm{Hom}_{\mathbf{Set}}(X, Z)\big) \to \mathrm{Hom}_{\mathbf{Set}}(Y \times X, Z),$$

$$\psi_{Y, Z}(g)(y, x) = \big(g(y)\big)(x),$$

for all $g \in \mathrm{Hom}_{\mathbf{Set}}\big(Y, \mathrm{Hom}_{\mathbf{Set}}(X, Z)\big)$, $y \in Y$ and $x \in X$.

Indeed, we have

$$\big(\psi_{Y,Z} \circ \theta_{Y,Z}\big)(f)(y,\, x) = \big(\theta_{Y,Z}(f)(y)\big)(x) = f(y,\, x),$$

$$\big(\theta_{Y,Z} \circ \psi_{Y,Z}\big)(g)(y)(x) = \psi_{Y,Z}(g)(y,\, x) = \big(g(y)\big)(x).$$

We are left to show the commutativity of diagrams (3.1) and (3.2). To start with, let $f \in \mathrm{Hom}_{\mathbf{Set}}(Y',\, Y)$; we need to show the commutativity of the following diagram:

$$
\begin{array}{ccc}
\mathrm{Hom}_{\mathbf{Set}}(Y \times X,\, Z) & \xrightarrow{\ \theta_{Y,Z}\ } & \mathrm{Hom}_{\mathbf{Set}}\big(Y,\, \mathrm{Hom}_{\mathbf{Set}}(X,\, Z)\big) \\
{\scriptstyle \mathrm{Hom}_{\mathbf{Set}}(f \times 1_X,\, Z)} \Big\downarrow & & \Big\downarrow {\scriptstyle \mathrm{Hom}_{\mathbf{Set}}\big(f,\, \mathrm{Hom}_{\mathbf{Set}}(X,\, Z)\big)} \\
\mathrm{Hom}_{\mathbf{Set}}(Y' \times X,\, Z) & \xrightarrow[\ \theta_{Y',Z}\ ]{} & \mathrm{Hom}_{\mathbf{Set}}\big(Y',\, \mathrm{Hom}_{\mathbf{Set}}(X,\, Z)\big)
\end{array}
$$

For any $t \in \mathrm{Hom}_{\mathbf{Set}}(Y \times X,\, Z)$, $y' \in Y'$, $x \in X$ we have

$$\Big(\mathrm{Hom}_{\mathbf{Set}}\big(f,\, \mathrm{Hom}_{\mathbf{Set}}(X,\, Z)\big) \circ \theta_{Y,Z}(t)\Big)(y')(x)$$

$$= \big(\theta_{Y,Z}(t) \circ f\big)(y')(x) = t\big(f(y'),\, x\big)$$

$$= t \circ (f \times 1_X)(y',\, x) = \theta_{Y',Z} \circ \big(t \circ (f \times 1_X)\big)(y')(x)$$

$$= \big(\theta_{Y',Z} \circ \mathrm{Hom}_{\mathbf{Set}}(f \times 1_X,\, Z)(t)\big)(y')(x)$$

and the commutativity of (3.1) is proved. Consider now $g \in \mathrm{Hom}_{\mathbf{Set}}(Z,\, Z')$; we are left to show the commutativity of the following diagram:

$$
\begin{array}{ccc}
\mathrm{Hom}_{\mathbf{Set}}(Y \times X,\, Z) & \xrightarrow{\ \theta_{Y,Z}\ } & \mathrm{Hom}_{\mathbf{Set}}\big(Y,\, \mathrm{Hom}_{\mathbf{Set}}(X,\, Z)\big) \\
{\scriptstyle \mathrm{Hom}_{\mathbf{Set}}(Y \times X,\, g)} \Big\downarrow & & \Big\downarrow {\scriptstyle \mathrm{Hom}_{\mathbf{Set}}\big(Y,\, \mathrm{Hom}_{\mathbf{Set}}(X,\, g)\big)} \\
\mathrm{Hom}_{\mathbf{Set}}(Y \times X,\, Z') & \xrightarrow[\ \theta_{Y,Z'}\ ]{} & \mathrm{Hom}_{\mathbf{Set}}\big(Y,\, \mathrm{Hom}_{\mathbf{Set}}(X,\, Z')\big)
\end{array}
$$

Indeed, if $t \in \mathrm{Hom}_{\mathbf{Set}}(Y \times X,\, Z)$, $y \in Y$, $x \in X$ we have

$$\big(\theta_{Y,Z'} \circ \mathrm{Hom}_{\mathbf{Set}}(Y \times X,\, g)(t)\big)(y)(x) = \theta_{Y,Z'}(g \circ t)(y)(x) = g \circ t(y,\, x)$$

$$= g \circ \theta_{Y,Z}(t)(y)(x) = \big(\mathrm{Hom}_{\mathbf{Set}}\big(Y,\, \mathrm{Hom}_{\mathbf{Set}}(X,\, g)\big) \circ \theta_{Y,Z}(t)\big)(y)(x),$$

which proves that (3.2) is also commutative.

(2) Let K be a field and denote the tensor product over K simply by $\otimes$ (i.e., $\otimes = \otimes_K$). If $X \in \mathrm{Ob}\,_K M$, then the tensor product functor $- \otimes X \colon {}_K M \to {}_K M$ (see

Example 1.5.3, (20)) has a right adjoint given by $\mathrm{Hom}_{K}\mathcal{M}(X, -)\colon {}_{K}M \to {}_{K}M$ (Example 1.5.3, (13)). Indeed, for all $Y, Z \in \mathrm{Ob}\, {}_{K}\mathcal{M}$ define

$$\theta_{Y,Z}\colon \mathrm{Hom}_{K}\mathcal{M}(Y \otimes X, Z) \to \mathrm{Hom}_{K}\mathcal{M}\big(Y, \mathrm{Hom}_{K}\mathcal{M}(X, Z)\big),$$

$$\theta_{Y,Z}(f)(y)(x) = f(y \otimes x) \text{ for all } f \in \mathrm{Hom}_{K}\mathcal{M}(Y \otimes X, Z),\ y \in Y,\ x \in X.$$

The inverse of $\theta_{Y,Z}$ is given as follows:

$$\psi_{Y,Z}\colon \mathrm{Hom}_{K}\mathcal{M}\big(Y, \mathrm{Hom}_{K}\mathcal{M}(X, Z)\big) \to \mathrm{Hom}_{K}\mathcal{M}(Y \otimes X, Z),$$

$$\psi_{Y,Z}(g)(y \otimes x) = \big(g(y)\big)(x),$$

for all $g \in \mathrm{Hom}_{K}\mathcal{M}\big(Y, \mathrm{Hom}_{K}\mathcal{M}(X, Z)\big)$, $y \in Y$ and $x \in X$.

Showing that the maps defined above are indeed K-linear is straightforward using the properties of the tensor product while proving the commutativity of the diagrams (3.1) and (3.2) goes very much along the lines of the previous example and is left to the reader.

(3) If X is a locally compact and Hausdorff topological space then the functor $- \times X\colon \mathbf{Top} \to \mathbf{Top}$ (see Example 1.5.3, (18)) is left adjoint to $\mathrm{Hom}_{\mathbf{Top}}(X, -)\colon \mathbf{Top} \to \mathbf{Top}$ (see Example 1.5.3, (11)) as proved, for instance, in [12, Chapter 5] in a more general setting. Recall that for any $Y \in \mathrm{Ob}\,\mathbf{Top}$ we consider on $\mathrm{Hom}_{\mathbf{Top}}(X, Y)$ the compact-open topology while $Y \times X$ is endowed with the product topology.

In the first example above we proved that for any $Y, Z \in \mathrm{Ob}\,\mathbf{Set}$ we have a set bijection between $\mathrm{Hom}_{\mathbf{Set}}(Y \times X, Z)$ and $\mathrm{Hom}_{\mathbf{Set}}\big(Y, \mathrm{Hom}_{\mathbf{Set}}(X, Z)\big)$. We will show that this bijection induces a continuous bijective map between $\mathrm{Hom}_{\mathbf{Top}}(Y \times X, Z)$ and $\mathrm{Hom}_{\mathbf{Top}}\big(Y, \mathrm{Hom}_{\mathbf{Top}}(X, Z)\big)$ with respect to the previously mentioned topologies. To this end, for all $Y, Z \in \mathrm{Ob}\,\mathbf{Top}$ define

$$\theta_{Y,Z}\colon \mathrm{Hom}_{\mathbf{Top}}(Y \times X, Z) \to \mathrm{Hom}_{\mathbf{Top}}\big(Y, \mathrm{Hom}_{\mathbf{Top}}(X, Z)\big),$$

$$\theta_{Y,Z}(f)(y)(x) = f(y, x) \text{ for all } f \in \mathrm{Hom}_{\mathbf{Top}}(Y \times X, Z),\ y \in Y,\ x \in X.$$

In order to show that $\theta_{Y,Z}$ is well-defined we need to prove that if $f \in \mathrm{Hom}_{\mathbf{Top}}(Y \times X, Z)$ then $\theta_{Y,Z}(f) = \overline{f} \in \mathrm{Hom}_{\mathbf{Top}}\big(Y, \mathrm{Hom}_{\mathbf{Top}}(X, Z)\big)$. Consider $W(K, V)$ to be a sub-basic open set of $\mathrm{Hom}_{\mathbf{Top}}(X, Z)$, i.e., $K \subseteq X$ is a compact subset, $V \subseteq Z$ is an open subset and

$$W(K, V) = \{f \in \mathrm{Hom}_{\mathbf{Top}}(X, Z) \mid f(K) \subseteq V\}.$$

If $\overline{f}^{-1}\big(W(K, V)\big) = \emptyset$ then $\overline{f}$ is obviously continuous, as desired. Assume now that $\overline{f}^{-1}\big(W(K, V)\big) \neq \emptyset$ and consider $y \in \overline{f}^{-1}\big(W(K, V)\big)$. Therefore, we have $\overline{f}(y) \in W(K, V)$ and we obtain $\overline{f}(y)(k) = f(y, k) \in V$ for all $k \in K$.

Since $f \colon Y \times X \to Z$ is continuous, it follows that $f^{-1}(V) = \{y\} \times K \subseteq Y \times X$ is an open subset. Now the tube lemma[2] implies that there exist open subsets $U_y \subseteq Y$ and $T_y \subseteq X$ such that

$$\{y\} \times K \subseteq U_y \times T_y \subseteq f^{-1}(V).$$

This shows that $y \in U_y \subseteq \overline{f}^{-1}\big(W(K, V)\big)$ and therefore we have

$$\overline{f}^{-1}\big(W(K, V)\big) = \bigcup_{y \in \overline{f}^{-1}\big(W(K,V)\big)} U_y.$$

We can now conclude that $\overline{f}^{-1}\big(W(K, V)\big)$ is an open set as a union of open sets. To summarize, we proved that if $f \colon Y \times X \to Z$ is continuous then $\overline{f} = \theta_{Y,Z}(f) \colon Y \to \mathrm{Hom}_{\mathbf{Top}}(X, Z)$ is continuous as well.

Consider now $\psi_{Y,Z}$, the inverse of $\theta_{Y,Z}$, given as follows:

$$\psi_{Y,Z} \colon \mathrm{Hom}_{\mathbf{Top}}\big(Y, \mathrm{Hom}_{\mathbf{Top}}(X, Z)\big) \to \mathrm{Hom}_{\mathbf{Top}}(Y \times X, Z),$$

$$\psi_{Y,Z}(g)(y, x) = \big(g(y)\big)(x),$$

for all $g \in \mathrm{Hom}_{\mathbf{Top}}\big(Y, \mathrm{Hom}_{\mathbf{Top}}(X, Z)\big)$, $y \in Y$ and $x \in X$.

We are left to show that $\psi_{Y,Z}$ is well-defined, i.e., if $g \colon Y \to \mathrm{Hom}_{\mathbf{Top}}(X, Z)$ is continuous then $\widehat{g} = \psi_{Y,Z}(g) \colon Y \times X \to Z$ is continuous as well. We start by showing that the evaluation map $\mathrm{ev} \colon \mathrm{Hom}_{\mathbf{Top}}(X, Z) \times X \to Z$ defined by $\mathrm{ev}(g, x) = g(x)$ is continuous at every point[3] $(g, x) \in \mathrm{Hom}_{\mathbf{Top}}(X, Z) \times X$. Indeed, if $V \subseteq Z$ is an open set such that $\mathrm{ev}(g, x) = g(x) \in V$ then the continuity of g implies that $g^{-1}(V) \subseteq X$ is an open subset such that $x \in g^{-1}(V)$. As X is Hausdorff and locally compact[4] we can find an open subset $U \subseteq X$ whose closure $\overline{U}$ is compact and $x \in U \subseteq \overline{U} \subseteq g^{-1}(V)$. Therefore, we have $g(x) \in g\big(\overline{U}\big) \subseteq V$, which shows that $W(\overline{U}, V) \times U \subseteq \mathrm{Hom}_{\mathbf{Top}}(X, Z) \times X$ is an open subset such that $(g, x) \in W(\overline{U}, V) \times U$ and $\mathrm{ev}\big(W(\overline{U}, V), U\big) \subseteq V$. Hence ev is a continuous map. Now, as we assumed $g \colon Y \to \mathrm{Hom}_{\mathbf{Top}}(X, Z)$ to be continuous, the

---

[2] The tube lemma: Let A and B compact subspaces of X and Y, respectively, and let N be an open set in $X \times Y$ containing $A \times B$. Then, there exist open subsets U and V in X and Y, respectively, such that $A \times B \subseteq U \times V \subseteq N$ (see [39, Lemma 26.8 and exercise 9 on page 171]).

[3] Let $f \colon A \to B$ be a map between two topological spaces. We say that f is *continuous at a point* $x \in A$ if, for each neighborhood V of $f(x)$, there is a neighborhood U of x whose closure $f(U) \subset V$. The map f is continuous if and only if is continuous at every point $x \in A$ ([39, Theorem 18.1]).

[4] Recall that if X is a Hausdorff space then X is *locally compact* if and only if given $x \in X$ and given a neighborhood U of x, there is a neighborhood V of x whose closure $\overline{V}$ is compact and $\overline{V} \subseteq U$ ([39, Theorem 29.2]).

following composition is also continuous:

$$Y \times X \xrightarrow{\ g \times 1_X\ } \text{Hom}_{\textbf{Top}}(X,\ Z) \times X \xrightarrow{\ \text{ev}\ } Z\ .$$

Furthermore, for all $(y,\ x) \in Y \times X$ we have

$$\text{ev} \circ (g \times 1_X)(y,\ x) = \text{ev}\big(g(y),\ x\big) = g(y)(x) = \psi_{Y,\ Z}(g)(y,\ x).$$

We can now conclude that $\psi_{Y,\ Z}(g) = \text{ev} \circ (g \times 1_X)$ is indeed continuous, as desired.

For a more general and comprehensive account of the various topologies that can be defined on a set of continuous maps and their behaviour with respect to the above adjunction, we refer to [12, Chapter 5].

(4) The inclusion functor $I \colon \textbf{Ab} \to \textbf{Grp}$ is right adjoint to the abelianization functor $F \colon \textbf{Grp} \to \textbf{Ab}$ defined in Example 1.5.3, (23). Indeed, for any $A \in \text{Ob}\,\textbf{Ab}$ and $G \in \text{Ob}\,\textbf{Grp}$, consider $\theta_{G,\,A} \colon \text{Hom}_{\textbf{Ab}}(F(G),\ A) \to \text{Hom}_{\textbf{Grp}}(G,\ I(A))$ defined by

$$\theta_{G,\,A}(f) = f \circ \pi_G,\ \text{ for all } f \in \text{Hom}_{\textbf{Ab}}(F(G),\ A),$$

where $\pi_G \colon G \to G_{ab}$ is the canonical projection. Furthermore, given a group homomorphism $g \in \text{Hom}_{\textbf{Grp}}(G,\ I(A))$, since A is an abelian group it can be easily seen that $[G,\ G] \subseteq \ker(g)$, where $[G,\ G]$ denotes the commutator subgroup. Therefore, the universal property of the quotient group G_{ab} yields a unique group homomorphism $h \in \text{Hom}_{\textbf{Ab}}(G_{ab},\ A)$ such that $h \circ \pi_G = g$. This shows that $\theta_{A,\,G}$ is bijective for all $A \in \text{Ob}\,\textbf{Ab}$ and $G \in \text{Ob}\,\textbf{Grp}$. We are left to show that θ defined above makes diagrams (3.1) and (3.2) commutative. To this end, let $f \in \text{Hom}_{\textbf{Grp}}(G',\ G)$ and $A \in \text{Ob}\,\textbf{Ab}$; we start by showing the commutativity of the following diagram:

$$
\begin{array}{ccc}
\text{Hom}_{\textbf{Ab}}(G_{ab},\ A) & \xrightarrow{\ \theta_{G,\,A}\ } & \text{Hom}_{\textbf{Grp}}(G,\ A) \\
{\scriptstyle \text{Hom}_{\textbf{Ab}}(f_{ab},\ A)}\big\downarrow & & \big\downarrow {\scriptstyle \text{Hom}_{\textbf{Grp}}(f,\ A)} \\
\text{Hom}_{\textbf{Ab}}(G'_{ab},\ A) & \xrightarrow[\ \theta_{G',\,A}\ ]{} & \text{Hom}_{\textbf{Grp}}(G',\ A)
\end{array}
$$

For any $t \in \text{Hom}_{\textbf{Ab}}(G_{ab},\ A)$ we have

$$\text{Hom}_{\textbf{Grp}}(f,\ A) \circ \theta_{G,\,A}(t) = \text{Hom}_{\textbf{Grp}}(f,\ A)(t \circ \pi_G)$$

$$= t \circ \pi_G \circ f \overset{(1.6)}{=} t \circ f_{ab} \circ \pi_{G'}$$

$$= \theta_{G',\,A}(t \circ f_{ab}) = \theta_{G',\,A} \circ \text{Hom}_{\textbf{Ab}}(f_{ab},\ A)(t),$$

which shows that (3.1) is commutative. Consider now $g \in \mathrm{Hom}_{\mathbf{Ab}}(A, A')$ and $G \in \mathrm{Ob}\,\mathbf{Grp}$. We are left to show the commutativity of the following diagram:

$$
\begin{array}{ccc}
\mathrm{Hom}_{\mathbf{Ab}}(G_{ab}, A) & \xrightarrow{\;\theta_{G,A}\;} & \mathrm{Hom}_{\mathbf{Grp}}(G, A) \\[2pt]
{\scriptstyle \mathrm{Hom}_{\mathbf{Ab}}(G_{ab}, g)} \Big\downarrow & & \Big\downarrow {\scriptstyle \mathrm{Hom}_{\mathbf{Grp}}(G, g)} \\[2pt]
\mathrm{Hom}_{\mathbf{Ab}}(G_{ab}, A') & \xrightarrow[\;\theta_{G,A'}\;]{} & \mathrm{Hom}_{\mathbf{Grp}}(G, A')
\end{array}
$$

Indeed, for any $t \in \mathrm{Hom}_{\mathbf{Ab}}(G_{ab}, A)$ we have

$$
\mathrm{Hom}_{\mathbf{Grp}}(G, g) \circ \theta_{G,A}(t) = \mathrm{Hom}_{\mathbf{Grp}}(G, g)(t \circ \pi_G) = g \circ t \circ \pi_G
$$

$$
= \theta_{G,A'}(g \circ t) = \theta_{G,A'} \circ \mathrm{Hom}_{\mathbf{Ab}}(G_{ab}, g)(t).
$$

Therefore, (3.2) is commutative and we have proved that F is left adjoint to I.

(5) Let $\mathbf{1}$ be the discrete category with one object denoted by $\star$. For any category C we can define a unique functor $T: C \to \mathbf{1}$. It can be easily seen that the functor T has a left (resp. right) adjoint if and only if C has an initial (resp. final) object.

Indeed, $L: \mathbf{1} \to C$ is a left adjoint for T if and only if for any $C \in \mathrm{Ob}\,C$ there exists a bijective map $\theta^l_{\star,C}: \mathrm{Hom}_C(L(\star), C) \to \mathrm{Hom}_{\mathbf{1}}(\star, T(C))$ which is natural in both variables. As $\mathrm{Hom}_{\mathbf{1}}(\star, T(C)) = \{1_\star\}$, $\theta^l_{\star,C}$ is a bijection for any $C \in \mathrm{Ob}\,C$ if and only if $\mathrm{Hom}_C(L(\star), C)$ has one element for any $C \in \mathrm{Ob}\,C$. This means precisely that $L(\star)$ is the initial object of C.

Similarly, $R: \mathbf{1} \to C$ is a right adjoint for T if and only if for any $C \in \mathrm{Ob}\,C$ there exists a bijective map $\theta^r_{C,\star}: \mathrm{Hom}_{\mathbf{1}}(T(C), \star) \to \mathrm{Hom}_C(C, R(\star))$ which is natural in both variables. As $\mathrm{Hom}_{\mathbf{1}}(T(C), \star) = \{1_\star\}$, $\theta^r_{C,\star}$ is a bijection for any $C \in \mathrm{Ob}\,C$ if and only if $\mathrm{Hom}_C(C, R(\star))$ has one element for any $C \in \mathrm{Ob}\,C$. This means precisely that $R(\star)$ is the final object of C.   □

We continue with further properties of adjoint functors. First we look at compositions of adjoint functors.

Proposition 3.4.2 *Consider the functors* $F: \mathcal{A} \to \mathcal{B}$, $G: \mathcal{B} \to \mathcal{A}$ *such that* $F \dashv G$ *and* $H: \mathcal{B} \to C$, $K: C \to \mathcal{B}$ *such that* $H \dashv K$. *Then* $HF \dashv GK$.

Proof We have the following natural isomorphisms for all $A \in \mathrm{Ob}\,\mathcal{A}$ and $C \in \mathrm{Ob}\,C$: $\mathrm{Hom}_C(HF(A), C) \approx \mathrm{Hom}_{\mathcal{B}}(F(A), K(C)) \approx \mathrm{Hom}_{\mathcal{A}}(A, GK(C))$.   □

Any pair of adjoint functors induces an adjunction between the opposite functors as follows:

Theorem 3.4.3 *Let* $F: C \to \mathcal{D}$ *and* $G: \mathcal{D} \to C$ *be two functors. Then* $F \dashv G$ *if and only if* $G^{op} \dashv F^{op}$.

Proof Assume that $F \dashv G$. Then, for any $Y \in \mathrm{Ob}\,C$ and $X \in \mathrm{Ob}\,\mathcal{D}$ we have a bijective map $\theta_{Y,X} \colon \mathrm{Hom}_{\mathcal{D}}(F(Y), X) \to \mathrm{Hom}_C(Y, G(X))$ which is natural in both variables. Consider now the map $\bar{\theta}_{X,Y} \colon \mathrm{Hom}_{C^{op}}(G^{op}(X), Y) \to \mathrm{Hom}_{\mathcal{D}^{op}}(X, F^{op}(Y))$ defined for all $t^{op} \in \mathrm{Hom}_{C^{op}}(G^{op}(X), Y)$ by

$$\bar{\theta}_{X,Y}\big(t^{op}\big) = \big(\theta^{-1}_{Y,X}(t)\big)^{op}. \tag{3.6}$$

The map defined above is bijective with inverse given by

$$\xi_{X,Y}\big(u^{op}\big) = \big(\theta_{Y,X}(u)\big)^{op} \tag{3.7}$$

for all $u^{op} \in \mathrm{Hom}_{\mathcal{D}^{op}}(X, F^{op}(Y))$ Indeed, for all $t^{op} \in \mathrm{Hom}_{C^{op}}(G^{op}(X), Y)$ and $u^{op} \in \mathrm{Hom}_{\mathcal{D}^{op}}(X, F^{op}(Y))$ we have

$$\bar{\theta}_{X,Y} \circ \xi_{X,Y}(u^{op}) = \bar{\theta}_{X,Y}\big(\theta_{Y,X}(u)^{op}\big) = \Big(\theta^{-1}_{Y,X}\big(\theta_{Y,X}(u)\big)\Big)^{op} = u^{op},$$

$$\xi_{X,Y} \circ \bar{\theta}_{X,Y}(t^{op}) = \xi_{X,Y}\Big(\big(\theta^{-1}_{Y,X}(t)\big)^{op}\Big) = \Big(\theta_{Y,X}\big(\theta^{-1}_{Y,X}(t)\big)\Big)^{op} = t^{op}.$$

We are left to show that $\bar{\theta}$ is natural in both variables. To this end, consider first $g^{op} \in \mathrm{Hom}_{\mathcal{D}^{op}}(D', D)$. The naturality of $\bar{\theta}$ in the first variable comes down to showing the commutativity of the following diagram:

$$
\begin{array}{ccc}
\mathrm{Hom}_{C^{op}}(G^{op}(D), C) & \xrightarrow{\ \bar{\theta}_{D,C}\ } & \mathrm{Hom}_{\mathcal{D}^{op}}(D, F^{op}(C)) \\[2pt]
{\scriptstyle \mathrm{Hom}_{C^{op}}(G^{op}(g^{op}), C)}\Big\downarrow & & \Big\downarrow{\scriptstyle \mathrm{Hom}_{\mathcal{D}^{op}}(g^{op}, F^{op}(C))} \\[2pt]
\mathrm{Hom}_{C^{op}}(G^{op}(D'), C) & \xrightarrow[\ \bar{\theta}_{D',C}\ ]{} & \mathrm{Hom}_{\mathcal{D}^{op}}(D', F^{op}(C))
\end{array}
$$

i.e., $\mathrm{Hom}_{\mathcal{D}^{op}}(g^{op}, F^{op}(C)) \circ \bar{\theta}_{D,C} = \bar{\theta}_{D',C} \circ \mathrm{Hom}_{C^{op}}(G^{op}(g^{op}), C).$ (3.8)

Since we have already proved that each $\bar{\theta}_{D,C}$ is invertible with inverse $\xi_{D,C}$ it will suffice to show that the following holds:

$$\xi_{D',C} \circ \mathrm{Hom}_{\mathcal{D}^{op}}(g^{op}, F^{op}(C)) = \mathrm{Hom}_{C^{op}}(G^{op}(g^{op}), C) \circ \xi_{D,C}.$$

To this end, given $u^{op} \in \mathrm{Hom}_{\mathcal{D}^{op}}(D, F^{op}(C))$, we have

$$\xi_{D',C} \circ \mathrm{Hom}_{\mathcal{D}^{op}}(g^{op}, F^{op}(C))(u^{op}) = \xi_{D',C}\big(u^{op} \circ^{op} g^{op}\big)$$
$$= \xi_{D',C}\big((g \circ u)^{op}\big)$$
$$= \big(\theta_{C,D'}(g \circ u)\big)^{op}$$

$$\overset{(3.2)}{=} \big(G(g) \circ \theta_{C, D}(u)\big)^{op}$$

$$= \big(\theta_{C, D}(u)\big)^{op} \circ^{op} G(g)^{op}$$

$$= \mathrm{Hom}_{C^{op}}(G^{op}(g^{op}), C)\Big(\big(\theta_{C, D}(u)\big)^{op}\Big)$$

$$= \mathrm{Hom}_{C^{op}}(G^{op}(g^{op}), C) \circ \xi_{D, C}(u^{op}),$$

which shows that (3.8) indeed holds.

Next we show that $\bar{\theta}$ is also natural in the second variable. Consider $f^{op} \in \mathrm{Hom}_{C^{op}}(C, C')$. The naturality of $\bar{\theta}$ in the second variable comes down to showing the commutativity of the following diagram:

$$
\begin{array}{ccc}
\mathrm{Hom}_{C^{op}}(G^{op}(D), C) & \xrightarrow{\ \bar{\theta}_{D, C}\ } & \mathrm{Hom}_{\mathcal{D}^{op}}(D, F^{op}(C)) \\[2mm]
{\scriptstyle \mathrm{Hom}_{C^{op}}(G^{op}(D), f^{op})} \Big\downarrow & & \Big\downarrow {\scriptstyle \mathrm{Hom}_{\mathcal{D}^{op}}(D, F^{op}(f^{op}))} \\[2mm]
\mathrm{Hom}_{C^{op}}(G^{op}(D), C') & \xrightarrow[\ \bar{\theta}_{D, C'}\ ]{} & \mathrm{Hom}_{\mathcal{D}^{op}}(D, F^{op}(C'))
\end{array}
$$

i.e., $\bar{\theta}_{D, C'} \circ \mathrm{Hom}_{C^{op}}(G^{op}(D), f^{op}) = \mathrm{Hom}_{\mathcal{D}^{op}}(D, F^{op}(f^{op})) \circ \bar{\theta}_{D, C}.$ (3.9)

Relying again on the fact that $\bar{\theta}_{D, C}$ is invertible with inverse $\xi_{D, C}$ it will suffice to show that the following holds:

$$\mathrm{Hom}_{C^{op}}(G^{op}(D), f^{op}) \circ \xi_{D, C} = \xi_{D, C'} \circ \mathrm{Hom}_{\mathcal{D}^{op}}(D, F^{op}(f^{op})).$$

Indeed, given $u^{op} \in \mathrm{Hom}_{\mathcal{D}^{op}}(D, F^{op}(C))$, we have

$$\mathrm{Hom}_{C^{op}}(G^{op}(D), f^{op}) \circ \xi_{D, C}(u^{op}) = \mathrm{Hom}_{C^{op}}(G^{op}(D), f^{op})\Big(\big(\theta_{C, D}(u)\big)^{op}\Big)$$

$$= f^{op} \circ^{op} \big(\theta_{C, D}(u)\big)^{op}$$

$$= \big(\theta_{C, D}(u) \circ f\big)^{op}$$

$$\overset{(3.1)}{=} \big(\theta_{C', D}(u \circ F(f))\big)^{op}$$

$$= \xi_{D, C'}\big((u \circ F(f))^{op}\big)$$

$$= \xi_{D, C'}\big(F(f)^{op} \circ^{op} u^{op}\big)$$

$$= \xi_{D, C'} \circ \mathrm{Hom}_{\mathcal{D}^{op}}(D, F^{op}(f^{op}))(u^{op}).$$

Hence (3.9) holds and this shows the naturality of $\bar{\theta}$ in the second variable.

Conversely, assume now that $G^{op} \dashv F^{op}$. Then, as proved above, we have $F^{op\,op} \dashv G^{op\,op}$, which comes down to $F \dashv G$, as desired.   □

The next result gives a necessary condition for the existence of adjoints: if a left (resp. right) adjoint of a functor F exists, then F has to preserve small limits (resp. colimits). This is, however, not a sufficient condition for a functor to admit an adjoint, as we will see in Sect. 3.11.

Theorem 3.4.4 *Consider the functors* $F: \mathcal{A} \to \mathcal{B}$, $G: \mathcal{B} \to \mathcal{A}$ *such that* $F \dashv G$. *Then* F *preserves small colimits while* G *preserves small limits.*

Proof We start by showing that G preserves all existing small limits of $\mathcal{B}$. Consider the natural isomorphism $\theta: \mathrm{Hom}_{\mathcal{B}}(F(-), -) \to \mathrm{Hom}_{\mathcal{A}}(-, G(-))$ corresponding to the adjunction $F \dashv G$. Let I be a small category and $H: I \to \mathcal{B}$ a functor whose limit we denote by $\left(L, (p_i: L \to H(i))_{i \in \mathrm{Ob}\,I}\right)$. We will prove that $\left(G(L), (G(p_i): G(L) \to GH(i))_{i \in \mathrm{Ob}\,I}\right)$ is the limit of $GH: I \to \mathcal{A}$. To start with, $\left(G(L), (G(p_i): G(L) \to GH(i))_{i \in \mathrm{Ob}\,I}\right)$ is a cone on GH by Lemma 2.5.5.

Consider now another cone $\left(A, (q_i: A \to GH(i))_{i \in \mathrm{Ob}\,I}\right)$ on GH. Since the map $\theta_{A, H(i)}: \mathrm{Hom}_{\mathcal{B}}(F(A), H(i)) \to \mathrm{Hom}_{\mathcal{A}}(A, GH(i))$ is a bijection, there exists a unique morphism $r_i \in \mathrm{Hom}_{\mathcal{B}}(F(A), H(i))$ such that $\theta_{A, H(i)}(r_i) = q_i$. We will prove that $\left(F(A), (r_i: F(A) \to H(i))_{i \in \mathrm{Ob}\,I}\right)$ is a cone on H, i.e., for any $d \in \mathrm{Hom}_I(i, j)$ we have $H(d) \circ r_i = r_j$. To this end, the naturality of θ renders the following diagram commutative:

$$
\begin{array}{ccc}
\mathrm{Hom}_{\mathcal{B}}(F(A), H(i)) & \xrightarrow{\;\theta_{A, H(i)}\;} & \mathrm{Hom}_{\mathcal{A}}(A, GH(i)) \\
{\scriptstyle \mathrm{Hom}_{\mathcal{B}}(F(A), H(d))} \downarrow & & \downarrow {\scriptstyle \mathrm{Hom}_{\mathcal{A}}(A, GH(d))} \\
\mathrm{Hom}_{\mathcal{B}}(F(A), H(j)) & \xrightarrow[\;\theta_{A, H(j)}\;]{} & \mathrm{Hom}_{\mathcal{A}}(A, GH(j))
\end{array}
$$

i.e., $\mathrm{Hom}_{\mathcal{A}}(A, GH(d)) \circ \theta_{A, H(i)} = \theta_{A, H(j)} \circ \mathrm{Hom}_{\mathcal{B}}(F(A), H(d))$.   (3.10)

Moreover, since $\left(A, (q_i: A \to GH(i))_{i \in \mathrm{Ob}\,I}\right)$ is a cone on GH the following diagram is commutative:

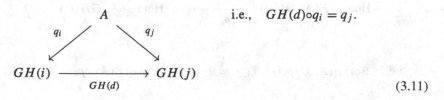

i.e., $GH(d) \circ q_i = q_j$.

$$
\begin{array}{ccc}
 & A & \\
{\scriptstyle q_i} \swarrow & & \searrow {\scriptstyle q_j} \\
GH(i) & \xrightarrow[GH(d)]{} & GH(j)
\end{array}
$$

(3.11)

Now by evaluating (3.10) at r_i we obtain

$$\mathrm{Hom}_{\mathcal{A}}(A,\ GH(d)) \circ \theta_{A,\ H(i)}(r_i) = \theta_{A,\ H(j)} \circ \mathrm{Hom}_{\mathcal{B}}(F(A),\ H(d))(r_i)$$

$$\Leftrightarrow \mathrm{Hom}_{\mathcal{A}}(A,\ GH(d))(q_i) = \theta_{A,\ H(j)}(H(d) \circ r_i)$$

$$\Leftrightarrow \underline{GH(d) \circ q_i} = \theta_{A,\ H(j)}(H(d) \circ r_i)$$

$$\overset{(3.11)}{\Leftrightarrow}\ q_j = \theta_{A,\ H(j)}(H(d) \circ r_i)\ \Leftrightarrow\ r_j = H(d) \circ r_i.$$

Thus $\big(F(A),\ (r_i)_{i \in \mathrm{Ob}\ I}\big)$ is a cone on H. Hence, there exists a unique morphism $f \in \mathrm{Hom}_{\mathcal{B}}(F(A),\ L)$ such that the following diagram is commutative for all $i \in \mathrm{Ob}\ I$:

$$
\begin{array}{ccc}
F(A) & \xrightarrow{\ r_i\ } & H(i) \\
\scriptstyle f \big\downarrow & \nearrow_{\ p_i} & \\
L & &
\end{array}
\qquad \text{i.e.,} \quad p_i \circ f = r_i.
$$

(3.12)

Denote $\theta_{A,L}(f) \in \mathrm{Hom}_{\mathcal{A}}(A,\ G(L))$ by g. We are left to prove that the following diagram is commutative for all $i \in \mathrm{Ob}\ I$:

$$
\begin{array}{ccc}
A & \xrightarrow{\ q_i\ } & GH(i) \\
\scriptstyle g \big\downarrow & \nearrow_{\ G(p_i)} & \\
G(L) & &
\end{array}
\qquad \text{i.e.,} \quad G(p_i) \circ g = q_i
$$

(3.13)

Using again the naturality of the bijection θ we obtain the following commutative diagram for all $i \in \mathrm{Ob}\ I$:

$$
\begin{array}{ccc}
\mathrm{Hom}_{\mathcal{B}}(F(A),\ L) & \xrightarrow{\ \theta_{A,\ L}\ } & \mathrm{Hom}_{\mathcal{A}}(A,\ G(L)) \\
{\scriptstyle \mathrm{Hom}_{\mathcal{B}}(F(A),\ p_i)} \big\downarrow & & \big\downarrow {\scriptstyle \mathrm{Hom}_{\mathcal{A}}(A,\ G(p_i))} \\
\mathrm{Hom}_{\mathcal{B}}(F(A),\ H(i)) & \xrightarrow[\ \theta_{A,\ H(i)}\ ]{} & \mathrm{Hom}_{\mathcal{A}}(A,\ GH(i))
\end{array}
$$

i.e., $\mathrm{Hom}_{\mathcal{A}}(A,\ G(p_i)) \circ \theta_{A,\ L} = \theta_{A,\ H(i)} \circ \mathrm{Hom}_{\mathcal{B}}(F(A),\ p_i)$. (3.14)

By evaluating (3.14) at $f \in \mathrm{Hom}_{\mathcal{B}}(F(A), L)$ we obtain

$$\mathrm{Hom}_{\mathcal{A}}(A, G(p_i)) \circ \theta_{A, L}(f) = \theta_{A, H(i)} \circ \mathrm{Hom}_{\mathcal{B}}(F(A), p_i)(f)$$

$$\Leftrightarrow G(p_i) \circ g = \theta_{A, H(i)}(\underline{p_i \circ f})$$

$$\overset{(3.12)}{\Leftrightarrow} G(p_i) \circ g = \theta_{A, H(i)}(r_i) \quad \Leftrightarrow \quad G(p_i) \circ g = q_i$$

for all $i \in \mathrm{Ob}\, I$. Hence the diagram (3.13) is indeed commutative.

Assume now that there exists another morphism $\overline{g} \in \mathrm{Hom}_{\mathcal{A}}(A, G(L))$ such that $G(p_i) \circ \overline{g} = q_i$ for all $i \in \mathrm{Ob}\, I$. Since $\theta_{A, L} \colon \mathrm{Hom}_{\mathcal{B}}(F(A), L) \to \mathrm{Hom}_{\mathcal{A}}(A, G(L))$ is bijective, there exists a unique morphism $\overline{f} \in \mathrm{Hom}_{\mathcal{B}}(F(A), L)$ such that $\theta_{A, L}(\overline{f}) = \overline{g}$. Now by evaluating (3.14) at $\overline{f}$ we arrive at

$$\mathrm{Hom}_{\mathcal{A}}(A, G(p_i)) \circ \theta_{A, L}(\overline{f}) = \theta_{A, H(i)} \circ \mathrm{Hom}_{\mathcal{B}}(F(A), p_i)(\overline{f})$$

$$\Leftrightarrow G(p_i) \circ \overline{g} = \theta_{A, H(i)}(p_i \circ \overline{f})$$

$$\Leftrightarrow q_i = \theta_{A, H(i)}(p_i \circ \overline{f})$$

for any $i \in \mathrm{Ob}\, I$. Therefore, we have $p_i \circ \overline{f} = r_i \overset{(3.12)}{=} p_i \circ f$ for all $i \in \mathrm{Ob}\, I$. By Proposition 2.2.14, (1) this implies $f = \overline{f}$ and consequently $g = \overline{g}$, as desired.

The second part of the theorem follows easily by duality. Indeed, if $F \dashv G$ then Theorem 3.4.3 implies that we also have $G^{op} \dashv F^{op}$. According to the above proof, F^{op} preserves all existing limits. Now using Lemma 2.5.2 we obtain that F preserves colimits, as desired. $\qquad\Box$

Theorem 3.4.4 can be very useful in ruling out the existence of left/right adjoints for certain functors, as shown in the following examples:

Examples 3.4.5

(1) The forgetful functor $F \colon \mathbf{Ab} \to \mathbf{Set}$ does not preserve coproducts. Therefore, by Theorem 3.4.4 it does not have a right adjoint.
(2) Consider now the inclusion functor $I \colon \mathbf{Ring} \to \mathbf{Rng}$. As $\mathbb{Z}$ is an initial object in $\mathbf{Ring}$ but not in $\mathbf{Rng}$ we can conclude by Theorem 3.4.4 that it does not admit a right adjoint.
(3) The forgetful functor $U \colon \mathbf{Field} \to \mathbf{Set}$ does not have a left adjoint. Indeed, if $F \colon \mathbf{Set} \to \mathbf{Field}$ is a left adjoint to U then by Theorem 3.4.4, F needs to preserve colimits. In particular, this would imply the existence of an initial object in $\mathbf{Field}$, which contradicts Example 1.3.10, (4). $\qquad\Box$

3.5   The Unit and Counit of an Adjunction

Our next result gives an important equivalent description of adjoint functors in terms of two natural transformations called the *unit* and the *counit* of the adjunction.

Theorem 3.5.1 *Let* $F: C \to \mathcal{D}$ *and* $G: \mathcal{D} \to C$ *be two functors. Then* F *is left adjoint to* G *if and only if there exist two natural transformations*

$$\eta: 1_C \to GF, \qquad \varepsilon: FG \to 1_{\mathcal{D}}$$

such that for all $C \in \mathrm{Ob}\,C$ *and* $D \in \mathrm{Ob}\,\mathcal{D}$ *we have*

$$1_{F(C)} = \varepsilon_{F(C)} \circ F\big(\eta_C\big) \tag{3.15}$$

$$1_{G(D)} = G\big(\varepsilon_D\big) \circ \eta_{G(D)} \tag{3.16}$$

In this case η *and* ε *are called the* unit *and the* counit *of the adjunction, respectively.*

Proof Suppose first that $F \dashv G$ and let $\theta: \mathrm{Hom}_{\mathcal{D}}(F(-), \, -) \to \mathrm{Hom}_C(-, \, G(-))$ be the corresponding natural isomorphism. For each $C \in \mathrm{Ob}\,C$ and $D \in \mathrm{Ob}\,\mathcal{D}$ we have the following bijective maps:

$$\theta_{C, \, F(C)}: \mathrm{Hom}_{\mathcal{D}}(F(C), \, F(C)) \to \mathrm{Hom}_C(C, \, GF(C)),$$

$$\theta_{G(D), \, D}: \mathrm{Hom}_{\mathcal{D}}(FG(D), \, D) \to \mathrm{Hom}_C(G(D), \, G(D)).$$

Now define $\eta_C = \theta_{C, \, F(C)}\big(1_{F(C)}\big): C \to GF(C)$ and $\varepsilon_D = \theta^{-1}_{G(D), \, D}\big(1_{G(D)}\big): FG(D) \to D$. We are left to prove (3.15) and (3.16) as well as the naturality of η and ε. We start by proving (3.15); indeed, if we consider the commutative diagram (3.1) for $X = GF(C)$, $X' = C$, $Y = F(C)$ and $f = \eta_C \in \mathrm{Hom}_C(C, \, GF(C))$ we obtain

$$
\begin{array}{ccc}
\mathrm{Hom}_{\mathcal{D}}(FGF(C), \, F(C)) & \xrightarrow{\;\theta_{GF(C), \, F(C)}\;} & \mathrm{Hom}_C(GF(C), \, GF(C)) \\
{\scriptstyle \mathrm{Hom}_{\mathcal{D}}(F(\eta_C), \, F(C))}\Big\downarrow & & \Big\downarrow{\scriptstyle \mathrm{Hom}_C(\eta_C, \, GF(C))} \\
\mathrm{Hom}_{\mathcal{D}}(F(C), \, F(C)) & \xrightarrow[\;\theta_{C, \, F(C)}\;]{} & \mathrm{Hom}_C(C, \, GF(C))
\end{array}
$$

From the commutativity of the above diagram applied to

$$\varepsilon_{F(C)} \in \mathrm{Hom}_{\mathcal{D}}(FGF(C), \, F(C))$$

we obtain

$$\text{Hom}_{\mathcal{C}}(\eta_C,\ GF(C)) \circ \theta_{GF(C),\ F(C)}(\varepsilon_{F(C)})$$

$$= \theta_{C,\ F(C)} \circ \text{Hom}_{\mathcal{D}}(F(\eta_C),\ F(C))(\varepsilon_{F(C)})$$

$$\Leftrightarrow \underline{\theta_{GF(C),\ F(C)}(\varepsilon_{F(C)})} \circ \eta_C = \theta_{C,\ F(C)}\big(\varepsilon_{F(C)} \circ F(\eta_C)\big)$$

$$\Leftrightarrow 1_{GF(C)} \circ \eta_C = \theta_{C,\ F(C)}\big(\varepsilon_{F(C)} \circ F(\eta_C)\big)$$

$$\Leftrightarrow \underline{\theta_{C,\ F(C)}^{-1}(\eta_C)} = \varepsilon_{F(C)} \circ F(\eta_C)$$

$$\Leftrightarrow 1_{F(C)} = \varepsilon_{F(C)} \circ F(\eta_C) \text{ i.e.,\ (3.15) holds.}$$

Similarly, by considering the commutative diagram (3.2) for $X = G(D)$, $Y = FG(D)$, $Y' = D$ and $g = \varepsilon_D \in \text{Hom}_{\mathcal{D}}(FG(D),\ D)$ we get

$$
\begin{array}{ccc}
\text{Hom}_{\mathcal{D}}(FG(D),\ FG(D)) & \xrightarrow{\ \ \theta_{G(D),\ FG(D)}\ \ } & \text{Hom}_{\mathcal{C}}(G(D),\ GFG(D)) \\
\Big\downarrow{\scriptstyle \text{Hom}_{\mathcal{D}}(FG(D),\ \varepsilon_D)} & & \Big\downarrow{\scriptstyle \text{Hom}_{\mathcal{C}}(G(D),\ G(\varepsilon_D))} \\
\text{Hom}_{\mathcal{D}}(FG(D),\ D) & \xrightarrow[\ \ \theta_{G(D),\ D}\ \ ]{} & \text{Hom}_{\mathcal{C}}(G(D),\ G(D))
\end{array}
$$

The commutativity of the above diagram applied to

$$1_{FG(D)} \in \text{Hom}_{\mathcal{D}}(FG(D),\ FG(D))$$

yields

$$\theta_{G(D),\ D} \circ \text{Hom}_{\mathcal{D}}(FG(D),\ \varepsilon_D)(1_{FG(D)})$$

$$= \text{Hom}_{\mathcal{C}}\big(G(D),\ G(\varepsilon_D)\big) \circ \underline{\theta_{G(D),\ FG(D)}(1_{FG(D)})}$$

$$\Leftrightarrow \theta_{G(D),\ D} \circ \text{Hom}_{\mathcal{D}}(FG(D),\ \varepsilon_D) \circ 1_{FG(D)} = G(\varepsilon_D) \circ \eta_{G(D)}$$

$$\Leftrightarrow \underline{\theta_{G(D),\ D}(\varepsilon_D)} = G(\varepsilon_D) \circ \eta_{G(D)}$$

$$\Leftrightarrow 1_{G(D)} = G(\varepsilon_D) \circ \eta_{G(D)} \text{ i.e.,\ (3.16) holds.}$$

Finally, we move on to proving that η and ε are natural transformations. First we will collect some compatibilities using the commutativity of the diagrams (3.1) and (3.2). Setting $X = C$, $X' = C'$ and $Y = F(C)$ in (3.1) yields the following commutative diagram for all $f \in \mathrm{Hom}_C(C', C)$:

$$\begin{array}{ccc}
\mathrm{Hom}_{\mathcal{D}}(F(C), F(C)) & \xrightarrow{\theta_{C, F(C)}} & \mathrm{Hom}_C(C, GF(C)) \\
{\scriptstyle \mathrm{Hom}_{\mathcal{D}}(F(f), F(C))}\Big\downarrow & & \Big\downarrow{\scriptstyle \mathrm{Hom}_C(f, GF(C))} \\
\mathrm{Hom}_{\mathcal{D}}(F(C'), F(C)) & \xrightarrow[\theta_{C', F(C)}]{} & \mathrm{Hom}_C(C', GF(C))
\end{array}$$

From the commutativity of the above diagram applied to $1_{F(C)}$ we get

$$\mathrm{Hom}_C(f, GF(C)) \circ \theta_{C, F(C)}(1_{F(C)}) = \theta_{C', F(C)} \circ \mathrm{Hom}_{\mathcal{D}}(F(f), F(C))(1_{F(C)})$$

$$\Leftrightarrow \mathrm{Hom}_C(f, GF(C)) \circ \eta_C = \theta_{C', F(C)}\big(F(f)\big)$$

$$\text{i.e., } \eta_C \circ f = \theta_{C', F(C)}\big(F(f)\big). \tag{3.17}$$

On the other hand, setting $X = C'$, $Y = F(C')$ and $Y' = F(C)$ in (3.2) yields the following commutative diagram for all $g \in \mathrm{Hom}_{\mathcal{D}}(F(C'), F(C))$:

$$\begin{array}{ccc}
\mathrm{Hom}_{\mathcal{D}}(F(C'), F(C')) & \xrightarrow{\theta_{C', F(C')}} & \mathrm{Hom}_C(C', GF(C')) \\
{\scriptstyle \mathrm{Hom}_{\mathcal{D}}(F(C'), g)}\Big\downarrow & & \Big\downarrow{\scriptstyle \mathrm{Hom}_C(C', G(g))} \\
\mathrm{Hom}_{\mathcal{D}}(F(C'), F(C)) & \xrightarrow[\theta_{C', F(C)}]{} & \mathrm{Hom}_C(C', GF(C))
\end{array}$$

By applying the commutativity of the above diagram to $1_{F(C')}$ we obtain

$$\mathrm{Hom}_C(C', G(g)) \circ \theta_{C', F(C')}(1_{F(C')}) = \theta_{C', F(C)} \circ \mathrm{Hom}_{\mathcal{D}}(F(C'), g)(1_{F(C')}),$$

$$\text{i.e., } G(g) \circ \eta_{C'} = \theta_{C', F(C)}(g). \tag{3.18}$$

Next we use the commutativity of the diagram (3.1) for $X = G(D)$, $X' = G(D')$ and $Y = D$. It comes down to the following commutative diagram for all $f \in \mathrm{Hom}_C(G(D'), G(D))$:

$$\begin{array}{ccc}
\mathrm{Hom}_{\mathcal{D}}(FG(D), D) & \xrightarrow{\theta_{G(D), D}} & \mathrm{Hom}_C(G(D), G(D)) \\
{\scriptstyle \mathrm{Hom}_{\mathcal{D}}(F(f), D)}\Big\downarrow & & \Big\downarrow{\scriptstyle \mathrm{Hom}_C(f, G(D))} \\
\mathrm{Hom}_{\mathcal{D}}(FG(D'), D) & \xrightarrow[\theta_{G(D'), D}]{} & \mathrm{Hom}_C(G(D'), G(D))
\end{array}$$

From the commutativity of the above diagram applied to ε_D we get

$$\theta_{G(D'), D} \circ \mathrm{Hom}_{\mathcal{D}}(F(f), D)(\varepsilon_D) = \mathrm{Hom}_{\mathcal{C}}(f, G(D)) \circ \theta_{G(D), D}(\varepsilon_D)$$

$$\Leftrightarrow \theta_{G(D'), D}(\varepsilon_D \circ F(f)) = 1_{G(D)} \circ f,$$

$$\text{i.e., } \varepsilon_D \circ F(f) = \theta^{-1}_{G(D'), D}(f). \tag{3.19}$$

Finally, we use the commutativity of the diagram (3.2) for $X = G(D')$, $Y = D'$ and $Y' = D$. It yields the following commutative diagram for all $g \in \mathrm{Hom}_{\mathcal{D}}(D', D)$:

$$
\begin{array}{ccc}
\mathrm{Hom}_{\mathcal{D}}(FG(D'), D') & \xrightarrow{\theta_{G(D'), D'}} & \mathrm{Hom}_{\mathcal{C}}(G(D'), G(D')) \\
{\scriptstyle \mathrm{Hom}_{\mathcal{D}}(FG(D'), g)} \downarrow & & \downarrow {\scriptstyle \mathrm{Hom}_{\mathcal{C}}(G(D'), G(g))} \\
\mathrm{Hom}_{\mathcal{D}}(FG(D'), D) & \xrightarrow{\theta_{G(D'), D}} & \mathrm{Hom}_{\mathcal{C}}(G(D'), G(D))
\end{array}
$$

The commutativity of the above diagram applied to $\varepsilon_{D'}$ gives

$$\theta_{G(D'), D} \circ \mathrm{Hom}_{\mathcal{D}}(FG(D'), g)(\varepsilon_{D'}) = \mathrm{Hom}_{\mathcal{C}}(G(D'), G(g)) \circ \theta_{G(D'), D'}(\varepsilon_{D'})$$

$$\Leftrightarrow \theta_{G(D'), D}(g \circ \varepsilon_{D'}) = G(g) \circ 1_{G(D')},$$

$$\text{i.e., } g \circ \varepsilon_{D'} = \theta^{-1}_{G(D'), D}(G(g)). \tag{3.20}$$

We are now in a position to prove that η and ε are natural transformations. Indeed, the naturality of η comes down to proving the commutativity of the following diagram for all $h \in \mathrm{Hom}_{\mathcal{C}}(C', C)$:

$$
\begin{array}{ccc}
C' & \xrightarrow{\eta_{C'}} & GF(C') \\
h \downarrow & & \downarrow GF(h) \\
C & \xrightarrow{\eta_C} & GF(C)
\end{array}
$$

To this end we have

$$\eta_C \circ h \overset{(3.17)}{=} \theta_{C', F(C)}(F(h)) \overset{(3.18)}{=} GF(h) \circ \eta_{C'},$$

where in the second equality we used (3.18) for $g = F(h)$. Thus η is a natural transformation.

The naturality of ε comes down to proving the commutativity of the following diagram for all $t \in \mathrm{Hom}_{\mathcal{D}}(D', D)$:

$$FG(D') \xrightarrow{\;\varepsilon_{D'}\;} D'$$

$$FG(t) \Big\downarrow \qquad\qquad \Big\downarrow t$$

$$FG(D) \xrightarrow[\varepsilon_D]{} D$$

which can be proved using (3.19) and respectively (3.20)

$$\varepsilon_D \circ FG(t) \overset{(3.19)}{=} \theta^{-1}_{G(D'),\,D}\big(G(t)\big) \overset{(3.20)}{=} g \circ \varepsilon_{D'}.$$

Note that the first equality follows by applying (3.19) for $f = G(t)$.

Assume now that there exist two natural transformations $\eta\colon 1_C \to GF$ and $\varepsilon\colon FG \to 1_{\mathcal{D}}$ such that (3.15) and (3.16) are fulfilled for any $C \in \mathrm{Ob}\,C$ and $D \in \mathrm{Ob}\,\mathcal{D}$. Define the following maps

$$\theta_{C,D}\colon \mathrm{Hom}_{\mathcal{D}}(F(C),\,D) \to \mathrm{Hom}_C(C,\,G(D)), \quad \theta_{C,D}(u) = G(u) \circ \eta_C,$$

$$\varphi_{C,D}\colon \mathrm{Hom}_C(C,\,G(D)) \to \mathrm{Hom}_{\mathcal{D}}(F(C),\,D), \quad \varphi_{C,D}(v) = \varepsilon_D \circ F(v),$$

for any $u \in \mathrm{Hom}_{\mathcal{D}}(F(C),\,D)$ and $v \in \mathrm{Hom}_C(C,\,G(D))$. First we will prove that $\theta_{C,D}$ and $\varphi_{C,D}$ are inverses to each other for any $C \in \mathrm{Ob}\,C$ and $D \in \mathrm{Ob}\,\mathcal{D}$. To start with, we note for further use that the naturality of η and ε imply the commutativity of the following diagrams for all $u \in \mathrm{Hom}_{\mathcal{D}}(F(C),\,D)$ and $v \in \mathrm{Hom}_C(C,\,G(D))$:

$$C \xrightarrow{\;\eta_C\;} GF(C) \qquad \text{i.e.,} \quad GF(v)\circ\eta_C = \eta_{G(D)}\circ v,$$

$$v \Big\downarrow \qquad\qquad \Big\downarrow GF(v)$$

$$G(D) \xrightarrow[\eta_{G(D)}]{} GFG(D) \tag{3.21}$$

$$FGF(C) \xrightarrow{\;\varepsilon_{F(C)}\;} F(C) \qquad \text{i.e.,} \quad \varepsilon_D\circ FG(u) = u\circ\varepsilon_{F(C)}.$$

$$FG(u) \Big\downarrow \qquad\qquad\quad \Big\downarrow u$$

$$FG(D) \xrightarrow[\varepsilon_D]{} D \tag{3.22}$$

Now, we have

$$\theta_{C,D} \circ \varphi_{C,D}(v) = \theta_{C,D}\big(\varepsilon_D \circ F(v)\big) = G\big(\varepsilon_D \circ F(v)\big) \circ \eta_C$$

$$= G(\varepsilon_D) \circ \underline{GF(v) \circ \eta_C}$$

$$\overset{(3.21)}{=} \underline{G(\varepsilon_D) \circ \eta_{G(D)}} \circ v$$

$$\overset{(3.16)}{=} v,$$

$$\varphi_{C,D} \circ \theta_{C,D}(u) = \varphi_{C,D}\big(G(u) \circ \eta_C\big) = \varepsilon_D \circ F\big(G(u) \circ \eta_C\big)$$

$$= \underline{\varepsilon_D \circ FG(u)} \circ F(\eta_C)$$

$$\overset{(3.22)}{=} u \circ \underline{\varepsilon_{F(C)} \circ F(\eta_C)}$$

$$\overset{(3.15)}{=} u.$$

Thus $\theta_{C,D}$ and $\varphi_{C,D}$ are inverses to each other for any $C \in \mathrm{Ob}\, C$ and $D \in \mathrm{Ob}\, \mathcal{D}$. We are left to prove that θ is natural in both variables, i.e., diagrams (3.1) and (3.2) are commutative. Indeed, let $f \in \mathrm{Hom}_C(C', C)$ and $u \in \mathrm{Hom}_{\mathcal{D}}(F(C), D)$; we have

$$\mathrm{Hom}_C(f, G(D)) \circ \theta_{C,D}(u) = \mathrm{Hom}_C(f, G(D)) \circ \big(G(u) \circ \eta_C\big)$$

$$= G(u) \circ \underline{\eta_C \circ f}$$

$$= G(u) \circ GF(f) \circ \eta_{C'}$$

$$= \theta_{C',D}\big(u \circ F(f)\big)$$

$$= \theta_{C',D} \circ \mathrm{Hom}_{\mathcal{D}}(F(f), D)(u),$$

where in the third equality we used the naturality of η applied to f. Thus, (3.1) holds.

Consider now $g \in \mathrm{Hom}_{\mathcal{D}}(D, D')$ and $u \in \mathrm{Hom}_{\mathcal{D}}(F(C), D)$. Then:

$$\mathrm{Hom}_C(C, G(g)) \circ \theta_{C,D}(u) = G(g) \circ G(u) \circ \eta_C$$

$$= G(g \circ u) \circ \eta_C$$

$$= \theta_{C,D'} \circ \mathrm{Hom}_{\mathcal{D}}(F(C), g)(u).$$

This proves that (3.2) also holds and the proof is now finished. $\qquad \square$

Examples 3.5.2

(1) Let $F : C \to \mathcal{D}$ be an isomorphism of categories with inverse $G : \mathcal{D} \to C$. Then (F, G) and (G, F) are pairs of adjoint functors with unit and counit given by the identity natural transformations. It is straightforward to check that the compatibility conditions (3.15) and (3.16) are trivially fulfilled.

(2) Consider the pair of adjoint functors $F = -\otimes X, G = \mathrm{Hom}_{K}\mathcal{M}(X, -) : {}_K\mathcal{M} \to {}_K\mathcal{M}$ from Example 3.4.1, (2) where $\otimes = \otimes_K$. The unit and counit of the

adjunction $F \dashv G$ are given as follows for any $Y, Z \in \mathrm{Ob}\,_K\mathcal{M}$:

$$\eta: 1_{_K\mathcal{M}} \to \mathrm{Hom}_{_K\mathcal{M}}(X, -\otimes X), \quad \varepsilon: \mathrm{Hom}_{_K\mathcal{M}}(X, -)\otimes X \to 1_{_K\mathcal{M}},$$

$$\eta_Y: Y \to \mathrm{Hom}_{_K\mathcal{M}}(X, Y\otimes X), \quad \eta_Y(y)(x) = y\otimes x, \; y \in Y, x \in X,$$

$$\varepsilon_Z: \mathrm{Hom}_{_K\mathcal{M}}(X, Z)\otimes X \to Z, \quad \varepsilon_Z(f\otimes x)$$

$$= f(x), \; f \in \mathrm{Hom}_{_K\mathcal{M}}(X, Z), x \in X.$$

Indeed, for all $x \in X$, $y \in Y$ and $f \in \mathrm{Hom}_{_K\mathcal{M}}(X, Y)$ we have

$$\varepsilon_{Y\otimes X} \circ (\eta_Y \otimes 1_X)(y\otimes x) = \varepsilon_{Y\otimes X}\big(\eta_Y(y)\otimes x\big)$$

$$= \eta_Y(y)(x) = y\otimes x,$$

$$\mathrm{Hom}_{_K\mathcal{M}}(X, \varepsilon_Y) \circ \eta_{\mathrm{Hom}_{_K\mathcal{M}}(X,Y)}(f)(x) = \mathrm{Hom}_{_K\mathcal{M}}(X, \varepsilon_Y)(f\otimes x)$$

$$= \varepsilon_Y(f\otimes x) = f(x),$$

which shows that (3.15) and (3.16) hold true.

(3) Similarly to the free group functor, we can construct the *free module functor* $\mathcal{R}: \mathbf{Set} \to \,_R\mathcal{M}$ as follows:

- for any $X \in \mathrm{Ob}\,\mathbf{Set}$, define $\mathcal{R}(X) = RX$, the free module generated by X;[5]
- given $f \in \mathrm{Hom}_{\mathbf{Set}}(X, Y)$, define $\mathcal{R}(f): RX \to RY$ by $\mathcal{R}(f) = \overline{f}$, where $\overline{f}$ is the unique homomorphism of R-modules which makes the following diagram commute:

$$\begin{array}{ccc} X & \xrightarrow{\;\;i_X\;\;} & RX \\ & \searrow_{\;i_Y \circ f} & \downarrow{\overline{f}} \\ & & RY \end{array} \qquad \text{i.e., } \overline{f}\circ i_X = i_Y\circ f,$$

(3.23)

where (RX, i_X) and (RY, i_Y) are the free R-modules generated by X and Y, respectively. $\mathcal{R}$ is the left adjoint of the forgetful functor $U: \,_R\mathcal{M} \to \mathbf{Set}$. Indeed, the unit $\eta: 1_{\mathbf{Set}} \to U\mathcal{R}$ is defined for all $X \in \mathrm{Ob}\,\mathbf{Set}$ by $\eta_X = i_X$, where $i_X: X \to RX$ is the map corresponding to the free R-module on X while the counit $\varepsilon: \mathcal{R}U \to 1_{_R\mathcal{M}}$ is defined for all $M \in \mathrm{Ob}\,_R\mathcal{M}$ as the unique homomorphism of R-modules $\varepsilon_M: R(U(M)) \to M$ such that the following

---

[5] The *free R-module* generated by X is a pair (RX, i_X), where RX is an R-module and $i_X: X \to RX$ is a map, such that for any R-module M and any map $f: X \to M$ there exists a unique R-module homomorphism $g: RX \to M$ such that $g \circ i_X = f$ (see [45, Definition 1.8]).

diagram is commutative:

$$U(M) \xrightarrow{\;i_{U(M)}\;} R(U(M)) \qquad \text{i.e.,} \quad \varepsilon_M \circ i_{U(M)} = 1_M.$$

with arrows 1_M and ε_M to M.

$$(3.24)$$

First, note that (3.23) implies in particular that for all $f \in \mathrm{Hom}_{\mathbf{Set}}(X, Y)$ the following diagram is commutative:

$$
\begin{array}{ccc}
X & \xrightarrow{\;\eta_X\;} & U\mathcal{R}(X) \\
{\scriptstyle f}\downarrow & & \downarrow{\scriptstyle U\mathcal{R}(f)} \\
Y & \xrightarrow{\;\eta_Y\;} & U\mathcal{R}(Y)
\end{array}
$$

In other words, η is a natural transformation. By applying U to (3.24) and having in mind that $\eta_{U(M)} = i_{U(M)}$ gives $U(\varepsilon_M) \circ \eta_{U(M)} = 1_{U(M)}$, which shows that condition (3.16) is fulfilled. We are left to show that ε is a natural transformation such that (3.15) also holds. To this end, we have

$$1_{RX} \circ i_X \overset{(3.24)}{=} \varepsilon_{RX} \circ i_{U(RX)} \circ i_X \overset{(3.23)}{=} \varepsilon_{RX} \circ R(i_X) \circ i_X.$$

This shows that $1_{RX} \circ i_X = \varepsilon_{RX} \circ R(i_X) \circ i_X$ and in light of [45, Definition 1.8] we can conclude that $1_{RX} = \varepsilon_{RX} \circ R(i_X)$. Hence, (3.15) also holds. Consider now $g \in \mathrm{Hom}_{R\mathcal{M}}(M, N)$; the proof will be finished once we show the commutativity of the following diagram:

$$
\begin{array}{ccc}
RU(M) & \xrightarrow{\;\varepsilon_M\;} & M \\
{\scriptstyle RU(g)}\downarrow & & \downarrow{\scriptstyle g} \\
RU(N) & \xrightarrow{\;\varepsilon_N\;} & N
\end{array}
$$

By the same argument used in the above paragraph, it will suffice to show that $g \circ \varepsilon_M \circ i_{U(M)} = \varepsilon_N \circ RU(g) \circ i_{U(M)}$. Indeed, we have

$$\varepsilon_N \circ RU(g) \circ i_{U(M)} \overset{(3.23)}{=} \varepsilon_N \circ i_{U(N)} \circ U(g)$$

$$\overset{(3.24)}{=} 1_N \circ g = g \circ 1_M \overset{(3.24)}{=} g \circ \varepsilon_M \circ i_{U(M)},$$

as desired.   □

We record here for further use the following slightly more general version of the compatibility conditions between the unit and counit of an adjunction:

Lemma 3.5.3 *Let $F: C \to \mathcal{D}$ and $G: \mathcal{D} \to C$ be two functors such that $F \dashv G$ and consider the corresponding natural isomorphism $\theta: \mathrm{Hom}_{\mathcal{D}}(F(-), -) \to \mathrm{Hom}_{C}(-, G(-))$. If η and ε are the unit and counit of this adjunction, then for all $u \in \mathrm{Hom}_{\mathcal{D}}(F(C), D)$ and $v \in \mathrm{Hom}_{C}(C, G(D))$ we have*

$$\varepsilon_D \circ F\big(\theta_{C, D}(u)\big) = u, \tag{3.25}$$

$$G\big(\theta_{C,D}^{-1}(v)\big) \circ \eta_C = v. \tag{3.26}$$

Proof To start with, if we consider the commutative diagram (3.1) for $X = G(D)$, $X' = C$, $Y = D$ and $f = \theta_{C,D}(u)$ we get

$$
\begin{array}{ccc}
\mathrm{Hom}_{\mathcal{D}}(FG(D), D) & \xrightarrow{\ \theta_{G(D), D}\ } & \mathrm{Hom}_{C}(G(D), G(D)) \\[2mm]
{\scriptstyle \mathrm{Hom}_{\mathcal{D}}\big(F(\theta_{C, D}(f)), D\big)} \downarrow & & \downarrow {\scriptstyle \mathrm{Hom}_{C}(\theta_{C, D}(u), G(D))} \\[2mm]
\mathrm{Hom}_{\mathcal{D}}(F(C), D) & \xrightarrow[\ \theta_{C, D}\ ]{} & \mathrm{Hom}_{C}(C, G(D))
\end{array}
$$

From the commutativity of the above diagram applied to $\varepsilon_D \in \mathrm{Hom}_{\mathcal{D}}(FG(D), D)$ we obtain

$$\mathrm{Hom}_{C}(\theta_{C, D}(u), G(D)) \circ \theta_{G(D), D}(\varepsilon_D) = \theta_{C, D} \circ \mathrm{Hom}_{\mathcal{D}}\Big(F\big(\theta_{C, D}(u)\big), D\Big)(\varepsilon_D)$$

$$\Leftrightarrow \underline{\theta_{G(D), D}(\varepsilon_D)} \circ \theta_{C, D}(u) = \theta_{C, D}\Big(\varepsilon_D \circ F\big(\theta_{C, D}(u)\big)\Big)$$

$$\Leftrightarrow 1_{G(D)} \circ \theta_{C, D}(u) = \theta_{C, D}\Big(\varepsilon_D \circ F\big(\theta_{C, D}(u)\big)\Big)$$

and the bijectivity of $\theta_{C, D}$ shows that (3.25) indeed holds.

For the second identity, consider the commutative diagram (3.2) for $X = C$, $Y = F(C)$, $Y' = D$ and $g = \theta_{C, D}^{-1}(v) \in \mathrm{Hom}_{\mathcal{D}}(F(C), D)$

$$
\begin{array}{ccc}
\mathrm{Hom}_{\mathcal{D}}(F(C), F(C)) & \xrightarrow{\ \theta_{C, F(C)}\ } & \mathrm{Hom}_{C}(C, GF(C)) \\[2mm]
{\scriptstyle \mathrm{Hom}_{\mathcal{D}}(F(C), \theta_{C, D}^{-1}(v))} \downarrow & & \downarrow {\scriptstyle \mathrm{Hom}_{C}\big(C, G(\theta_{C, D}^{-1}(v))\big)} \\[2mm]
\mathrm{Hom}_{\mathcal{D}}(F(C), D) & \xrightarrow[\ \theta_{C, D}\ ]{} & \mathrm{Hom}_{C}(C, G(D))
\end{array}
$$

The commutativity of the above diagram applied to $1_{F(C)} \in \mathrm{Hom}_{\mathcal{D}}(F(C), F(C))$ yields

$$\theta_{C,D} \circ \mathrm{Hom}_{\mathcal{D}}(F(C), \theta_{C,D}^{-1}(v))(1_{F(C)})$$

$$= \mathrm{Hom}_C\left(C, G\left(\theta_{C,D}^{-1}(v)\right)\right) \circ \underline{\theta_{C,F(C)}(1_{F(C)})}$$

$$\Leftrightarrow v = G\left(\theta_{C,D}^{-1}(v)\right) \circ \eta_C,$$

which shows that (3.26) also holds.   □

Corollary 3.5.4 *Let $F: C \to \mathcal{D}$ and $G: \mathcal{D} \to C$ be two functors such that $F \dashv G$ and consider η and ε to be the unit and respectively the counit of this adjunction. Then ε^{op} and η^{op} are the unit and respectively the counit of the adjunction $G^{op} \dashv F^{op}$.*

Proof Indeed, by Theorem 3.5.1 the unit and counit of the adjunction $F \dashv G$ are defined as follows for all $C \in \mathrm{Ob}\,C$ and $D \in \mathrm{Ob}\,\mathcal{D}$:

$$\eta_C = \theta_{C,F(C)}\left(1_{F(C)}\right), \quad \varepsilon_D = \theta_{G(D),D}^{-1}\left(1_{G(D)}\right),$$

where θ denotes the natural isomorphism induced by the adjunction. Similarly, if $\overline{\eta}$, $\overline{\varepsilon}$, $\overline{\theta}$ denote the unit, the counit and respectively the natural isomorphism induced by the adjunction $G^{op} \dashv F^{op}$, we have

$$\overline{\eta}_D = \overline{\theta}_{D,G^{op}(D)}\left(1_{G^{op}(D)}^{op}\right) \overset{(3.6)}{=} \left(\theta_{G(D),D}^{-1}\left(1_{G(D)}\right)\right)^{op} = \varepsilon_D^{op},$$

$$\overline{\varepsilon}_C = \overline{\theta}_{F^{op}(C),C}^{-1}\left(1_{F^{op}(C)}^{op}\right) \overset{(3.7)}{=} \left(\theta_{C,F(C)}\left(1_{F(C)}\right)\right)^{op} = \eta_C^{op},$$

and the proof is now finished.   □

Our next result provides a way of inducing natural transformations between two pairs of adjoint functors.

Theorem 3.5.5 *Let $F_i: C \to \mathcal{D}$ and $G_i: \mathcal{D} \to C$ be functors such that $F_i \dashv G_i$ and denote by θ^i the corresponding natural isomorphism for all $i = 1, 2$. Then, for any natural transformation $\alpha: G_1 \to G_2$ there exists a unique natural transformation $\overline{\alpha}: F_2 \to F_1$ such that the following diagram is commutative for all*

$C \in \mathrm{Ob}\,C$ *and* $D \in \mathrm{Ob}\,\mathcal{D}$:

$$
\begin{array}{ccc}
\mathrm{Hom}_{\mathcal{D}}(F_1(C),\, D) & \xrightarrow{\;\theta^1_{C,D}\;} & \mathrm{Hom}_C(C,\, G_1(D)) \\[2pt]
{\scriptstyle \mathrm{Hom}_{\mathcal{D}}(\overline{\alpha}_C,\, D)}\Big\downarrow & & \Big\downarrow {\scriptstyle \mathrm{Hom}_C(C,\, \alpha_D)} \\[2pt]
\mathrm{Hom}_{\mathcal{D}}(F_2(C),\, D) & \xrightarrow[\;\theta^2_{C,D}\;]{} & \mathrm{Hom}_C(C,\, G_2(D))
\end{array}
$$

i.e., $\quad \theta^2_{C,D} \circ \mathrm{Hom}_{\mathcal{D}}(\overline{\alpha}_C,\, D)(f) = \mathrm{Hom}_C(C,\, \alpha_D) \circ \theta^1_{C,D}(f),$ \hfill (3.27)

for all $f \in \mathrm{Hom}_{\mathcal{D}}(F_1(C),\, D).$

Proof For all $C \in \mathrm{Ob}\,C$, we denote by $\psi^C \colon \mathrm{Hom}_{\mathcal{D}}(F_1(C),\, -) \to \mathrm{Hom}_{\mathcal{D}}(F_2(C),\, -)$ the natural transformation defined for all $D \in \mathrm{Ob}\,\mathcal{D}$ as the following composition:

$$
\psi^C_D = \left(\theta^2_{C,D}\right)^{-1} \circ \mathrm{Hom}_C(C,\, \alpha_D) \circ \theta^1_{C,D}. \tag{3.28}
$$

We will show first that each ψ^C is indeed a natural transformation between $\mathrm{Hom}_{\mathcal{D}}(F_1(C),\, -)$ and $\mathrm{Hom}_{\mathcal{D}}(F_2(C),\, -)$. To this end, we will show that the following diagram is commutative for all $u \in \mathrm{Hom}_{\mathcal{D}}(D,\, D')$:

$$
\begin{array}{ccc}
\mathrm{Hom}_{\mathcal{D}}(F_1(C),\, D) & \xrightarrow{\;\psi^C_D\;} & \mathrm{Hom}_C(F_2(C),\, D) \\[2pt]
{\scriptstyle \mathrm{Hom}_{\mathcal{D}}(F_1(C),\, u)}\Big\downarrow & & \Big\downarrow {\scriptstyle \mathrm{Hom}_C(F_2(C),\, u)} \\[2pt]
\mathrm{Hom}_{\mathcal{D}}(F_1(C),\, D') & \xrightarrow[\;\psi^C_{D'}\;]{} & \mathrm{Hom}_{\mathcal{D}}(F_2(C),\, D')
\end{array}
$$

i.e., $\quad u \circ \psi^C_D(v) = \psi^C_{D'}(u \circ v)$ for all $v \in \mathrm{Hom}_{\mathcal{D}}(F_1(C),\, D).$

Indeed, this can be written equivalently as follows:

$$
u \circ \left(\theta^2_{C,D}\right)^{-1}\!\left(\alpha_D \circ \theta^1_{C,D}(v)\right) = \left(\theta^2_{C,D'}\right)^{-1}\!\left(\alpha_{D'} \circ \theta^1_{C,D'}(u \circ v)\right)
$$

$$
\Leftrightarrow \theta^2_{C,D'}\!\left(u \circ \left(\theta^2_{C,D}\right)^{-1}\!\left(\alpha_D \circ \theta^1_{C,D}(v)\right)\right) = \alpha_{D'} \circ \theta^1_{C,D'}(u \circ v)
$$

$$
\overset{(3.2)}{\Leftrightarrow} G_2(u) \circ \theta^2_{C,D}\!\left(\left(\theta^2_{C,D}\right)^{-1}\!\left(\alpha_D \circ \theta^1_{C,D}(v)\right)\right) = \alpha_{D'} \circ \theta^1_{C,D'}(u \circ v)
$$

$$
\Leftrightarrow G_2(u) \circ \alpha_D \circ \theta^1_{C,D}(v) = \alpha_{D'} \circ \theta^1_{C,D'}(u \circ v)
$$

$$
\overset{(3.2)}{\Leftrightarrow} G_2(u) \circ \alpha_D \circ \theta^1_{C,D}(v) = \alpha_{D'} \circ G_1(u) \circ \theta^1_{C,D}(v),
$$

and the last equality holds because $\alpha \colon G_1 \to G_2$ is a natural transformation, i.e., we have $G_2(u) \circ \alpha_D = \alpha_{D'} \circ G_1(u)$ for all $u \in \mathrm{Hom}_{\mathcal{D}}(D, D')$.

This shows that, for each $C \in \mathrm{Ob}\, C$, ψ^C is indeed a natural transformation and by Corollary 1.10.4, (1) there exists a unique morphism $\bar{\alpha}_C \in \mathrm{Hom}_{\mathcal{D}}(F_2(C), F_1(C))$ such that $\psi^C_D(f) = f \circ \bar{\alpha}_C$ for all $f \in \mathrm{Hom}_{\mathcal{D}}(F_1(C), D)$. In light of (3.28), $\bar{\alpha}_C \in \mathrm{Hom}_{\mathcal{D}}(F_2(C), F_1(C))$ is the unique morphism such that

$$\theta^2_{C, D}(f \circ \bar{\alpha}_C) = \alpha_D \circ \theta^1_{C, D}(f), \quad \text{for all } f \in \mathrm{Hom}_{\mathcal{D}}(F_1(C), D), \qquad (3.29)$$

which proves that diagram (3.27) is commutative.

Next we show that $\bar{\alpha} \colon F_2 \to F_1$ defined by (3.27) is in fact a natural transformation, i.e., the following diagram is commutative for all $g \in \mathrm{Hom}_C(C, C')$:

$$
\begin{array}{ccc}
F_2(C) & \xrightarrow{\;\;F_2(g)\;\;} & F_2(C') \\
\bar{\alpha}_C \downarrow & & \downarrow \bar{\alpha}_{C'} \\
F_1(C) & \xrightarrow[\;\;F_1(g)\;\;]{} & F_1(C')
\end{array}
\qquad \text{i.e., } \bar{\alpha}_{C'} \circ F_2(g) = F_1(g) \circ \bar{\alpha}_C.
$$

Given the bijectivity of each $\theta^2_{C, F_1(C')}$ it will suffice to prove that the following holds:

$$\theta^2_{C, F_1(C')}\big(\bar{\alpha}_{C'} \circ F_2(g)\big) = \theta^2_{C, F_1(C')}\big(F_1(g) \circ \bar{\alpha}_C\big).$$

To this end, we have

$$\theta^2_{C, F_1(C')}\big(F_1(g) \circ \bar{\alpha}_C\big) \overset{(3.29)}{=} \alpha_{F_1(C')} \circ \theta^1_{C, F_1(C')}\big(F_1(g)\big)$$

$$\overset{(3.17)}{=} \alpha_{F_1(C')} \circ \eta^1_{C'} \circ g$$

$$= \alpha_{F_1(C')} \circ \theta^1_{C', F_1(C')}\big(1_{F_1(C')}\big) \circ g$$

$$\overset{(3.29)}{=} \theta^2_{C', F_1(C')}\big(\bar{\alpha}_{C'}\big) \circ g$$

$$\overset{(3.1)}{=} \theta^2_{C, F_1(C')}\big(\bar{\alpha}_{C'} \circ F_2(g)\big),$$

where η^1 and θ^1 denote the unit and respectively the natural isomorphism induced by the adjunction $F_1 \dashv G_1$. $\qquad \square$

Recall that if I is a small category then any functor $F \colon C \to \mathcal{D}$ induces a functor between the corresponding functor categories $F_* \colon \mathrm{Fun}\,(I, C) \to \mathrm{Fun}\,(I, \mathcal{D})$ as in (1.36). Our next result shows that any adjunction can be lifted to an adjunction between the corresponding induced functors.

Proposition 3.5.6 *Let* $F : C \to \mathcal{D}$, $G : \mathcal{D} \to C$ *be two functors such that* $F \dashv G$ *and* I *a small category. Then* $F_* \dashv G_*$, *where* $F_* : \mathrm{Fun}\,(\mathrm{I}, C) \to \mathrm{Fun}\,(\mathrm{I}, \mathcal{D})$ *and* $G_* : \mathrm{Fun}\,(\mathrm{I}, \mathcal{D}) \to \mathrm{Fun}\,(\mathrm{I}, C)$ *are the corresponding induced functors.*

Proof Let $\eta : 1_C \to GF$ and $\varepsilon : FG \to 1_{\mathcal{D}}$ denote the unit and respectively the counit of the adjunction $F \dashv G$. Consider now the natural transformations $\overline{\eta} : 1_{\mathrm{Fun}\,(\mathrm{I}, C)} \to G_* F_*$ and $\overline{\varepsilon} : F_* G_* \to 1_{\mathrm{Fun}\,(\mathrm{I}, \mathcal{D})}$ defined for all functors $H : I \to C$, $K : I \to \mathcal{D}$ by the whiskering of η on the left by H and respectively the whiskering of ε on the left by K (see Example 1.7.2, (7)). More precisely, we have

$$\overline{\eta}_H = \eta H, \quad \overline{\varepsilon}_K = \varepsilon K.$$

In order to show that $\overline{\eta}$ and $\overline{\varepsilon}$ are indeed natural transformations, consider two natural transformations $\alpha : H \to H'$ and $\beta : K \to K'$, where H, $H' : I \to C$ and K, $K' : I \to \mathcal{D}$ are functors. First, by the naturality of η and ε, the following diagrams are commutative for all $i \in \mathrm{Ob}\,I$:

$$
\begin{array}{ccc}
H(i) \xrightarrow{\;\eta_{H(i)}\;} GF(H(i)) & \qquad & FG(K(i)) \xrightarrow{\;\varepsilon_{K(i)}\;} K(i) \\
\alpha_i \downarrow \qquad\qquad \downarrow GF(\alpha_i) & & FG(\beta_i) \downarrow \qquad\qquad \downarrow \beta_i \\
H'(i) \xrightarrow[\;\eta_{H'(i)}\;]{} GF(H'(i)) & & FG(K'(i)) \xrightarrow[\;\varepsilon_{K'(i)}\;]{} K'(i)
\end{array}
$$

$$(3.30)$$

To summarize, for all $i \in \mathrm{Ob}\,I$ we obtain

$$\left(G_* F_*(\alpha) \circ \overline{\eta}_H\right)_i \overset{(1.36)}{=} GF(\alpha_i) \circ \eta_{H(i)} \overset{(3.30)}{=} \eta_{H'(i)} \circ \alpha_i = (\overline{\eta}_{H'} \circ \alpha)_i,$$

$$\left(\overline{\varepsilon}_{K'} \circ F_* G_*(\beta)\right)_i \overset{(1.36)}{=} \varepsilon_{K'(i)} \circ FG(\beta_i) \overset{(3.30)}{=} \beta_i \circ \varepsilon_{K(i)} = (\beta \circ \overline{\varepsilon}_K)_i,$$

which shows that $\overline{\eta}$ and $\overline{\varepsilon}$ are natural transformations. The proof will be finished once we show that $\overline{\eta}$ and $\overline{\varepsilon}$ fulfill (3.15) and (3.16). To start with, note that since η and ε fulfill (3.15) and (3.16), in particular the following hold for all $i \in \mathrm{Ob}\,I$:

$$1_{FH(i)} = \varepsilon_{FH(i)} \circ F\left(\eta_{H(i)}\right), \qquad 1_{GK(i)} = G\left(\varepsilon_{K(i)}\right) \circ \eta_{GK(i)}.$$

The above identities come down to the following:

$$1_{F_*(H)} = \overline{\varepsilon}_{F_*(H)} \circ F_*\left(\overline{\eta}_H\right), \qquad 1_{G_*(K)} = G_*\left(\overline{\varepsilon}_K\right) \circ \overline{\eta}_{G_*(K)},$$

which shows precisely that $\bar{\eta}$ and $\bar{\varepsilon}$ fulfill (3.15) and (3.16) and the proof is now finished. $\qquad\qquad\qquad\qquad\qquad\qquad\qquad\qquad\qquad\qquad\qquad\qquad\qquad$ □

3.6 Another Characterisation of Adjoint Functors

A very useful characterization of an adjunction involving only the (co)unit is the following:

Theorem 3.6.1 *Let $F : C \to \mathcal{D}$ and $G : \mathcal{D} \to C$ be two functors. The following are equivalent:*

(1) $F \dashv G$;
(2) *there exists a natural transformation $\eta : 1_C \to GF$ such that for any morphism $f \in \mathrm{Hom}_C(C, G(D))$ there exists a unique morphism $g \in \mathrm{Hom}_{\mathcal{D}}(F(C), D)$ which makes the following diagram commutative:*

$$
\begin{array}{ccc}
C & \xrightarrow{\ \eta_C\ } & GF(C) \\
 & \searrow{\scriptstyle f} & \downarrow{\scriptstyle G(g)} \\
 & & G(D)
\end{array}
\qquad \text{i.e.,}\quad G(g) \circ \eta_C = f;
$$
$$\text{(3.31)}$$

(3) *there exists a natural transformation $\varepsilon : FG \to 1_{\mathcal{D}}$ such that for any morphism $f \in \mathrm{Hom}_{\mathcal{D}}(F(C), D)$ there exists a unique morphism $g \in \mathrm{Hom}_C(C, G(D))$ which makes the following diagram commutative:*

$$
\begin{array}{ccc}
F(C) & & \\
\downarrow{\scriptstyle F(g)} & \searrow{\scriptstyle f} & \\
FG(D) & \xrightarrow[\varepsilon_D]{} & D
\end{array}
\qquad \text{i.e.,}\quad \varepsilon_D \circ F(g) = f.
$$
$$\text{(3.32)}$$

As the notation suggests, the natural transformations η and ε are precisely the unit and the counit, respectively, of the adjunction $F \dashv G$.

Proof We start by proving the equivalence between (1) and (2). Suppose first that $F \dashv G$ and let θ be the corresponding natural isomorphism. We define the natural transformation $\eta : 1_C \to GF$ as in the proof of Theorem 3.5.1, namely by $\eta_C = \theta_{C,F(C)}(1_{F(C)})$ for any $C \in \mathrm{Ob}\,C$. Let $f \in \mathrm{Hom}_C(C, G(D))$; we will prove that $g = \theta_{C,D}^{-1}(f) \in \mathrm{Hom}_{\mathcal{D}}(F(C), D)$ is the unique morphism in $\mathcal{D}$ which makes diagram (3.31) commutative. Indeed, setting $X = C$, $Y = F(C)$ and $Y' = D$ in (3.2) gives the following commutative diagram for all $u \in \mathrm{Hom}_C(F(C), D)$:

$$
\begin{array}{ccc}
\mathrm{Hom}_{\mathcal{D}}(F(C), F(C)) & \xrightarrow{\ \theta_{C, F(C)}\ } & \mathrm{Hom}_C(C, GF(C)) \\
{\scriptstyle \mathrm{Hom}_{\mathcal{D}}(F(C), u)} \Big\downarrow & & \Big\downarrow {\scriptstyle \mathrm{Hom}_C(C, G(u))} \\
\mathrm{Hom}_{\mathcal{D}}(F(C), D) & \xrightarrow[\ \theta_{C, D}\ ]{} & \mathrm{Hom}_C(C, G(D))
\end{array}
$$

By applying the commutativity of the above diagram to $1_{F(C)}$ we obtain

$$
\mathrm{Hom}_C(C, G(u)) \circ \theta_{C, F(C)}(1_{F(C)}) = \theta_{C, D} \circ \mathrm{Hom}_{\mathcal{D}}(F(C), u)(1_{F(C)}),
$$

$$
\text{i.e.,} \quad G(u) \circ \eta_C = \theta_{C, D}(u). \tag{3.33}
$$

Thus, we have

$$
G(g) \circ \eta_C \overset{(3.33)}{=} \theta_{C,D}(g) = \theta_{C,D} \circ \theta_{C,D}^{-1}(f) = f,
$$

which shows that $g = \theta_{C,D}^{-1}(f) \in \mathrm{Hom}_{\mathcal{D}}(F(C), D)$ makes diagram (3.31) commutative. Assume now that there exists another morphism $g' \in \mathrm{Hom}_{\mathcal{D}}(F(C), D)$ such that $G(g') \circ \eta_C = f$ and let $f' = \theta_{C,D}(g')$. Following the same steps as in the argument above it can be easily seen that $G(g') \circ \eta_C = \theta_{C,D}(g') = f'$. Our assumption now implies that $f = f'$ and therefore, since $\theta_{C, D}$ is a bijection, we obtain $g = g'$.

Assume now that 2) holds, i.e., for any $f \in \mathrm{Hom}_C(C, G(D))$ there exists a unique morphism $g \in \mathrm{Hom}_{\mathcal{D}}(F(C), D)$ such that (3.31) is fulfilled. Given $C \in \mathrm{Ob}\,C$ and $D \in \mathrm{Ob}\,\mathcal{D}$ we define the following map:

$$
\theta_{C, D} : \mathrm{Hom}_{\mathcal{D}}(F(C), D) \to \mathrm{Hom}_C(C, G(D)), \quad \theta_{C, D}(u) = G(u) \circ \eta_C \tag{3.34}
$$

for any $u \in \mathrm{Hom}_{\mathcal{D}}(F(C), D)$. Obviously, our assumption implies that $\theta_{C, D}$ is a set bijection for all $C \in \mathrm{Ob}\,C$ and $D \in \mathrm{Ob}\,\mathcal{D}$. The fact that θ defined in (3.34) is natural in both variables follows exactly as in the proof of Theorem 3.5.1.

Finally, we are left to show the equivalence between (1) and (3). Indeed, by Theorem 3.4.3, $F \dashv G$ if and only if $G^{op} \dashv F^{op}$. By applying the equivalence between (1) and (2) we obtain that $G^{op} \dashv F^{op}$ if and only if there exists a natural transformation $\overline{\eta} : 1_{\mathcal{D}^{op}} \to F^{op} G^{op}$ with the property that for any $f^{op} \in \mathrm{Hom}_{\mathcal{D}^{op}}(D, F^{op}(C))$ there exists a unique $g^{op} \in \mathrm{Hom}_{C^{op}}(G^{op}(D), C)$ such that the following holds:

$$F^{op}(g^{op}) \circ^{op} \overline{\eta}_D = f^{op}. \tag{3.35}$$

By Proposition 1.8.3, there exists a natural transformation $\varepsilon \colon FG \to 1_{\mathcal{D}}$ such that $\overline{\eta} = \varepsilon^{op}$. Now it can be easily seen that (3.35) comes down to $\varepsilon_D \circ F(g) = f$. Putting everything together, we obtain that $F \dashv G$ if and only if there exists a natural transformation $\varepsilon \colon FG \to 1_{\mathcal{D}}$ such that for any $f \in \mathrm{Hom}_{\mathcal{D}}(D, F(C))$ there exists a unique $g \in \mathrm{Hom}_C(G(D), C)$ such that $\varepsilon_D \circ F(g) = f$. The proof is now complete. $\qquad\square$

As a straightforward consequence of Theorem 3.6.1 we have:

Corollary 3.6.2 *Suppose $F \colon C \to \mathcal{D}$ and $G \colon \mathcal{D} \to C$ are two functors such that $F \dashv G$ and let $\eta \colon 1_C \to GF$, $\varepsilon \colon FG \to 1_{\mathcal{D}}$ be the unit, respectively the counit of the adjunction.*

(1) If $g, g' \in \mathrm{Hom}_{\mathcal{D}}(F(C), D)$ such that $G(g) \circ \eta_C = G(g') \circ \eta_C$ then $g = g'$.
(2) If $h, h' \in \mathrm{Hom}_C(C, G(D))$ such that $\varepsilon_D \circ F(h) = \varepsilon_D \circ F(h')$ then $h = h'$.

Proof

(1) Follows trivially from Theorem 3.6.1, (2) by considering $f = G(g') \circ \eta_C$. Then both morphisms g and g' make diagram (3.31) commutative, which implies $g = g'$. The second part follows in a similar manner by using Theorem 3.6.1, (3).

$\qquad\square$

Examples 3.6.3

(1) The forgetful functor $U \colon \mathbf{Top} \to \mathbf{Set}$ has both a left and a right adjoint. We start by constructing the left adjoint functor $F \colon \mathbf{Set} \to \mathbf{Top}$ which endows each $X \in \mathrm{Ob}\,\mathbf{Set}$ with the discrete topology. We define a natural transformation $\eta \colon 1_{\mathbf{Set}} \to UF$ by $\eta_X(x) = x$ for any $X \in \mathrm{Ob}\,\mathbf{Set}$ and $x \in X$. Consider now $f \in \mathrm{Hom}_{\mathbf{Set}}(X, U(Y))$, where $Y \in \mathrm{Ob}\,\mathbf{Top}$. According to Theorem 3.6.1, (2) in order to prove that $F \dashv U$ we need to find a unique morphism $g \in \mathrm{Hom}_{\mathbf{Top}}(F(X), Y)$ such that the following diagram commutes:

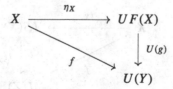

To this end, it is enough to consider $g = f$. Note that f is obviously continuous since $F(X)$ is endowed with the discrete topology.

On the other hand, the right adjoint $G \colon \mathbf{Set} \to \mathbf{Top}$ endows each $X \in \mathrm{Ob}\,\mathbf{Set}$ with the indiscrete topology. We define a natural transformation $\eta \colon 1_{\mathbf{Top}} \to$

GU by $\eta_X(x) = x$ for all $X \in \text{Ob } \textbf{Top}$ and $x \in X$. Now since $\eta_X^{-1}(\emptyset) = \emptyset$ and $\eta_X^{-1}(G(X)) = G(X) = X$ we obtain that each η_X is continuous. Consider now $f \in \text{Hom}_{\textbf{Top}}(X, G(Y))$, where $Y \in \text{Ob } \textbf{Set}$. We aim to find a unique morphism $g \in \text{Hom}_{\textbf{Set}}(U(X), Y)$ such that the following diagram is commutative:

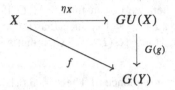

As before, we set $g = f$.

(2) The Stone–Čech compactification functor $\mathcal{S}\colon \textbf{Top} \to \textbf{KHaus}$ defined in Example 1.5.3, (24) is left adjoint to the inclusion functor $I\colon \textbf{KHaus} \to \textbf{Top}$. Indeed, let $i\colon 1_{\textbf{Top}} \to I\mathcal{S}$ be the natural transformation defined for all topological spaces X by the continuous map $i_X\colon X \to \mathcal{S}(X)$ associated with the Stone–Čech compactification of X. If $f \in \text{Hom}_{\textbf{Top}}(X, Y)$ then $\mathcal{S}(f)$ is defined by (1.7) as the unique morphism in $\textbf{KHaus}$ such that $I\mathcal{S}(f) \circ i_X = i_Y \circ f$. Therefore, the following diagram is commutative for all $f \in \text{Hom}_{\textbf{Top}}(X, Y)$:

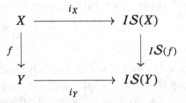

This shows that i is indeed a natural transformation. Now recall that by the universal property of the Stone–Čech compactification, for any $f \in \text{Hom}_{\textbf{Top}}(X, I(Z))$, where $Z \in \text{Ob } \textbf{KHaus}$, there exists a unique $g \in \text{Hom}_{\textbf{KHaus}}(\mathcal{S}(X), Z)$ such that the following diagram is commutative:

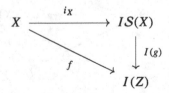

Using Theorem 3.6.1, (2) we can conclude that $\mathcal{S}$ is left adjoint to I, as desired.

(3) The Grothendieck group functor $\mathcal{G}\colon \textbf{Mon} \to \textbf{Grp}$ defined in Example 1.5.3, (25) is left adjoint to the inclusion functor $I\colon \textbf{Grp} \to \textbf{Mon}$. Indeed, let $i\colon 1_{\textbf{Mon}} \to I\mathcal{G}$ be the natural transformation defined for all monoids M

by the homomorphism of monoids $i_M \colon M \to IG(M)$ associated with the Grothendieck group of M. Indeed, if $f \in \mathrm{Hom}_{\mathbf{Mon}}(M, N)$ then $G(f)$ is defined by (1.8) as the unique morphism in **Grp** such that $IG(f) \circ i_M = i_N \circ f$. Therefore, the following diagram is commutative for all $f \in \mathrm{Hom}_{\mathbf{Mon}}(M, N)$:

$$
\begin{array}{ccc}
M & \xrightarrow{\;\;i_M\;\;} & IG(M) \\[2pt]
\big\downarrow{\scriptstyle f} & & \big\downarrow{\scriptstyle IG(f)} \\[2pt]
N & \xrightarrow[\;i_N\;]{} & IG(YN
\end{array}
$$

In particular, this shows that i is indeed a natural transformation. Now using the universal property of the Grothendieck group, for any $f \in \mathrm{Hom}_{\mathbf{Mon}}(M, I(N))$, where $N \in \mathrm{Ob}\,\mathbf{Grp}$, there exists a unique $g \in \mathrm{Hom}_{\mathbf{Grp}}(G(M), N)$ such that the following diagram is commutative:

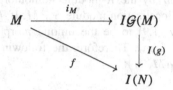

Using Theorem 3.6.1, (2) we can conclude that G is left adjoint to I, as desired.

(4) The functor $\mathcal{U} \colon \mathbf{Mon} \to \mathbf{Grp}$ defined in Example 1.5.3, (26) which assigns to each monoid its group of invertible elements, is right adjoint to the inclusion functor $I \colon \mathbf{Grp} \to \mathbf{Mon}$. Indeed, let $\varepsilon \colon I\mathcal{U} \to 1_{\mathbf{Mon}}$ be the natural transformation defined for all $M \in \mathrm{Ob}\,\mathbf{Mon}$ by $\varepsilon_M \colon U(M) \to M$, $\varepsilon_M = i_M$, where i_M denotes the inclusion map. If $f \in \mathrm{Hom}_{\mathbf{Mon}}(M, N)$, then for all $m \in U(M)$ we have

$$
(f \circ i_M)(m) = f(m) = \big(i_N \circ f_{|U(M)}\big)(m) = \big(i_N \circ I\mathcal{U}(f)\big)(m).
$$

Therefore, the following diagram is commutative for all $f \in \mathrm{Hom}_{\mathbf{Mon}}(M, N)$:

$$
\begin{array}{ccc}
I\mathcal{U}(M) & \xrightarrow{\;\;\varepsilon_M\;\;} & M \\[2pt]
\big\downarrow{\scriptstyle I\mathcal{U}(f)} & & \big\downarrow{\scriptstyle f} \\[2pt]
I\mathcal{U}(N) & \xrightarrow[\;\varepsilon_N\;]{} & N
\end{array}
$$

which shows that ε is a natural transformation. Now, if $f \in \mathrm{Hom}_{\mathbf{Mon}}(I(G), M)$, where $G \in \mathrm{Ob}\,\mathbf{Grp}$, then $f(G)$ is also a group and therefore $f(G) \subseteq U(M)$. Hence $f \in \mathrm{Hom}_{\mathbf{Grp}}(G, U(M))$ and f is the unique morphism which makes the following diagram commutative:

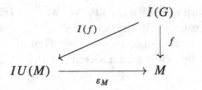

Using Theorem 3.6.1, (3) we can conclude that $\mathcal{U}$ is right adjoint to I, as desired.

(5) Let R be a commutative ring with unity and $(S^{-1}R, j)$ its localization with respect to the multiplicative set $S \subset R$, where $j: R \rightarrow S^{-1}R$ is the ring homomorphism defined by $j(r) = \frac{r}{1}$ for all $r \in R$. Then the localization functor $\mathcal{L}: {}_R\mathcal{M} \rightarrow {}_{S^{-1}R}\mathcal{M}$ defined in Example 1.5.3, (29) is left adjoint to the restriction of scalars functor $F_j: {}_{S^{-1}R}\mathcal{M} \rightarrow {}_R\mathcal{M}$ induced by the ring homomorphism j (see Example 1.5.3, (32)).

Throughout, if M is an R-module we denote by $(S^{-1}M, \varphi_M)$ its localization module at S. Consider the natural transformation $\varphi: 1_{{}_R\mathcal{M}} \rightarrow F_j\mathcal{L}$ defined for all R-modules M by the R-module homomorphism $\varphi_M: M \rightarrow S^{-1}M$ associated with the localization module $S^{-1}M$. If $f \in \mathrm{Hom}_{{}_R\mathcal{M}}(M, N)$ then $\mathcal{L}(f)$ is defined by (1.9) to be the unique morphism in ${}_{S^{-1}R}\mathcal{M}$ such that $\varphi_N \circ f = F_j\mathcal{L}(f) \circ \varphi_M$. Therefore, the following diagram is commutative for all $f \in \mathrm{Hom}_{{}_R\mathcal{M}}(M, N)$:

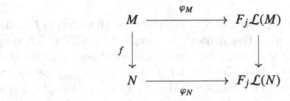

In particular, this shows that φ is indeed a natural transformation. Now recall that by the universal property of the localization module ([3, Theorem 12.3]), for any $f \in \mathrm{Hom}_{{}_R\mathcal{M}}(M, F_j(N))$, with $N \in \mathrm{Ob}\,{}_{S^{-1}R}\mathcal{M}$, there exists a unique $g \in \mathrm{Hom}_{{}_{S^{-1}R}\mathcal{M}}(\mathcal{L}(M), N)$ such that the following diagram is commutative:

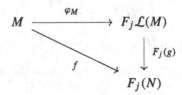

Using Theorem 3.6.1, (2) we can now conclude that $\mathcal{L}$ is left adjoint to F_j, as desired.

(6) The Hausdorff quotient functor $\mathcal{H}\colon \mathbf{Top} \to \mathbf{Haus}$ defined in Example 1.5.3, (30) is left adjoint to the inclusion functor $I\colon \mathbf{Haus} \to \mathbf{Top}$. Indeed, let $q\colon 1_{\mathbf{Top}} \to I\mathcal{H}$ be the natural transformation defined for all topological spaces X by the continuous map $q_X\colon X \to H(X)$ associated with the Hausdorff quotient of X. If $f \in \mathrm{Hom}_{\mathbf{Top}}(X, Y)$ then $\mathcal{H}(f)$ is defined by (1.10) as the unique morphism in $\mathbf{Haus}$ such that $q_Y \circ f = I\mathcal{H}(f) \circ q_X$. Therefore, the following diagram is commutative for all $f \in \mathrm{Hom}_{\mathbf{Top}}(X, Y)$:

$$
\begin{array}{ccc}
X & \xrightarrow{\;q_X\;} & I\mathcal{H}(X) \\
{\scriptstyle f}\downarrow & & \downarrow{\scriptstyle I\mathcal{H}(f)} \\
Y & \xrightarrow[\;q_Y\;]{} & I\mathcal{H}(Y)
\end{array}
$$

This shows that q is indeed a natural transformation. Now recall that by the universal property of the Hausdorff quotient, for any $f \in \mathrm{Hom}_{\mathbf{Top}}(X, I(Z))$, where $Z \in \mathrm{Ob}\,\mathbf{Haus}$, there exists a unique $g \in \mathrm{Hom}_{\mathbf{Haus}}(H(X),\ Z)$ such that the following diagram is commutative:

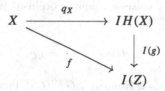

Using Theorem 3.6.1, (2) we can conclude that $\mathcal{H}$ is left adjoint to I, as desired.

(7) The Dorroh extension functor $\mathcal{D}\colon \mathbf{Rng} \to \mathbf{Ring}$ defined in Example 1.5.3, (31) is left adjoint to the inclusion functor $I\colon \mathbf{Ring} \to \mathbf{Rng}$. Indeed, let $j\colon 1_{\mathbf{Rng}} \to I\mathcal{D}$ be the natural transformation defined for all rings R by the ring homomorphism $j_R\colon R \to D(R)$ associated with the Dorroh extension of R. If $f \in \mathrm{Hom}_{\mathbf{Rng}}(R, S)$ then $\mathcal{D}(f)$ is defined by (1.11) as the unique morphism in $\mathbf{Ring}$ such that $j_S \circ f = I\mathcal{D}(f) \circ j_R$. Therefore, the following diagram is commutative for all $f \in \mathrm{Hom}_{\mathbf{Rng}}(R, S)$:

$$
\begin{array}{ccc}
R & \xrightarrow{\;q_R\;} & I\mathcal{D}(R) \\
{\scriptstyle f}\downarrow & & \downarrow{\scriptstyle I\mathcal{D}(f)} \\
S & \xrightarrow[\;q_S\;]{} & I\mathcal{D}(S)
\end{array}
$$

In particular, this shows that j is a natural transformation. Now recall that by the universal property of the Dorroh extension, for any $f \in \mathrm{Hom}_{\mathbf{Rng}}(R, I(T))$, where $T \in$ Ob **Ring**, there exists a unique $g \in \mathrm{Hom}_{\mathbf{Ring}}(D(R), T)$ such that the following diagram is commutative:

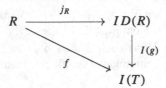

Using Theorem 3.6.1, (2) we can conclude that $\mathcal{D}$ is left adjoint to I, as desired. $\qquad\qquad\qquad\qquad\qquad\qquad\qquad\qquad\qquad\qquad\qquad\qquad\qquad\quad\square$

As another application of Theorem 3.6.1 we will show that, when they exist, left/right adjoints are unique up to natural isomorphism.

Theorem 3.6.4 *Any two left (right) adjoints of a given functor are naturally isomorphic.*

Proof Assume F, $F': C \to \mathcal{D}$ are both left adjoint functors of $G: \mathcal{D} \to C$. Then there exist natural transformations $\eta: 1_C \to GF$ and $\eta': 1_C \to GF'$ satisfying the conditions in Theorem 3.6.1, (2). Given $C \in \mathrm{Ob}\,C$, as $F' \dashv G$ and $\eta_C \in \mathrm{Hom}_C(C, GF(C))$, there exists a unique morphism $\gamma_C \in \mathrm{Hom}_{\mathcal{D}}(F'(C), F(C))$ such that

$$G(\gamma_C) \circ \eta'_C = \eta_C. \qquad\qquad (3.36)$$

Similarly, as $F \dashv G$ and $\eta'_C \in \mathrm{Hom}_C(C, GF'(C))$, there exists a unique morphism $\gamma'_C \in \mathrm{Hom}_{\mathcal{D}}(F(C), F'(C))$ such that

$$G(\gamma'_C) \circ \eta_C = \eta'_C. \qquad\qquad (3.37)$$

We will see that each γ_C is an isomorphism with the inverse given precisely by γ'_C. Indeed, using (3.36) and (3.37) we can easily see that $G(\gamma'_C \circ \gamma_C) \circ \eta'_C = \eta'_C$ and since we obviously also have $G(1_{F'(C)}) \circ \eta'_C = \eta'_C$ it follows by Corollary 3.6.2, (1) that $\gamma'_C \circ \gamma_C = 1_{F'(C)}$. Similarly, one can prove that $\gamma_C \circ \gamma'_C = 1_{F(C)}$.

We are left to prove that $\gamma: F' \to F$ is a natural transformation, i.e., for any $f \in \mathrm{Hom}_C(C, C')$ the following diagram is commutative:

$$
\begin{array}{ccc}
F'(C) & \xrightarrow{\ \gamma_C\ } & F(C) \\
{\scriptstyle F'(f)}\downarrow & & \downarrow{\scriptstyle F(f)} \\
F'(C') & \xrightarrow{\ \gamma_{C'}\ } & F(C')
\end{array}
\qquad\text{i.e.,}\quad F(f) \circ \gamma_C = \gamma_{C'} \circ F'(f).
$$

Using Corollary 3.6.2, (1) it is enough to prove that the following holds:

$$G(F(f) \circ \gamma_C) \circ \eta'_C = G(\gamma_{C'} \circ F'(f)) \circ \eta'_C. \tag{3.38}$$

To this end, we use the naturality of η and respectively η'; that is, the commutativity of the following diagrams:

$$
\begin{array}{ccc}
C & \xrightarrow{\;\eta_C\;} & GF(C) \\
{\scriptstyle f}\downarrow & & \downarrow{\scriptstyle GF(f)} \\
C' & \xrightarrow{\;\eta_{C'}\;} & GF(C')
\end{array}
\qquad \text{i.e.,}\quad GF(f)\circ\eta_C = \eta_{C'}\circ f,
\tag{3.39}
$$

$$
\begin{array}{ccc}
C & \xrightarrow{\;\eta'_C\;} & GF'(C) \\
{\scriptstyle f}\downarrow & & \downarrow{\scriptstyle GF'(f)} \\
C' & \xrightarrow{\;\eta'_{C'}\;} & GF'(C')
\end{array}
\qquad \text{i.e.,}\quad GF'(f)\circ\eta'_C = \eta'_{C'}\circ f.
\tag{3.40}
$$

Then, we have

$$
\begin{aligned}
GF(f) \circ G(\gamma_C) \circ \eta'_C &\overset{(3.36)}{=} GF(f) \circ \eta_C \\
&\overset{(3.39)}{=} \eta_{C'} \circ f \\
&\overset{(3.36)}{=} G(\gamma_{C'}) \circ \eta'_{C'} \circ f \\
&\overset{(3.40)}{=} G(\gamma_{C'}) \circ GF'(f) \circ \eta'_C.
\end{aligned}
$$

Therefore, (3.38) indeed holds. To summarize, we have proved that there exists a natural isomorphism $\gamma : F' \to F$ and the proof is now finished. $\qquad\square$

Adjunctions can also be used to easily derive important properties of certain functorial constructions, as the following examples show. This includes, for instance, the commutation of tensor products or localizations with direct sums of modules. All of these are obtained by applying Theorem 3.4.4.

Example 3.6.5 Given a commutative ring R, for any $X \in \mathrm{Ob}_R\mathcal{M}$ and any family $(M_i)_{i \in I}$ of R-modules we have the following isomorphisms of R-modules:

$$\left(\oplus_{i \in I} M_i \right) \otimes X \simeq \left(\oplus_{i \in I} M_i \otimes X \right),$$
$$S^{-1}\left(\oplus_{i \in I} M_i \right) \simeq \oplus_{i \in I} S^{-1}M_i,$$

where $\otimes = \otimes_R$. Indeed, both statements are consequences of the fact that both the tensor product functor $- \otimes X: {}_RM \rightarrow {}_RM$ and the localization functor $\mathcal{L}: {}_RM \rightarrow {}_{S^{-1}R}M$ are left adjoints (see Example 3.4.1, (2) and Example 3.6.3, (5)) and therefore they preserve coproducts (see Example 2.1.5, (12)). $\square$

We end this section with the following useful result:

Lemma 3.6.6 *Let* $F: C \rightarrow \mathcal{D}$, $G: \mathcal{D} \rightarrow C$ *be two functors such that* $F \dashv G$ *and let* $\eta: 1_C \rightarrow GF$ *and* $\varepsilon: FG \rightarrow 1_{\mathcal{D}}$ *be the unit and counit of the adjunction, respectively.*

(1) F is faithful if and only if η_C is a monomorphism for all $C \in \mathrm{Ob}\, C$; G is faithful if and only if ε_D is an epimorphism for all $D \in \mathrm{Ob}\, \mathcal{D}$;

(2) F is full if and only if η_C is a split epimorphism for all $C \in \mathrm{Ob}\, C$; G is full if and only if ε_D is a split monomorphism for all $D \in \mathrm{Ob}\, \mathcal{D}$;

(3) F is fully faithful if and only if the unit of the adjunction is a natural isomorphism; G is fully faithful if and only if the counit of the adjunction is a natural isomorphism.

Proof

(1) Assume first that F is faithful and let f_1, $f_2 \in \mathrm{Hom}_C(C', C)$ such that $\eta_C \circ f_1 = \eta_C \circ f_2$. Applying F to the last identity yields $F(\eta_C) \circ F(f_1) = F(\eta_C) \circ F(f_2)$ and after composing on the left with $\varepsilon_{F(C)}$ and using (3.15) we obtain $F(f_1) = F(f_2)$. As F is faithful we arrive at $f_1 = f_2$, which shows that η_C is indeed a monomorphism.

Conversely, assume η_C is a monomorphism for all $C \in \mathrm{Ob}\, C$ and let f_1, $f_2 \in \mathrm{Hom}_C(C', C)$ such that $F(f_1) = F(f_2)$. Using (3.15) we obtain $\varepsilon_{F(C)} \circ F(\eta_C \circ f_1) = \varepsilon_{F(C)} \circ F(\eta_C \circ f_2)$. Now Corollary 3.6.2, (2) implies $\eta_C \circ f_1 = \eta_C \circ f_2$ and since η_C is a monomorphism we obtain $f_1 = f_2$. Therefore, F is faithful.

The result concerning the functor G follows by duality. Indeed, by Theorem 3.4.3 we have $G^{op} \dashv F^{op}$ and moreover, Corollary 3.5.4 shows that the unit of this adjunction is precisely ε^{op}. Since G is faithful if and only if G^{op} is faithful, the desired conclusion follows the first part of the proof and Proposition 1.4.3, (1).

(2) Let F be a full functor and $C \in \mathrm{Ob}\, C$. Then $\varepsilon_{F(C)} \in \mathrm{Hom}_{\mathcal{D}}(FGF(C), F(C))$ and since F is full there exists a morphism $u_C \in \mathrm{Hom}_C(GF(C), C)$ such that $\varepsilon_{F(C)} = F(u_C)$. We obtain

$$\varepsilon_{F(C)} \circ F(\eta_C \circ u_C) = \underline{\varepsilon_{F(C)} \circ F(\eta_C)} \circ F(u_C) \overset{(3.15)}{=} F(u_C)$$

$$= \varepsilon_{F(C)} = \varepsilon_{F(C)} \circ F(1_{GF(C)})$$

Now Corollary 3.6.2, (2) implies $\eta_C \circ u_C = 1_{GF(C)}$, as desired.

Conversely, assume η_C is a split epimorphism for all $C \in \mathrm{Ob}\, C$. Thus, there exists a morphism $v_C \in \mathrm{Hom}_C(GF(C), C)$ such that $\eta_C \circ v_C = 1_{GF(C)}$.

Let $g \in \mathrm{Hom}_{\mathcal{D}}(F(C), F(C'))$. The naturality of η applied to υ_C renders the following diagram commutative:

$$
\begin{array}{ccc}
GF(C) & \xrightarrow{\;\eta_{GF(C)}\;} & GFGF(C) \\
{\scriptstyle \upsilon_C}\big\downarrow & & \big\downarrow{\scriptstyle GF(\upsilon_C)} \\
C & \xrightarrow[\;\eta_C\;]{} & GF(C)
\end{array}
\qquad \text{i.e.,} \quad GF(\upsilon_C) \circ \eta_{GF(C)} = \eta_C \circ \upsilon_C.
$$

$$(3.41)$$

Therefore, for all $C \in \mathrm{Ob}\,C$, we obtain

$$
GF(\upsilon_C) \circ \eta_{GF(C)} \stackrel{(3.41)}{=} \eta_C \circ \upsilon_C = 1_{GF(C)} \stackrel{(3.16)}{=} G(\varepsilon_{F(C)}) \circ \eta_{GF(C)}.
$$

Now Corollary 3.6.2, (1) implies $F(\upsilon_C) = \varepsilon_{F(C)}$. Finally, the naturality of ε applied to g yields

$$
\begin{array}{ccc}
FGF(C) & \xrightarrow{\;\varepsilon_{F(C)}\;} & F(C) \\
{\scriptstyle FG(g)}\big\downarrow & & \big\downarrow{\scriptstyle GF(\alpha)} \\
FGF(C') & \xrightarrow[\;\varepsilon_{F(C')}\;]{} & F(C')
\end{array}
\qquad \text{i.e.,} \quad g \circ \varepsilon_{F(C)} = \varepsilon_{F(C')} \circ FG(g).
$$

$$(3.42)$$

Putting all this together we obtain

$$
g = g \circ 1_{F(C)} \stackrel{(3.15)}{=} g \circ \varepsilon_{F(C)} \circ F(\eta_C)
$$
$$
\stackrel{(3.42)}{=} \varepsilon_{F(C')} \circ FG(g) \circ F(\eta_C) = F\big(\upsilon_{C'} \circ G(g) \circ \eta_C\big),
$$

which shows that F is full.

The result concerning the functor G follows again by duality. Indeed, by Theorem 3.4.3 we have $G^{op} \dashv F^{op}$ and, moreover, Corollary 3.5.4 shows that the unit of this adjunction is precisely ε^{op}. Since G is full if and only if G^{op} is full, the desired conclusion follows from Exercise 1.6.

(3) Using (1) and (2), F is fully faithful if and only if η_C is both a monomorphism and a split epimorphism for all $C \in \mathrm{Ob}\,C$. Similarly, G is fully faithful if and only if ε_D is both an epimorphism and a split monomorphism and for all $D \in \mathrm{Ob}\,\mathcal{D}$. The conclusion now follows from Exercise 1.6, (b).

$\square$

3.7   (Co)reflective Subcategories

This section is devoted to a special kind of adjunction, namely those for which one of the functors involved is an inclusion.

Definition 3.7.1 A full subcategory $\mathcal{A}$ of $\mathcal{B}$ is called *reflective* if the inclusion functor $I: \mathcal{A} \to \mathcal{B}$ admits a left adjoint, called a *reflector*. Dually, a full subcategory $\mathcal{A}$ of $\mathcal{B}$ is called *coreflective* if the inclusion functor admits a right adjoint, called a *coreflector*.

We have already encountered many examples of such subcategories:

Examples 3.7.2

(1) **Ab** is a reflective subcategory of **Grp**, as shown in Example 3.4.1, (4).
(2) **Grp** is both a reflective and a coreflective subcategory of **Mon** as shown in Example 3.6.3, (3) and (4).
(3) **KHaus** is a reflective subcategory of **Top**. The Stone–Čech compactification provides the reflector, as shown in Example 3.6.3, (2).
(4) **Haus** is a reflective subcategory of **Top**. The Hausdorff quotient provides the reflector, as shown in Example 3.6.3, (6).
(5) **Ring** is not a reflective subcategory of **Rng**. Indeed, note that although the inclusion functor $I: \mathbf{Ring} \to \mathbf{Rng}$ has a left adjoint, as shown in Example 3.6.3, (7) the category **Ring** is not a full subcategory of **Rng**.   □

One important feature of (co)reflective subcategories is that they behave well with respect to (co)limits. The remaining of this section will be devoted to studying their (co)completeness and providing explicit descriptions of the (co)limits.

Proposition 3.7.3 *Let $\mathcal{A}$ be a subcategory of $\mathcal{B}$.*

(1) If $\mathcal{A}$ is a reflective subcategory of the complete category $\mathcal{B}$ then $\mathcal{A}$ is itself complete.
(2) If $\mathcal{A}$ is a coreflective subcategory of the cocomplete category $\mathcal{B}$ then $\mathcal{A}$ is itself cocomplete.

Proof

(1) Let $R: \mathcal{B} \to \mathcal{A}$ be the reflector of the inclusion functor $I: \mathcal{A} \to \mathcal{B}$, i.e., $R \dashv I$. Let J be a small category and $F: J \to \mathcal{A}$ a functor. Since $\mathcal{B}$ is complete, the functor $IF: J \to \mathcal{B}$ has a limit, which we denote by $\big(L, (p_j: L \to IF(j))_{j \in \mathrm{Ob}\,J}\big)$. Since $\big(L, (p_j: L \to IF(j))_{j \in \mathrm{Ob}\,J}\big)$ is in particular a cone on

IF, the following diagram is commutative for all $d \in \mathrm{Hom}_J(j,\, l)$:

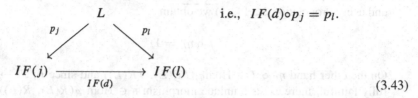

i.e., $IF(d)\circ p_j = p_l$.

$$(3.43)$$

Let $\eta\colon 1_{\mathcal{B}} \to IR$ be the unit of the adjunction $R \dashv I$ (see Theorem 3.5.1). By Theorem 3.6.1, (2) for any $p_j \in \mathrm{Hom}_{\mathcal{B}}(L,\, IF(j))$ there exists a unique morphism $q_j \in \mathrm{Hom}_{\mathcal{A}}(R(L),\, F(j))$ such that the following diagram commutes for all $j \in \mathrm{Ob}\, J$:

$$L \xrightarrow{\;\eta_L\;} IR(L) \qquad \text{i.e.,} \quad I(q_j)\circ\eta_L = p_j.$$

$$p_j \searrow \quad \downarrow I(q_j)$$

$$IF(j)$$

$$(3.44)$$

We will prove first that $\big(R(L),\, (q_j\colon R(L) \to F(j))_{j\in\mathrm{Ob}\,J}\big)$ is a cone on F. Indeed, for all $d \in \mathrm{Hom}_J(j,\, l)$ we have

$$I\big(F(d)\circ q_j\big)\circ\eta_L = IF(d)\circ I(q_j)\circ\eta_L \overset{(3.44)}{=} IF(d)\circ p_j \overset{(3.43)}{=} p_l \overset{(3.44)}{=} I(q_l)\circ\eta_L.$$

Now by Corollary 3.6.2, (1) we get $F(d) \circ q_j = q_l$, i.e.,

$$\big(R(L),\, (q_j\colon R(L) \to F(j))_{j\in\mathrm{Ob}\,J}\big)$$

is a cone on F. We will show that $\big(R(L),\, (q_j)_{j\in\mathrm{Ob}\,J}\big)$ is in fact the limit of F.

We start by showing that $\eta_L\colon L \to IR(L)$ is an isomorphism. Indeed, since $\big(IR(L),\, (I(q_j)\colon IR(L) \to IF(j))_{j\in\mathrm{Ob}\,J}\big)$ is a cone on IF by Lemma 2.5.5 and $\big(L,\, (p_j\colon L \to IF(j))_{j\in\mathrm{Ob}\,J}\big)$ is its limit, there exists a unique morphism $f \in \mathrm{Hom}_{\mathcal{B}}(IR(L),\, L)$ such that the following diagram is commutative for all $j \in \mathrm{Ob}\, J$:

$$L \xrightarrow{\;p_j\;} IF(j) \qquad \text{i.e.,} \quad p_j\circ f = I(q_j).$$

$$f \uparrow \quad \nearrow I(q_j)$$

$$IR(L)$$

$$(3.45)$$

Thus, for any $j \in \mathrm{Ob}\, J$ we have $p_j \circ f \circ \eta_L \overset{(3.45)}{=} I(q_j) \circ \eta_L \overset{(3.44)}{=} p_j = p_j \circ 1_L$ and using Proposition 2.2.14, (1) we obtain

$$f \circ \eta_L = 1_L. \tag{3.46}$$

On the other hand $\eta_L \circ f \in \mathrm{Hom}_{\mathcal{B}}(IR(L), IR(L))$ and since $I : \mathcal{A} \to \mathcal{B}$ is fully faithful, there exists a unique morphism $t \in \mathrm{Hom}_{\mathcal{A}}(R(L), R(L))$ such that

$$\eta_L \circ f = I(t). \tag{3.47}$$

Moreover, we have

$$I(t) \circ \eta_L \overset{(3.47)}{=} \eta_L \circ f \circ \eta_L \overset{(3.46)}{=} \eta_L = I(1_{R(L)}) \circ \eta_L.$$

Using again Corollary 3.6.2, (1) we get $t = 1_{R(L)}$ and hence $\eta_L \circ f = 1_{IR(L)}$, so η_L is an isomorphism, as desired.

Consider now another cone $\left(L', (t_j : L' \to F(j))_{j \in \mathrm{Ob}\, J}\right)$ on F. Then $\left(I(L'), (I(t_j)_{j \in \mathrm{Ob}\, J}\right)$ is a cone on IF. Therefore, there exists a unique morphism $g \in \mathrm{Hom}_{\mathcal{B}}(I(L'), L)$ such that the following diagram is commutative for all $j \in \mathrm{Ob}\, J$:

$$\begin{array}{ccc} IR(L) & & \text{i.e.,} \quad p_j \circ g = I(t_j). \\ \eta_L \uparrow \quad \searrow^{I(q_j)} & & \\ \quad \quad p_j \searrow & & \\ L \xrightarrow{\quad} IF(j) & & \\ g \uparrow \quad \nearrow & & \\ \quad \nearrow_{I(t_j)} & & \\ I(L') & & \end{array} \tag{3.48}$$

Since we also have $p_j = I(q_j) \circ \eta_L$ for all $j \in \mathrm{Ob}\, J$, we obtain

$$I(q_j) \circ \eta_L \circ g = I(t_j). \tag{3.49}$$

As $\eta_L \circ g \in \mathrm{Hom}_{\mathcal{B}}(I(L'), IR(L))$ and I is fully faithful, there exists a unique morphism $h \in \mathrm{Hom}_{\mathcal{A}}(L', R(L))$ such that $I(h) = \eta_L \circ g$. Then (3.49) becomes $I(q_j \circ h) = I(t_j)$ and since I is fully faithful we get $q_j \circ h = t_j$ for all $j \in \mathrm{Ob}\, J$, i.e., the following diagram is commutative:

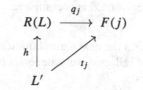

The proof will be finished once we show that h is the unique morphism which makes the above diagram commutative. Indeed, suppose there exists an $\overline{h} \in \mathrm{Hom}_{\mathcal{A}}(L', R(L))$ such that $q_j \circ \overline{h} = t_j$ for all $j \in \mathrm{Ob}\, J$. Then we also have $I(q_j) \circ I(\overline{h}) = I(t_j)$ and using (3.44) and respectively (3.48) we get

$$p_j \circ \eta_L^{-1} \circ I(\overline{h}) = p_j \circ g$$

for all $j \in \mathrm{Ob}\, J$. Proposition 2.2.14, (1) implies $\eta_L^{-1} \circ I(\overline{h}) = g$ and thus $I(\overline{h}) = \eta_L \circ g$. Since h is the unique morphism such that $I(h) = \eta_L \circ g$, we get $\overline{h} = h$. This shows that $\big(R(L), (q_j)_{j \in \mathrm{Ob}\, J}\big)$ is indeed the limit of F and therefore $\mathcal{A}$ is a complete category.

(2) Let $\mathcal{A}$ be a coreflective subcategory of a cocomplete category $\mathcal{B}$ and denote by $C : \mathcal{B} \to \mathcal{A}$ the right adjoint of the inclusion functor $I : \mathcal{A} \to \mathcal{B}$ and by ε the counit of this adjunction. Let J be a small category and $F : J \to \mathcal{A}$ a functor.

Then $\mathcal{A}^{op}$ is obviously a subcategory of $\mathcal{B}^{op}$ and the inclusion functor is precisely I^{op}. Theorem 3.4.3 implies that $C^{op} \dashv I^{op}$ and, moreover, as proved in Corollary 3.5.4, ε^{op} is the unit of this adjunction. Using the first part of the proof, the limit of the functor $F^{op} : J^{op} \to \mathcal{A}^{op}$ is given by the pair

$$\Big(C(L), \ (q_j^{op} \in \mathrm{Hom}_{\mathcal{A}^{op}}(C(L), F(j^{op})))_{i \in \mathrm{Ob}\, J}\Big),$$

where q_j^{op} is the unique morphism which makes the following diagram commutative:

$$L \xrightarrow{\ \varepsilon_L^{op}\ } I^{op} C^{op}(L) \qquad \text{i.e.,} \quad I^{op}(q_j^{op}) \circ^{op} \varepsilon_L^{op} = p_j^{op},$$

with arrows p_j^{op} and $I^{op}(q_j^{op})$ pointing to $I^{op} F^{op}(j)$

and $\Big(L, \ (p_j^{op} \in \mathrm{Hom}_{\mathcal{B}^{op}}(L, I^{op} F^{op}(j)))_{i \in \mathrm{Ob}\, J}\Big)$ is the limit of the functor $I^{op} F^{op} : J^{op} \to \mathcal{B}^{op}$. In other words, for all $j \in \mathrm{Ob}\, J$, q_j is the unique morphism in $\mathcal{A}$ such that $\varepsilon_L \circ I(q_j) = p_j$. Now we can conclude by

Lemma 2.2.12 that $\left(C(L), (q_j \in \mathrm{Hom}_{\mathcal{A}}(F(j), C(L)))_{i \in \mathrm{Ob}\, J} \right)$ is the colimit of the functor F and therefore $\mathcal{A}$ is cocomplete.

$\square$

In light of Proposition 3.7.3 the next natural question we are led to consider concerns the cocompleteness of reflective subcategories (and, dually, the completeness of coreflective subcategories).

Proposition 3.7.4 *Let $\mathcal{A}$ be a subcategory of $\mathcal{B}$.*

(1) If $\mathcal{A}$ is a reflective subcategory of a cocomplete category $\mathcal{B}$ then $\mathcal{A}$ is also cocomplete.

(2) If $\mathcal{A}$ is a coreflective subcategory of a complete category $\mathcal{B}$ then $\mathcal{A}$ is also complete.

Proof

(1) Let $I: \mathcal{A} \to \mathcal{B}$ be the inclusion functor and $R: \mathcal{B} \to \mathcal{A}$ the reflector. Let $F: J \to \mathcal{A}$ be a functor, where J is a small category. Since $\mathcal{B}$ is cocomplete, the functor $IF: J \to \mathcal{B}$ has a colimit, which we denote by

$$\left(D, (q_j: IF(j) \to D)_{j \in \mathrm{Ob}\, J} \right).$$

R is left adjoint to I and by Theorem 3.4.4 it preserves colimits, so

$$\left(R(D), (R(q_j): RIF(j) \to R(D))_{j \in \mathrm{Ob}\, J} \right)$$

is the colimit of the functor $RIF: J \to \mathcal{A}$. By Lemma 3.6.6, (3) we know that the counit $\varepsilon: RI \to 1_{\mathcal{A}}$ of the adjunction $R \dashv I$ is a natural isomorphism. Therefore, the natural transformation $\varepsilon_F: RIF \to F$ defined by

$$(\varepsilon_F)_j = \varepsilon_{F(j)} \text{ for all } j \in \mathrm{Ob}\, J$$

is also a natural isomorphism (Example 1.7.2, (7)). Now, in light of Lemma 2.2.15, (2) we can conclude that $\left(R(D), (R(q_j) \circ \varepsilon_{F(j)}^{-1}: F(j) \to R(D))_{j \in \mathrm{Ob}\, J} \right)$ is the colimit of F.

(2) Let $I: \mathcal{A} \to \mathcal{B}$ to be the inclusion functor and denote by $C: \mathcal{B} \to \mathcal{A}$ the coreflector and by η the unit of this adjunction, which is a natural isomorphism by Lemma 3.6.6, (3). Let $F: J \to \mathcal{A}$ be a functor, where J is a small category.

By Theorem 3.4.3 we also have an adjunction $C^{op} \dashv I^{op}$ whose counit is precisely η^{op}. This shows, in particular, that $\mathcal{A}^{op}$ is a reflective subcategory of the cocomplete category $\mathcal{B}^{op}$ with inclusion functor I^{op}. By part 1) proved above, the colimit of the functor F^{op} is given by the pair

$$\left(C^{op}(D), (C^{op}(q_j^{op}) \circ^{op} (\eta_{F^{op}(j)}^{-1})^{op} \in \mathrm{Hom}_{\mathcal{A}^{op}}(F^{op}(j), C^{op}(D)))_{j \in \mathrm{Ob}\, J} \right),$$

where $\left(D,\ (q_j^{op})_{j \in \mathrm{Ob}\,J}\right)$ is the colimit of the functor $I^{op} F^{op} \colon J^{op} \to \mathcal{B}^{op}$. Now Lemma 2.2.12 implies that $\left(C(D),\ (\eta_{F(j)}^{-1} \circ C(q_j))_{j \in \mathrm{Ob}\,J}\right)$ is the limit of the functor F.

$\square$

3.8 Equivalence of Categories

When studying categories which are practically *the same*, the first notion we usually encounter is that of an isomorphism of categories, as introduced in Definition 1.6.6. However, this concept turns out to be too strict, as there are many examples of categories with similar properties (such as completeness, cocompleteness etc.) which are not isomorphic. To express that two categories share many of the same properties, a more suitable notion than isomorphism is the following:

Definition 3.8.1 A functor $F \colon C \to \mathcal{D}$ is called an *equivalence of categories* and the category C is said to be *equivalent* to $\mathcal{D}$ if there exists another functor $G \colon \mathcal{D} \to C$ such that we have natural isomorphisms $GF \cong 1_C$ and $FG \cong 1_{\mathcal{D}}$. A contravariant functor $F \colon C \to \mathcal{D}$ for which $F \colon C^{op} \to \mathcal{D}$ is an equivalence of categories is called a *duality of categories*.

Example 3.8.2 Given a field K, the category $\mathbf{Mat}_K$ defined in Example 1.2.2, (19) is equivalent to the category of finite-dimensional K-vector spaces $_K \mathcal{M}^{fd}$. Indeed, the functor $F \colon \mathbf{Mat}_K \to {}_K \mathcal{M}^{fd}$ defined below is an equivalence of categories:

$F(n) = K^n$, the n-dimensional space of column vectors over K for all $n \in \mathbb{N}$,

$F(A) = M_A$, for all morphisms $A \colon m \to n$ in $\mathbf{Mat}_K$, where $M_A \colon K^m \to K^n$ is

given by $M_A(v) = Av$ for all $v \in K^m$.

To this end, we choose a basis[6] $\mathcal{B}_V$ for each finite dimensional vector space V and we define a functor $G \colon {}_K \mathcal{M}^{fd} \to \mathbf{Mat}_K$ as follows:

$G(V) = \dim(V)$, for all finite-dimensional vector spaces V,

$G(\alpha) = U_\alpha$, where U_α is the matrix of the linear map $\alpha \colon V \to W$ with respect to the chosen bases $\mathcal{B}_V$ and $\mathcal{B}_W$ of V and W respectively.

Throughout this example, by convention, the basis we consider on K^n will be the standard basis. We start by showing that $GF = 1_{\mathbf{Mat}_K}$. First, for any $n \in \mathbb{N}$ we

---

[6] Note that this is always possible due to the axiom of choice.

have

$$GF(n) = G(K^n) = \dim(K^n) = n = 1_{\mathbf{Mat}_K}(n).$$

Moreover, if $A\colon m \to n$ is a morphism in $\mathbf{Mat}_K$, we have $GF(A) = G(M_A) = U_{M_A}$, where U_{M_A} is the matrix of the linear map $M_A\colon K^m \to K^n$ given by $M_A(v) = Av$ with respect to the standard bases $\{e_1, e_2, \ldots, e_m\}$ and $\{f_1, f_2, \ldots, f_n\}$ of K^m and K^n, respectively. Having in mind that the element e_i (resp. f_j) of the standard basis is the column vector in K^m (resp. in K^n) with 1 on the i-th (resp. j-th) position and zeros elsewhere for all $i = 1, 2, \ldots, m$ (resp. $j = 1, 2, \ldots, n$), we obtain $M_A(e_i) = Ae_i = \sum_{j=1}^n a_{ji} f_j$, where $A = (a_{kl})_{k=\overline{1,n}, \, l=\overline{1,m}}$. This proves that $U_{M_A} = A$, i.e., $GF(A) = A$, as desired. Hence, we have proved that $GF = 1_{\mathbf{Mat}_K}$, which shows that, in particular, GF is naturally isomorphic to $1_{\mathbf{Mat}_K}$.

We are left to show that FG is naturally isomorphic to $1_{K\mathcal{M}^{fd}}$. Consider $\eta\colon 1_{K\mathcal{M}^{fd}} \to FG$ defined for any vector space V by $\eta_V\colon V \to K^{\dim(V)}$, $\eta_V(v) = [v]$, where we denote by $[v]$ the (column) coordinate vector of v with respect to the chosen basis of V. We claim that η is a natural isomorphism. To start with, each η_V is clearly a linear bijection. We are left to check the naturality condition. To this end, let $\alpha\colon V \to W$ be a morphism in $K\mathcal{M}^{fd}$ and consider the chosen bases $\mathcal{B}_V = \{t_1, t_2, \ldots, t_m\}$ and $\mathcal{B}_W = \{w_1, w_2, \ldots, w_n\}$ in V and W, respectively, where $m = \dim(V)$, $n = \dim(W)$. The proof will be finished once we show that the following diagram is commutative:

$$
\begin{array}{ccc}
V & \xrightarrow{\ \eta_V\ } & K^m \\[2pt]
{\scriptstyle \alpha}\Big\downarrow & & \Big\downarrow{\scriptstyle FG(\alpha)} \\[2pt]
W & \xrightarrow[\ \eta_W\ ]{} & K^n
\end{array}
\tag{3.50}
$$

If we define $U_\alpha = (u_{kl})_{k=\overline{1,n}, \, l=\overline{1,m}}$, then for any $v = \sum_{i=1}^m v_i t_i \in V$ we have $\alpha(v) = \sum_{i=1}^m v_i \, \alpha(t_i) = \sum_{i=1}^m v_i \left(\sum_{j=1}^n u_{ji} w_j \right) = \sum_{j=1}^n \left(\sum_{i=1}^m v_i u_{ji} \right) w_j$. This shows that the j-th component of the column vector $[\alpha(v)]$ is $\sum_{i=1}^m v_i u_{ji}$, for all $j = 1, 2, \ldots, n$. Moreover, a similar straightforward computation shows that the j-th component of $U_\alpha[v]$ is $\sum_{i=1}^m u_{ji} v_i$ for all $j = 1, 2, \ldots, n$. Putting everything together we have

$$[\alpha(v)] = U_\alpha[v]. \tag{3.51}$$

Therefore, for any $v \in V$, we have

$$FG(\alpha) \circ \eta_V(v) = FG(\alpha)([v]) = M_{U_\alpha}([v]) = U_\alpha[v] \overset{(3.51)}{=} [\alpha(v)] = \eta_W \circ \alpha(v),$$

which proves the commutativity of diagram (3.50).   □

Remark 3.8.3 The categories $\mathbf{Mat}_K$ and $_K\mathcal{M}^{fd}$ from the previous example are equivalent but not isomorphic. Indeed, this follows easily by noticing that $\mathbf{Mat}_K$ is a small category while $_K\mathcal{M}^{fd}$ has a class of objects.

Proposition 3.8.4 *Let $\mathcal{A}$, $\mathcal{B}$ and C be three categories. The following hold:*

(1) any category is equivalent to itself;
(2) if $\mathcal{A}$ is equivalent to $\mathcal{B}$ then $\mathcal{B}$ is equivalent to $\mathcal{A}$;
(3) if $\mathcal{A}$ is equivalent to $\mathcal{B}$ and $\mathcal{B}$ is equivalent to C then $\mathcal{A}$ is equivalent to C.

Proof

(1) Any category $\mathcal{A}$ is equivalent to itself as the identity functor $1_{\mathcal{A}} : \mathcal{A} \to \mathcal{A}$ is obviously an equivalence of categories.
(2) Assume that the category $\mathcal{A}$ is equivalent to $\mathcal{B}$ and $F : \mathcal{A} \to \mathcal{B}$ is the equivalence functor. Then there exists another functor $G : \mathcal{B} \to \mathcal{A}$ and two natural isomorphisms $GF \cong 1_{\mathcal{A}}$ and $FG \cong 1_{\mathcal{B}}$. This shows that G is also an equivalence of categories and therefore $\mathcal{B}$ is equivalent to $\mathcal{A}$.
(3) Assume that $\mathcal{A}$ is equivalent to $\mathcal{B}$ and $\mathcal{B}$ is equivalent to C. Then, we have two pairs of functors and their corresponding natural isomorphisms

$$F : \mathcal{A} \to \mathcal{B}, \quad G : \mathcal{B} \to \mathcal{A}, \quad \alpha : FG \to 1_{\mathcal{B}}, \quad \beta : GF \to 1_{\mathcal{A}},$$

$$H : \mathcal{B} \to C, \quad T : C \to \mathcal{B}, \quad \gamma : HT \to 1_C, \quad \sigma : TH \to 1_{\mathcal{B}}.$$

Consider now the functors $HF : \mathcal{A} \to C$ and $GT : C \to \mathcal{A}$ and the following natural transformations obtained by whiskering α and σ as in Example 1.7.2, (7) both on the left and on the right:

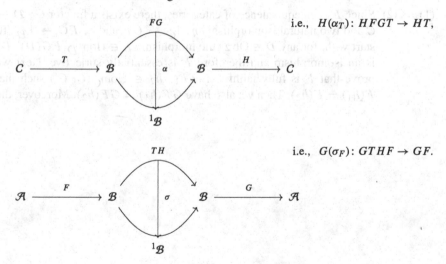

i.e., $H(\alpha_T) : HFGT \to HT$,

i.e., $G(\sigma_F) : GTHF \to GF$.

Note that the above natural transformations are in fact natural isomorphisms since α_T and σ_F are natural isomorphisms (Example 1.7.2, (7)) and all functors preserve isomorphisms (Proposition 1.6.9, (1)). This gives rise to the following natural isomorphisms:

$$HFGT \xrightarrow{\;\;H(\alpha_T)\;\;} HT \xrightarrow{\;\;\gamma\;\;} 1_C \;,\;\; \gamma \circ H(\alpha_T) \colon HFGT \to 1_C,$$

$$GTHF \xrightarrow{\;\;G(\sigma_F)\;\;} GF \xrightarrow{\;\;\beta\;\;} 1_{\mathcal{A}} \;,\;\; \beta \circ G(\sigma_F) \colon GTHF \to 1_{\mathcal{A}},$$

and we can now conclude that $\mathcal{A}$ is equivalent to C.

$\square$

We have the following very useful characterization of equivalences of categories:

Theorem 3.8.5 *Let $F \colon C \to \mathcal{D}$ be a functor. The following are equivalent:*

(1) F is an equivalence of categories;
(2) F is fully faithful and essentially surjective;
(3) there exists a right adjoint $G \colon \mathcal{D} \to C$ of F such that the unit and counit of the adjunction are natural isomorphisms;
(4) there exists a left adjoint $G \colon \mathcal{D} \to C$ of F such that the unit and counit of the adjunction are natural isomorphisms.

In particular, as the notation suggests, an equivalence of categories has a left adjoint which is also a right adjoint and the unit and counit of these adjunctions are natural isomorphisms.

Proof

$(1) \Rightarrow (2)$ Since F is an equivalence of categories, there exists a functor $G \colon \mathcal{D} \to C$ and two natural isomorphisms $\eta \colon 1_C \to GF$ and $\varepsilon \colon FG \to 1_{\mathcal{D}}$. To start with, for any $D \in \mathrm{Ob}\,\mathcal{D}$ the morphism $\varepsilon_D \in \mathrm{Hom}_{\mathcal{D}}(FG(D), D)$ is an isomorphism and therefore F is essentially surjective. Next we prove that F is fully faithful. Let $h_1, h_2 \in \mathrm{Hom}_C(C, C')$ such that $F(h_1) = F(h_2)$. Then we also have $GF(h_1) = GF(h_2)$. Moreover, the naturality of η renders the following diagrams commutative:

$$
\begin{CD}
C @>{\eta_C}>> GF(C) \\
@V{h_1}VV @VV{GF(h_1)}V \\
C' @>>{\eta_{C'}}> GF(C')
\end{CD}
\qquad \text{i.e.,} \quad GF(h_1) \circ \eta_C = \eta_{C'} \circ h_1,
$$

$$
\begin{CD}
C @>{\eta_C}>> GF(C) \\
@V{h_2}VV @VV{GF(h_2)}V \\
C' @>>{\eta_{C'}}> GF(C')
\end{CD}
\qquad \text{i.e.,} \quad GF(h_2) \circ \eta_C = \eta_{C'} \circ h_2.
$$

(3.52)

(3.53)

From (3.52) and (3.53) we obtain $\eta_{C'} \circ h_1 = \eta_{C'} \circ h_2$ and since $\eta_{C'}$ is an isomorphism we get $h_1 = h_2$, as desired. Similarly, using the naturality of ε it follows that G is faithful as well.

Consider now C, $C' \in \mathrm{Ob}\, C$ and $g \in \mathrm{Hom}_{\mathcal{D}}(F(C), F(C'))$. Now define

$$ f = \eta_{C'}^{-1} \circ G(g) \circ \eta_C \in \mathrm{Hom}_C(C, C'). \qquad (3.54) $$

We will prove that $F(f) = g$. Indeed, using the naturality of η applied to f, we obtain

$$ \eta_{C'}^{-1} \circ GF(f) \circ \eta_C \overset{(3.52)}{=} f \overset{(3.54)}{=} \eta_{C'}^{-1} \circ G(g) \circ \eta_C. $$

Since η_C and $\eta_{C'}$ are isomorphisms, we get $GF(f) = G(g)$. As G is faithful, the above equality comes down to $F(f) = g$, which proves that F is full as well.

(2) $\Rightarrow$ (1)   Assume now that F is fully faithful and essentially surjective. First note that since F is fully faithful it reflects isomorphisms (see Proposition 1.6.9, (2)) and therefore two objects C and C' are isomorphic in C if and only if $F(C)$ and $F(C')$ are isomorphic in $\mathcal{D}$. Therefore, as F is essentially surjective, for any $D \in \mathrm{Ob}\,\mathcal{D}$ there exists a unique (up to isomorphism) $C \in \mathrm{Ob}\, C$ such that $F(C) \simeq D$. Thus, for any $D \in \mathrm{Ob}\,\mathcal{D}$ we can choose[7] an object $G(D) \in \mathrm{Ob}\, C$ and an isomorphism $\varepsilon_D \colon FG(D) \to D$ in $\mathcal{D}$. Now if $g \in \mathrm{Hom}_{\mathcal{D}}(D, D')$ we have the

---

[7] We assume that the axiom of choice holds.

following morphism in $\mathcal{D}$:

$$\varepsilon_{D'}^{-1} \circ g \circ \varepsilon_D \colon FG(D) \to FG(D').$$

Since F is fully faithful, there exists a unique morphism $G(g) \in \mathrm{Hom}_C(G(D), G(D'))$ such that $FG(g) = \varepsilon_{D'}^{-1} \circ g \circ \varepsilon_D$. The last equality implies the commutativity of the following diagram:

$$
\begin{array}{ccc}
FG(D) & \xrightarrow{\ \varepsilon_D\ } & D \\
{\scriptstyle FG(g)}\downarrow & & \downarrow{\scriptstyle g} \\
FG(D') & \xrightarrow[\ \varepsilon_{D'}\ ]{} & D'
\end{array}
\tag{3.55}
$$

We will prove that G defined above is in fact a functor and $\varepsilon \colon FG \to 1_{\mathcal{D}}$ is a natural transformation. Indeed, setting $D = D'$ and $g = 1_D$ yields

$$\varepsilon_D^{-1} \circ 1_D \circ \varepsilon_D = 1_{FG(D)} \colon FG(D) \to FG(D)$$

and there exists a unique morphism $G(1_D) \in \mathrm{Hom}_C(G(D), G(D))$ such that $FG(1_D) = 1_{FG(D)} = F(1_{G(D)})$ where the last equality holds because F is a functor. Since F is faithful we get $G(1_D) = 1_{G(D)}$. Consider now $g \in \mathrm{Hom}_{\mathcal{D}}(D, D')$, $g' \in \mathrm{Hom}_{\mathcal{D}}(D', D'')$ and the unique morphisms $G(g) \in \mathrm{Hom}_C(G(D), G(D'))$, respectively $G(g') \in \mathrm{Hom}_C(G(D'), G(D''))$ such that

$$FG(g) = \varepsilon_{D'}^{-1} \circ g \circ \varepsilon_D \quad \text{and} \quad FG(g') = \varepsilon_{D''}^{-1} \circ g' \circ \varepsilon_{D'}.$$

This yields

$$F\big(G(g') \circ G(g)\big) = \varepsilon_{D''}^{-1} \circ g' \circ g \circ \varepsilon_D. \tag{3.56}$$

Now having in mind that there exists a unique morphism $G(g' \circ g) \in \mathrm{Hom}_C(G(D), G(D''))$ such that $FG(g' \circ g) = \varepsilon_{D''}^{-1} \circ (g' \circ g) \circ \varepsilon_D$, it follows from (3.56) that $G(g' \circ g) = G(g') \circ G(g)$. Therefore, G is indeed a functor. Now the commutativity of diagram (3.55) implies that ε is a natural transformation. Recall that every ε_D is an isomorphism and thus ε is in fact a natural isomorphism. We are left to construct a natural isomorphism $\eta \colon 1_C \to GF$. For any $C \in \mathrm{Ob}\,C$ we have $\varepsilon_{F(C)}^{-1} \in \mathrm{Hom}_{\mathcal{D}}(F(C), FGF(C))$, an isomorphism in $\mathcal{D}$. Since F is fully faithful, there exists a unique $\eta_C \in \mathrm{Hom}_C(C, GF(C))$ such that $F(\eta_C) = \varepsilon_{F(C)}^{-1}$. Obviously, η_C is an isomorphism for all $C \in$

Ob C since F reflects isomorphisms. We prove now that η is a natural transformation. To this end, let $f \in \mathrm{Hom}_C(C, C')$; we need to prove the commutativity of the following diagram:

$$
\begin{array}{ccc}
C & \xrightarrow{\ \eta_C\ } & GF(C) \\
{\scriptstyle f}\downarrow & & \downarrow{\scriptstyle GF(f)} \\
C' & \xrightarrow[\ \eta_{C'}\ ]{} & C'
\end{array}
\qquad \text{i.e.,} \quad \eta_{C'}\circ f = GF(f)\circ\eta_C.
$$

$$(3.57)$$

By naturality of ε applied to $F(f)$ we have the following commutative diagram:

$$
\begin{array}{ccc}
FGF(C) & \xrightarrow{\ \varepsilon_{F(C)}\ } & F(C) \\
{\scriptstyle FGF(f)}\downarrow & & \downarrow{\scriptstyle F(f)} \\
FGF(C') & \xrightarrow[\ \varepsilon_{F(C')}\ ]{} & F(C')
\end{array}
$$

i.e., $\quad \varepsilon_{F(C')} \circ FGF(f) = F(f) \circ \varepsilon_{F(C)}$

$$\Leftrightarrow FGF(f) \circ \varepsilon_{F(C)}^{-1} = \varepsilon_{F(C')}^{-1} \circ F(f)$$

$$\Leftrightarrow FGF(f) \circ F(\eta_C) = F(\eta_{C'}) \circ F(f)$$

$$\Leftrightarrow F\big(GF(f) \circ \eta_C\big) = F(\eta_{C'} \circ f).$$

Since F is faithful we get $GF(f) \circ \eta_C = \eta_{C'} \circ f$, i.e., (3.57) is commutative, as desired.

$(3) \Rightarrow (1)$ Obvious.

$(1) \Rightarrow (3)$ Suppose F is an equivalence of categories and let $G: \mathcal{D} \to C$ such that $GF \cong 1_C$ and $FG \cong 1_{\mathcal{D}}$. We will prove that G is right adjoint to F. Denote by $\varepsilon: FG \to 1_{\mathcal{D}}$ the natural isomorphism arising from the above equivalence. Thus, for any $C \in \mathrm{Ob}\,C$ the morphism $\varepsilon_{F(C)} \in \mathrm{Hom}_{\mathcal{D}}(FGF(C), F(C))$ is an isomorphism. Therefore, $\varepsilon_{F(C)}^{-1} \in \mathrm{Hom}_{\mathcal{D}}(F(C), FGF(C))$ and since F is fully faithful (see $(1) \Rightarrow (2)$) there exists a unique morphism $\eta_C \in \mathrm{Hom}_C(C, GF(C))$ such that $F(\eta_C) = \varepsilon_{F(C)}^{-1}$. Since F is fully faithful, it reflects isomorphisms; thus η_C is also an isomorphism. Furthermore, one can show exactly as in the proof of $(2) \Rightarrow (1)$ that η is also a natural transformation. In light of Theorem 3.5.1, the proof will be finished once we show that (3.15) and (3.16) hold. To start with, for all $C \in \mathrm{Ob}\,C$ we

have

$$\varepsilon_{F(C)} \circ F(\eta_C) = \varepsilon_{F(C)} \circ \varepsilon_{F(C)}^{-1} = 1_{F(C)}, \quad \text{i.e.,} \quad (3.15) \text{ is fulfilled.}$$

Consider now $D \in \mathrm{Ob}\,\mathcal{D}$ and $\varepsilon_D^{-1}: D \to FG(D)$. From the naturality of ε applied to the morphism ε_D^{-1} we obtain the following commutative diagram:

$$
\begin{array}{ccc}
FG(D) & \xrightarrow{\ \varepsilon_D\ } & D \\
\ \downarrow{\scriptstyle FG(\varepsilon_D^{-1})} & & \ \downarrow{\scriptstyle \varepsilon_D^{-1}} \\
FGFG(D) & \xrightarrow[\ \varepsilon_{FG(D)}\ ]{} & FG(D)
\end{array}
\qquad \text{i.e.,} \quad \varepsilon_{FG(D)} \circ FG(\varepsilon_D^{-1}) = 1_{FG(D)}
$$

Therefore, since F is faithful, we have

$$FG(\varepsilon_D) \circ \varepsilon_{FG(D)}^{-1} = 1_{FG(D)}$$

$$\Leftrightarrow FG(\varepsilon_D) \circ F(\eta_{G(D)}) = 1_{FG(D)}$$

$$\Leftrightarrow F(G(\varepsilon_D) \circ \eta_{G(D)}) = F\big(1_{G(D)}\big)$$

$$\Leftrightarrow G(\varepsilon_D) \circ \eta_{G(D)} = 1_{G(D)}, \quad \text{i.e.,} \quad (3.16) \text{ holds as well.}$$

$(3) \Rightarrow (4)$ Assume now that $G: \mathcal{D} \to \mathcal{C}$ is a functor such that $F \dashv G$ and the unit $\eta: 1_{\mathcal{C}} \to GF$ and counit $\varepsilon: FG \to 1_{\mathcal{D}}$ of this adjunction are natural isomorphisms. Then $G \dashv F$ with unit $\varepsilon^{-1}: 1_{\mathcal{D}} \to FG$ and counit $\eta^{-1}: GF \to 1_{\mathcal{C}}$. Indeed, as η and ε are natural isomorphisms, the compatibility conditions (3.15) and (3.16) imply that for all $C \in \mathrm{Ob}\,\mathcal{C}$ and $D \in \mathrm{Ob}\,\mathcal{D}$ we have

$$\varepsilon_{F(C)}^{-1} = F(\eta_C), \qquad \eta_{G(D)}^{-1} = G(\varepsilon_D).$$

Therefore, for all $C \in \mathrm{Ob}\,\mathcal{C}$ and $D \in \mathrm{Ob}\,\mathcal{D}$ we have

$$1_{G(D)} = \eta_{G(D)}^{-1} \circ G\big(\varepsilon_D^{-1}\big),$$

$$1_{F(C)} = F\big(\eta_C^{-1}\big) \circ \varepsilon_{F(C)}^{-1},$$

which shows that the compatibility conditions (3.15) and (3.16) are fulfilled for ε^{-1} and η^{-1}.

$(4) \Rightarrow (3)$ Follows in the same fashion as $3) \Rightarrow 4)$. The proof is now finished.

$\square$

As an application of the previous theorem we will highlight an equivalence of categories involving ring localizations.

Example 3.8.6 Let R be a commutative ring with unity and $(S^{-1}R, \; j)$ its localization at the multiplicative set $S \subset R$. We will show, using Theorem 3.8.5, (2) that the category $_{S^{-1}R}\mathcal{M}$ of modules over the localization ring $S^{-1}R$ is equivalent to the category $_R\mathcal{M}^{S-\text{aut}}$ of modules over R on which S acts as automorphisms (see Example 1.2.2, (14)).

Indeed, consider the restriction of scalars functor $F_j \colon {}_{S^{-1}R}\mathcal{M} \to {}_R\mathcal{M}$ induced by the ring homomorphism $j \colon R \to S^{-1}R$ as defined in Example 1.5.3, (32). First note that since $j \colon R \to S^{-1}R$ is an epimorphism in $\mathbf{Ring}^c$ (see Example 1.3.2, (5)), the corresponding restriction of scalars functor F_j is fully faithful, as proved in Example 1.6.2, (6).

Furthermore, one can easily show that S acts as automorphisms on $F_j(M)$, for any $S^{-1}R$-module M. Indeed, it is straightforward to see that for all $s \in S$ the inverse of the multiplication map $\mu_s \colon M \to M$, $\mu_s(m) = sm$ is given by the R-linear homomorphism $\mu_{\frac{1}{s}}$, where the juxtaposition denotes the R-module structure on $F_j(M) = M$. This proves that the image of the restriction of scalars functor F_j is contained in the category $_R\mathcal{M}^{S-\text{aut}}$.

Therefore, we have a fully faithful functor $F_j \colon {}_{S^{-1}R}\mathcal{M} \to {}_R\mathcal{M}^{S-\text{aut}}$. We are left to show that F_j is essentially surjective as well. To this end, let $M \in \text{Ob}\, {}_R\mathcal{M}^{S-\text{aut}}$. Then M admits an $S^{-1}R$-module structure defined for all $r \in R$, $s \in S$ and $m \in M$ as follows:

$$\frac{r}{s} \star m = rn, \tag{3.58}$$

where the juxtaposition denotes the R-module structure on M and n is the unique element of M such that $sn = m$. Note that the existence and uniqueness of the element n with this property is a consequence of M being an R-module on which S acts as an automorphism. Moreover, if $M \in \text{Ob}\, {}_{S^{-1}R}\mathcal{M}$ with the $S^{-1}R$-module structure given in (3.58) then $F_j(M)$ has the R-module structure defined as follows for all $r \in R$ and $m \in M$:

$$j(r) \star m = \frac{r}{1} \star m = rm,$$

i.e., it coincides with the initial R-module structure on M. This finishes the proof. $\square$

As stated in the beginning, equivalent categories share most of the important properties:

Proposition 3.8.7 *Let C and $\mathcal{D}$ be two equivalent categories. Then C is (co)complete if and only if $\mathcal{D}$ is (co)complete.*

Proof Given an equivalence of categories $F \colon C \to \mathcal{D}$, by Theorem 3.8.5, (3) there exists a right adjoint $G \colon \mathcal{D} \to C$ of F such that the unit $\eta \colon 1_C \to GF$ and the counit $\varepsilon \colon FG \to 1_{\mathcal{D}}$ of the adjunction are natural isomorphisms. Assume first that $\mathcal{D}$ is complete and let $H \colon J \to C$ be a functor, where J is a small category. As $\mathcal{D}$ is a complete category, the functor $FH\colon J \to \mathcal{D}$ has a limit, say $\big(L,\, (p_j\colon L \to FH(j))_{j\in\mathrm{Ob}\,J}\big)$. Moreover, as G is right adjoint to F, Theorem 3.4.4 implies that $\big(G(L),\, (G(p_j)\colon G(L) \to GFH(j))_{j\in\mathrm{Ob}\,J}\big)$ is the limit of the functor $GFH\colon J \to C$. Now $\eta^{-1}\colon GF \to 1_C$ (as defined in Example 1.7.2, (6)) is a natural isomorphism and consequently $\eta_H^{-1}\colon GFH \to H$ (as defined in Example 1.7.2, (7)) is also a natural isomorphism. Now Lemma 2.2.15, (1) implies that $\big(G(L),\, (\eta_{H(j)}^{-1} \circ G(p_j)\colon G(L) \to H(j))_{j\in\mathrm{Ob}\,J}\big)$ is the limit of H and therefore the category C is complete.

The statement concerning cocompleteness follows similarly using Theorem 3.8.5, (4).   □

Definition 3.8.8 A *skeleton* of a category C is a full subcategory C_0 of C such that each object of C is isomorphic to exactly one object of C_0.

Example 3.8.9 A skeleton of a given category C always exists; indeed, it can be constructed by choosing[8] an object from each isomorphism class of objects in C and considering the full subcategory of C with this objects class.   □

Moreover, as an easy consequence of Theorem 3.8.5 we obtain the following:

Corollary 3.8.10 *A category is equivalent to any of its skeletons.*

Proof Let C_0 be a skeleton of a given category C. Then the inclusion functor $I\colon C_0 \to C$ is fully faithful and essentially surjective. The conclusion now follows from Theorem 3.8.5, (2).   □

Proposition 3.8.11 *Let C_0 and $\mathcal{D}_0$ be two skeletons of the categories C and $\mathcal{D}$, respectively. Then C and $\mathcal{D}$ are equivalent if and only if C_0 and $\mathcal{D}_0$ are isomorphic.*

Proof Assume first that the categories C_0 and $\mathcal{D}_0$ are isomorphic; in particular, the two categories are also equivalent. Moreover, recall that the categories C_0 and C, respectively $\mathcal{D}_0$ and $\mathcal{D}$, are equivalent. Then Proposition 3.8.4, (2) and (3) show that C is equivalent to $\mathcal{D}$.

Conversely, assume now that C and $\mathcal{D}$ are equivalent. As noted before, C_0 and C, respectively $\mathcal{D}_0$ and $\mathcal{D}$, are also equivalent and using again Proposition 3.8.4, (2) and (3) we obtain that C_0 and $\mathcal{D}_0$ are equivalent too. Thus, there exist two functors $F\colon C_0 \to \mathcal{D}_0$, $G\colon \mathcal{D}_0 \to C_0$ and two natural isomorphisms $\psi\colon FG \to 1_{\mathcal{D}_0}$, $\varphi\colon GF \to 1_{C_0}$. Hence, for each $C \in \mathrm{Ob}\,C_0$, we have an isomorphism $\varphi_C\colon GF(C) \to C$. Since C_0 is a skeleton of C, such an isomorphism can only be an identity morphism and therefore $GF(C) = C$. Similarly, we obtain $FG(D) = D$ for all $D \in \mathrm{Ob}\,\mathcal{D}_0$. Furthermore, if $f \in \mathrm{Hom}_{\mathcal{D}_0}(D_1, D_2) = \mathrm{Hom}_{\mathcal{D}_0}(FG(D_1), FG(D_2))$, since F is in particular fully faithful by Theorem 3.8.5, (2) there exists a unique $f' \in \mathrm{Hom}_{C_0}(G(D_1), G(D_2))$ such that $f = F(f')$. We can now define a functor $H\colon \mathcal{D}_0 \to C_0$ as follows:

---

[8] Recall that the axiom of choice is assumed to hold.

$$H(D) = G(D), \quad D \in \mathrm{Ob}\,\mathcal{D}_0, \tag{3.59}$$

$$H(f) = f', \quad f \in \mathrm{Hom}_{\mathcal{D}_0}(D_1, D_2). \tag{3.60}$$

It can be easily seen that H is a functor. Indeed, for all $D \in \mathrm{Ob}\,\mathcal{D}_0$ we have $1_D = 1_{FG(D)} = F(1_{G(D)})$, which implies that $H(1_D) = 1_{G(D)} = 1_{H(D)}$. Furthermore, consider $f \in \mathrm{Hom}_{\mathcal{D}_0}(D_1, D_2)$, $g \in \mathrm{Hom}_{\mathcal{D}_0}(D_2, D_3)$ and let $f' \in \mathrm{Hom}_{C_0}(G(D_1), G(D_2))$, $g' \in \mathrm{Hom}_{C_0}(G(D_2), G(D_3))$ be the unique morphisms such that $F(f') = f$ and $F(g') = g$. Then $g' \circ f' \in \mathrm{Hom}_{C_0}(G(D_1), G(D_3))$ is the unique morphism such that $g \circ f = F(g' \circ f')$ and we obtain

$$H(g) \circ H(f) = g' \circ f' = H(g \circ f).$$

This shows that H is a functor. The proof will be finished once we show that H is the inverse of F. Indeed, for all $C \in \mathrm{Ob}\,C_0$ and $D \in \mathrm{Ob}\,\mathcal{D}_0$ we have

$$FH(D) \overset{(3.59)}{=} FG(D) = D, \qquad HF(C) \overset{(3.59)}{=} GF(C) = C.$$

Furthermore, if $f \in \mathrm{Hom}_{\mathcal{D}_0}(D_1, D_2)$ and $f' \in \mathrm{Hom}_{C_0}(G(D_1), G(D_2))$ is the unique morphism such that $F(f') = f$, we obtain

$$F\big(H(f)\big) \overset{(3.60)}{=} F(f') = f.$$

Similarly, if we have $t \in \mathrm{Hom}_{C_0}(C_1, C_2) = \mathrm{Hom}_{C_0}(GF(C_1), GF(C_2))$, then $H\big(F(t)\big) \overset{(3.60)}{=} t$ and the proof is now finished. $\qquad\square$

In light of our previous result, loosely speaking, we can conclude that two equivalent categories might differ only by the numbers of isomorphic copies of the same object. Another important consequence is the following:

Corollary 3.8.12 *The skeleton of a category is unique up to isomorphism.*

Proof As proved in Proposition 3.8.4, (1) any category C is trivially equivalent to itself. Now any two skeletons of C are isomorphic by Proposition 3.8.11. $\qquad\square$

Categories with a small skeleton have been mentioned in passing in Example 1.2.2, (5). We discuss them here in more detail.

Definition 3.8.13 A category is called *essentially small* if its skeleton is a small category.

A useful characterization of essentially small categories is the following:

Proposition 3.8.14 *A category is essentially small if and only if it is equivalent to a small category.*

Proof Consider C to be a category equivalent to a small category $\mathcal{D}$. If C_0 and $\mathcal{D}_0$ denote the skeleton of C and $\mathcal{D}$, respectively, then in particular $\mathcal{D}_0$ is also a small category. Now Proposition 3.8.11 implies that C_0 and $\mathcal{D}_0$ are isomorphic categories and therefore C_0 is small. This shows that C is essentially small.

Conversely, if C is essentially small then its skeleton C_0 is a small category and the conclusion follows from Corollary 3.8.10.   □

Examples 3.8.15

(1) The categories **FinSet** and $_K\mathcal{M}^{fd}$ are essentially small. Indeed, a skeleton of **FinSet** is given by its full subcategory whose objects are the sets $\bar{n}$, for all $n \in \mathbb{N}$, where $\bar{n} = \begin{cases} \varnothing \text{ if } n = 0 \\ \{1, \ldots, n\} \text{ if } n \in \mathbb{N}\backslash\{0\} \end{cases}$. The latter category is obviously small.

Furthermore, Example 3.8.2 shows that the category $_K\mathcal{M}^{fd}$ is equivalent to the small category **Mat**$_K$ defined in Example 1.2.2, (19). Hence, $_K\mathcal{M}^{fd}$ is essentially small by virtue of Proposition 3.8.14.

(2) The categories **Set**, **Grp**, **Ring**, **Top** and $_K\mathcal{M}$ are not essentially small. We only prove the assertion regarding the category **Set** and leave the others to the reader. To this end, assume there exists a small skeleton C of **Set**. Given that Ob C is a set we can consider Ob $C = \{X_i \mid i \in I\}$, where I is a set and $X_i \in$ Ob **Set** for all $i \in I$. Now let $X = \mathcal{P}(\coprod_{i \in I} X_i)$ be the power set of the coproduct of the family of objects $(X_i)_{i \in I}$ in **Set**. As C is assumed to be the skeleton of **Set**, there exists some $i_0 \in I$ and an isomorphism in **Set** (i.e., a set bijection) between X and X_{i_0}. From Cantor's theorem[9] we have $|\coprod_{i \in I} X_i| < |X|$. Furthermore, as $\coprod_{i \in I} X_i$ is the union of the sets $X_i' = X_i \times \{i\}$ (see Example 2.1.5, (9)) we also have $|X_i| \leq |\coprod_{i \in I} X_i|$ for all $i \in I$. Putting all this together leads in particular to $|X_{i_0}| < |X|$ and we have reached a contradiction as X was assumed to be isomorphic to X_{i_0}. Therefore, **Set** cannot have a small skeleton, as desired.   □

We end this section with a generic example of a duality of categories. Let C and $\mathcal{D}$ be two small categories. We denote by Fun$_L(C, \mathcal{D})$ the full subcategory of Fun$(C, \mathcal{D})$ consisting of all functors from C to $\mathcal{D}$ which admit a left adjoint. Similarly, Fun$_R(C, \mathcal{D})$ denotes the full subcategory of Fun$(C, \mathcal{D})$ consisting of all functors from C to $\mathcal{D}$ which admit a right adjoint.

Theorem 3.8.16 *For all small categories C and $\mathcal{D}$, we have a duality of categories between* Fun$_R(C, \mathcal{D})$ *and* Fun$_L(\mathcal{D}, C)$.

Proof We will define an equivalence of categories

$$H: \text{Fun}_L(\mathcal{D}, C) \to \text{Fun}_R(C, \mathcal{D})^{op}$$

as follows. Given $G \in \text{Fun}_L(\mathcal{D}, C)$, we choose a functor $F: C \to \mathcal{D}$ such that $F \dashv G$ and define $H(G) = F$.

---

[9] Cantor's theorem: For any set X we have $|X| < |\mathcal{P}(X)|$ ([37, Theorem 2.21]).

Furthermore, if G_1, $G_2 \in \mathrm{Fun}_L(\mathcal{D}, C)$ with corresponding left adjoint functors F_1, $F_2: C \to \mathcal{D}$ and $\alpha: G_1 \to G_2$ is a natural transformation then we define $H(\alpha) = \overline{\alpha}$, where $\overline{\alpha}: F_2 \to F_1$ is the unique natural transformation which makes diagram (3.27) commutative, as constructed in (the proof of) Theorem 3.5.5.

We show first that H is indeed a functor. To start with, let $G: \mathcal{D} \to C$ be a functor whose chosen left adjoint we denote by F and consider α to be the identity natural transformation on G, i.e., $\alpha = \left(1_{G(D)}\right)_{D \in \mathrm{Ob}\,\mathcal{D}}$. Then, for all $C \in \mathrm{Ob}\,C$, $\overline{\alpha}_C$ is the unique morphism such that $\theta_{C,D}(f \circ \overline{\alpha}_C) = \theta_{C,D}(f)$ holds for all morphisms $f: F(C) \to D$. Obviously, this implies $\overline{\alpha}_C = 1_{F(C)}$, which shows that H respects identities.

Consider now functors $G_i: \mathcal{D} \to C$ and choose $F_i: C \to \mathcal{D}$ to be the corresponding left adjoints, $i = 1, 2, 3$. Moreover, denote by θ^i the natural isomorphism induced by the adjunction $F_i \dashv G_i$, $i = 1, 2, 3$. If $\alpha: G_1 \to G_2$ and $\beta: G_2 \to G_3$ are natural transformations and $H(\alpha) = \overline{\alpha}$, $H(\beta) = \overline{\beta}$ we aim to show that $H(\beta \circ \alpha) = \overline{\beta} \circ^{op} \overline{\alpha}$. This comes down to showing that for each $C \in \mathrm{Ob}\,C$, the morphism $\left(\overline{\beta} \circ^{op} \overline{\alpha}\right)_C = \overline{\alpha}_C \circ \overline{\beta}_C$ is the unique one such that the following holds:

$$\theta^3_{C,D}(h \circ \overline{\alpha}_C \circ \overline{\beta}_C) = \beta_D \circ \alpha_D \circ \theta^1_{C,D}(h) \qquad (3.61)$$

for all $h \in \mathrm{Hom}_{\mathcal{D}}(F_1(C), D)$. To this end, recall that $\overline{\alpha}_C$ and $\overline{\beta}_C$ are the unique morphisms such that

$$\theta^2_{C,D}(f \circ \overline{\alpha}_C) = \alpha_D \circ \theta^1_{C,D}(f), \qquad (3.62)$$

$$\theta^3_{C,D}(g \circ \overline{\beta}_C) = \beta_D \circ \theta^2_{C,D}(g) \qquad (3.63)$$

for all $f \in \mathrm{Hom}_{\mathcal{D}}(F_1(C), D)$, $g \in \mathrm{Hom}_{\mathcal{D}}(F_2(C), D)$. Putting all the above together yields

$$\beta_D \circ \alpha_D \circ \underline{\theta^1_{C,D}(h)} \overset{(3.62)}{=} \beta_D \circ \underline{\theta^2_{C,D}(h \circ \overline{\alpha}_C)} \overset{(3.63)}{=} \theta^3_{C,D}(h \circ \overline{\alpha}_C \circ \overline{\beta}_C)$$

for all $h \in \mathrm{Hom}_{\mathcal{D}}(F_1(C), D)$. Therefore, H is indeed a functor.

The proof will be finished once we show that H is essentially surjective and fully faithful. To this end, let $F \in \mathrm{Ob}\left(\mathrm{Fun}_R(C, \mathcal{D})^{op}\right)$ and consider a right adjoint $G: \mathcal{D} \to C$ of F. Then $H(G) = F'$ for some $F': C \to \mathcal{D}$ such that $F' \dashv G$. By Theorem 3.6.4 we have a natural isomorphism $F \simeq F'$. Therefore $F \simeq F' = H(G)$, which shows that H is essentially surjective.

Next we show that H is fully faithful. To this end, consider G_1, $G_2 \in \mathrm{Ob}\left(\mathrm{Fun}_L(\mathcal{D}, C)\right)$ and the map

$$\mathcal{H}_{G_1, G_2}: \mathrm{Hom}_{\mathrm{Fun}_L(\mathcal{D}, C)}(G_1, G_2) \to \mathrm{Hom}_{\mathrm{Fun}_R(C, \mathcal{D})^{op}}(F_1, F_2)$$

defined by

$$\mathcal{H}_{G_1, G_2}(\gamma) = H(\gamma) = \overline{\gamma} \text{ for all natural transformations } \gamma \colon G_1 \to G_2,$$

where $F_i \colon C \to \mathcal{D}$ is the left adjoint of G_i and denote by θ^i the natural isomorphism corresponding to the adjunction $F_i \dashv G_i$, $i = 1, 2$.

Assume α, $\beta \colon G_1 \to G_2$ are natural transformations such that $\mathcal{H}_{G_1, G_2}(\alpha) = \mathcal{H}_{G_1, G_2}(\beta)$, i.e., $\overline{\alpha} = \overline{\beta}$. This implies $\theta^2_{C, D}(f \circ \overline{\alpha}_C) = \theta^2_{C, D}(f \circ \overline{\beta}_C)$ and consequently we have

$$\alpha_D \circ \theta^1_{C, D}(f) = \beta_D \circ \theta^1_{C, D}(f), \tag{3.64}$$

for all $f \in \mathrm{Hom}_{\mathcal{D}}(F_1(C), D)$. Considering

$$C = G_1(D) \text{ and } f = \left(\theta^1_{G_1(D), D}\right)^{-1}(1_{G(D)})$$

in (3.64) yields $\alpha_D = \beta_D$ for all $D \in \mathrm{Ob}\,\mathcal{D}$. Hence $\alpha = \beta$ and H is faithful.

We are left to show that H is full. Consider a natural transformation $\mu \colon F_2 \to F_1$, i.e., $\mu \in \mathrm{Hom}_{\mathrm{Fun}_R(C, \mathcal{D})^{op}}(F_1, F_2)$. For all $D \in \mathrm{Ob}\,\mathcal{D}$ define

$$\alpha_D := \theta^2_{G_1(D), D}(\varepsilon^1_D \circ \mu_{G_1(D)}) \in \mathrm{Hom}_{\mathcal{D}}(G_1(D), G_2(D)),$$

where ε^1 denotes the counit of the adjunction $F_1 \dashv G_1$. We show first that the morphisms α_D, $D \in \mathrm{Ob}\,\mathcal{D}$, form a natural transformation $\alpha \colon G_1 \to G_2$, i.e., for all $r \in \mathrm{Hom}_{\mathcal{D}}(D, D')$ the following diagram is commutative:

$$
\begin{array}{ccc}
G_1(D) & \xrightarrow{\;\;\alpha_D\;\;} & G_2(D) \\
{\scriptstyle G_1(r)}\big\downarrow & & \big\downarrow{\scriptstyle G_2(r)} \\
G_1(D') & \xrightarrow[\;\;\alpha_{D'}\;\;]{} & G_2(D')
\end{array}
\qquad \text{i.e., } \alpha_{D'} \circ G_1(r) = G_2(r) \circ \alpha_D.
$$

$$\tag{3.65}$$

To start with, note that $\varepsilon^1 \circ \mu_{G_1} \colon F_2 G_1 \to 1_{\mathcal{D}}$ is a natural transformation and therefore the following diagram is commutative:

$$
\begin{array}{ccc}
F_2 G_1(D) & \xrightarrow{\;\;\varepsilon^1_D \circ \mu_{G_1(D)}\;\;} & D \\
{\scriptstyle F_2 G_1(r)}\big\downarrow & & \big\downarrow{\scriptstyle r} \\
F_2 G_1(D') & \xrightarrow[\;\;\varepsilon^1_{D'} \circ \mu_{G_1(D')}\;\;]{} & D'
\end{array}
\qquad \text{i.e., } r \circ \varepsilon^1_D \circ \mu_{G_1(D)} = \varepsilon^1_{D'} \circ \mu_{G_1(D')} \circ F_2 G_1(r).
$$

$$\tag{3.66}$$

Therefore, we have

$$G_2(r) \circ \alpha_D = G_2(r) \circ \theta^2_{G_1(D), D}\left(\varepsilon^1_D \circ \mu_{G_1(D)}\right)$$

$$\overset{(3.2)}{=} \theta^2_{G_1(D), D'}\left(r \circ \varepsilon^1_D \circ \mu_{G_1(D)}\right)$$

$$\overset{(3.66)}{=} \theta^2_{G_1(D), D'}\left(\varepsilon^1_{D'} \circ \mu_{G_1(D')} \circ F_2 G_1(r)\right)$$

$$\overset{(3.1)}{=} \theta^2_{G_1(D'), D'}\left(\varepsilon^1_{D'} \circ \mu_{G_1(D')}\right) \circ G_1(r)$$

$$= \alpha_{D'} \circ G_1(r),$$

which shows that (3.65) indeed holds and hence α is a natural transformation. The proof will be finished if we show that $\mathcal{H}_{G_1, G_2}(\alpha) = \mu$ or, equivalently, that the following holds for all $f \in \mathrm{Hom}_{\mathcal{D}}(F_1(C), D)$:

$$\theta^2_{C, D}(f \circ \mu_C) = \alpha_D \circ \theta^1_{C, D}(f).$$

To start with, recall that $\mu \colon F_2 \to F_1$ is a natural transformation and therefore the following diagram is commutative:

$$
\begin{array}{ccc}
F_2(C) & \xrightarrow{\;F_2\left(\theta^1_{C, D}(f)\right)\;} & F_2 G_1(D) \\[4pt]
\Big\downarrow{\mu_C} & & \Big\downarrow{\mu_{G_1(D)}} \\[4pt]
F_1(C) & \xrightarrow[\;F_1\left(\theta^1_{C, D}(f)\right)\;]{} & F_1 G_1(D)
\end{array}
$$

(3.67)

i.e., $F_1\left(\theta^1_{C, D}(f)\right) \circ \mu_C = \mu_{G_1(D)} \circ F_2\left(\theta^1_{C, D}(f)\right)$.

Then, by the way we defined α, we have

$$\alpha_D \circ \theta^1_{C, D}(f) = \theta^2_{G_1(D), D}\left(\varepsilon^1_D \circ \mu_{G_1(D)}\right) \circ \theta^1_{C, D}(f)$$

$$\overset{(3.1)}{=} \theta^2_{C, D}\left(\varepsilon^1_D \circ \mu_{G_1(D)} \circ F_2\left(\theta^1_{C, D}(f)\right)\right)$$

$$\overset{(3.67)}{=} \theta^2_{C, D}\left(\varepsilon^1_D \circ F_1\left(\theta^1_{C, D}(f)\right) \circ \mu_C\right)$$

$$\overset{(3.25)}{=} \theta^2_{C, D}(f \circ \mu_C),$$

as desired. This concludes the proof. $\qquad\square$

Examples of duality theorems abound in the mathematical landscape and are often used to build bridges between different fields. We only mention here some of the most important ones: the category of compact topological abelian groups is dual to the category of abelian groups (*Pontryagin duality*); the category of commutative unital C^*-algebras is dual to the category of compact Hausdorff topological spaces (*Gelfand-Naimark duality*); the category of compact and totally disconnected topological spaces[10] is dual to the category of Boolean algebras (*Stone duality*). For further details we refer the reader to [14, 43].

3.9   Localization

The idea of formally adjoining inverses in a systematic way, called *localization*, exists for many algebraic structures such as rings or modules. A similar construction can be performed in the general setting of category theory. Indeed, consider S to be a class of morphisms in a category C. The purpose of localization as first introduced in [26] is to construct a new category C_S in which all morphisms in S became *invertible*, while approximating the original category as closely as possible. The precise definition is the following:

Definition 3.9.1 A *localization of a category* C (or *category of fractions* as referred to in ([8, Section 5.2]) by a class of morphisms S of C is a category C_S together with a functor $F: C \to C_S$ such that

(1) for any $s \in S$, $F(s)$ is an isomorphism in C_S;
(2) if $G: C \to \mathcal{D}$ is a functor such that for all $s \in S$, $G(s)$ is an isomorphism in $\mathcal{D}$, there exists a unique functor $H: C_S \to \mathcal{D}$ such that the following diagram is commutative:

$$C \xrightarrow{\ F\ } C_S \qquad \text{i.e.,} \quad H \circ F = G.$$

$$G \searrow \quad \downarrow H$$

$$\mathcal{D} \tag{3.68}$$

Theorem 3.9.2 *Let C be a category. Then there exists a localization of C by any set of morphisms S of C.*

Proof In order to construct the localization of C by the set S we start by defining an oriented graph Γ as follows:

- the vertices of Γ are the objects of C;

---

[10] A topological space that is compact and totally disconnected is called a *Stone space*.

- the edges of Γ are the morphisms of C (any morphism $f \in \mathrm{Hom}_C(X, Y)$ is seen as an oriented edge $X \xrightarrow{\ f\ } Y$) together with the set $\{x_s \mid s \in S\}$, where x_s is an edge having the same vertices as s but the opposite orientation (i.e., if $s \in \mathrm{Hom}_C(X, Y)$ then $Y \xrightarrow{\ x_s\ } X$).

Two paths in the above graph will be called *equivalent* if one can be transformed into the other by applying the following elementary operations a finite number of times:

- if $f \in \mathrm{Hom}_C(X, Y)$ and $g \in \mathrm{Hom}_C(Y, Z)$ then the path $X \xrightarrow{\ f\ } Y \xrightarrow{\ g\ } Z$ can be replaced by the composition path $X \xrightarrow{\ g \circ f\ } Z$;
- if $s \in S$, $s \in \mathrm{Hom}_C(X, Y)$ then the path $X \xrightarrow{\ s\ } Y \xrightarrow{\ x_s\ } X$ can be replaced by the path $X \xrightarrow{\ 1_X\ } X$; similarly, the path $Y \xrightarrow{\ x_s\ } X \xrightarrow{\ s\ } Y$ can be replaced by the path $Y \xrightarrow{\ 1_Y\ } Y$.

It is straightforward to see that this is an equivalence relation on the class of paths of Γ. We denote by $\widehat{\gamma}$ the equivalence class of the path γ. The localization category C_S is defined as follows:

$\mathrm{Ob}\, C_S = \mathrm{Ob}\, C$;
$\mathrm{Hom}_{C_S}(X, Y) = \{\widehat{\gamma} \mid \gamma \text{ is a path in } \Gamma \text{ from } X \text{ to } Y\}$,[11] for all $X, Y \in \mathrm{Ob}\, C$,

with the composition of morphisms in C_S induced by the concatenation of paths and the identity maps given by the trivial paths. The functor $F: C \to C_S$ is defined as follows:

$F(X) = X$, for all $X \in \mathrm{Ob}\, C$;
$F(f) = \widehat{f}$, for all $f \in \mathrm{Hom}_C(X, Y)$.

Note that if $s \in S$, $s \in \mathrm{Hom}_C(X, Y)$, then $F(s) = \widehat{s}$ has an inverse in C_S, namely $\widehat{x_s}$, where $Y \xrightarrow{\ x_s\ } X$. We are left to show that the pair (C_S, F) satisfies the second condition in Definition 3.9.1 as well. To this end, let $\mathcal{D}$ be a category and $G: C \to \mathcal{D}$ a functor such that $G(s)$ is an isomorphism for any $s \in S$. Consider the functor $H: C_S \to \mathcal{D}$ defined as follows:

$H(X) = G(X)$, for all $X \in \mathrm{Ob}\, C_s = \mathrm{Ob}\, C$;
$H(\widehat{f}) = G(f)$, for all $f \in \mathrm{Hom}_C(X, Y)$;
$H\left(W \xrightarrow{\ \widehat{x_s}\ } Z\right) = G(s)^{-1}$, for all $s \in S, s \in \mathrm{Hom}_C(Z, W)$.

The way we defined the functor above ensures the commutativity of diagram (3.68) as well as the uniqueness of H with this property. Indeed, if a functor H makes

---

[11] This is obviously a set as a consequence of S being a set.

diagram (3.68) commutative, then we have $H(X) = G(X)$ and $H(\widehat{f}) = G(f)$, for all $X \in \mathrm{Ob}\, C_s = \mathrm{Ob}\, C$ and $f \in \mathrm{Hom}_C(X, Y)$; furthermore, in order for H to be a functor and to respect compositions and identities, it should satisfy $H(\widehat{x_s}) = G(s)^{-1}$, for all $s \in S$.

We are left to prove that H is well-defined. To this end, consider two paths u and v in Γ such that $\widehat{u} = \widehat{v}$. Since the paths u and v are equivalent, we can turn u into v after a finite number of elementary operations. Thus, it suffice to prove that by applying H to each of these elementary operations we obtain equalities in $\mathcal{D}$. Indeed, whenever $X \xrightarrow{\widehat{f}} Y \xrightarrow{\widehat{g}} Z \;=\; X \xrightarrow{\widehat{g \circ f}} Z$ in C_S we obviously have

$$H(\widehat{g \circ f}) = G(g \circ f) = G(g) \circ G(f) = H(\widehat{g}) \circ H(\widehat{f}) = H(\widehat{g} \circ \widehat{f}).$$

Analogously, whenever $X \xrightarrow{\widehat{s}} Y \xrightarrow{\widehat{x_s}} X \;=\; X \xrightarrow{\widehat{1_X}} X$ in C_S it follows that we have

$$H(\widehat{x_s} \circ \widehat{s}) = H(\widehat{x_s}) \circ H(\widehat{s}) = G(s)^{-1} \circ G(s) = 1_{G(X)} = G(1_X) = H(\widehat{1_X}).$$

Therefore H is well-defined and the proof is now finished. $\qquad\qquad\qquad\square$

Proposition 3.9.3 *When it exists, the localization of a category C by a class of morphisms S of C is unique up to isomorphism.*

Proof Suppose (C_S, F) and $(\overline{C_S}, \overline{F})$ are two localizations of C by S. Thus, there exists a unique functor $G: C_S \to \overline{C_S}$ such that

$$G \circ F = \overline{F}. \tag{3.69}$$

Similarly, as $(\overline{C_S}, \overline{F})$ is also a localization of C by S, there exists a unique functor $G': \overline{C_S} \to C_S$ such that

$$G' \circ \overline{F} = F. \tag{3.70}$$

By putting all this together we obtain

$$F \overset{(3.70)}{=} G' \circ \overline{F} \overset{(3.69)}{=} G' \circ G \circ F = (G' \circ G) \circ F. \tag{3.71}$$

Applying Definition 3.9.1 to the pair (C_S, F), seen both as a localization and as the other pair, yields a unique functor $H: C_S \to C_S$ such that $H \circ F = F$. By the uniqueness of H we must have $H = 1_{C_S}$. Moreover, since by (3.71) the functor

$G' \circ G$ makes the same diagram commutative, we obtain $G' \circ G = 1_{C_S}$.

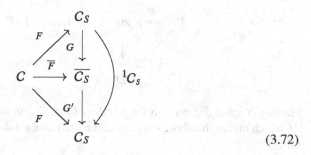

$$(3.72)$$

Similarly one can prove that $G \circ G' = 1_{\overline{C_S}}$ and therefore the categories C_S and $\overline{C_S}$ are isomorphic, as desired. The proof is now finished. □

One of the situations when the localization of a category can be described, up to equivalence of categories, even without assuming the localizing class of morphisms to be a set, is that of reflective subcategories.

Theorem 3.9.4 *Let* $I\colon \mathcal{A} \to \mathcal{B}$ *be a reflective subcategory inclusion with reflector* $R\colon \mathcal{B} \to \mathcal{A}$ *and denote by* S *the class of all morphisms* s *of* $\mathcal{B}$ *such that* $R(s)$ *is an isomorphism in* $\mathcal{A}$. *Then the localization of* $\mathcal{B}$ *by* S *is equivalent to* $\mathcal{A}$.

Proof Let $\eta\colon 1_{\mathcal{B}} \to IR$ and $\varepsilon\colon RI \to 1_{\mathcal{A}}$ be the unit and respectively the counit of the adjunction $R \dashv I$. To start with, recall that by Lemma 3.6.6, (3) the counit $\varepsilon\colon RI \to 1_{\mathcal{A}}$ is a natural isomorphism. Moreover, for all $B \in \mathrm{Ob}\mathcal{B}$ we have $1_{R(B)} \overset{(3.15)}{=} \varepsilon_{R(B)} \circ R(\eta_B)$ and given that $\varepsilon_{R(B)}$ is an isomorphism it follows that $R(\eta_B) \in \mathrm{Hom}_{\mathcal{A}}\big(R(B),\, RIR(B)\big)$ is an isomorphism as well. Therefore, $\eta_B \in S$ for all $B \in \mathrm{Ob}\,\mathcal{B}$.

Define a category $\mathcal{B}_S$ as follows:

$\mathrm{Ob}\,\mathcal{B}_S = \mathrm{Ob}\,\mathcal{B}$;
$\mathrm{Hom}_{\mathcal{B}_S}(B,\, B') = \mathrm{Hom}_{\mathcal{A}}(R(B),\, R(B'))$ for all $B,\, B' \in \mathrm{Ob}\,\mathcal{B}_S$,

with the composition of morphisms and identities given by those of $\mathcal{A}$.

First we prove that $(\mathcal{B}_S, F)$ is the localization of $\mathcal{B}$ with respect to S, where $F\colon \mathcal{B} \to \mathcal{B}_S$ is the functor defined as follows:

$F(B) = B$, for all $B \in \mathrm{Ob}\,\mathcal{B}$;
$F(f) = R(f)$, for all $f \in \mathrm{Hom}_{\mathcal{B}}(B,\, B')$.

Recall that S is the class of all morphisms s of $\mathcal{B}$ such that $R(s)$ is an isomorphism and therefore $F(s)$ is obviously an isomorphism for any $s \in S$.

Consider now another functor $G\colon \mathcal{B} \to \mathcal{D}$ such that $G(s)$ is an isomorphism for all $s \in S$. We need to find a functor $H\colon \mathcal{B}_S \to \mathcal{D}$ which makes the following diagram commutative:

$$
\begin{array}{ccc}
\mathcal{B} & \xrightarrow{\;F\;} & \mathcal{B}_S \\
 & \diagdown{G} \quad & \downarrow{H} \\
 & & \mathcal{D}
\end{array}
\qquad \text{i.e.,}\quad H \circ F = G.
$$

$$(3.73)$$

Having in mind that $\eta_B \in S$ for any $B \in \mathrm{Ob}\mathcal{B}$, it can be easily seen that a functor H which makes the above diagram commute has the following property for all $B \in \mathrm{Ob}\mathcal{B}$:

$$
H(\varepsilon_{R(B)}) \overset{(3.15)}{=} H\big(R(\eta_B)^{-1}\big) = \Big(H\big(R(\eta_B)\big)\Big)^{-1}
$$

$$
= \Big(H\big(F(\eta_B)\big)\Big)^{-1} \overset{(3.73)}{=} G(\eta_B)^{-1}. \qquad (3.74)
$$

Furthermore, for any morphism $f \in \mathrm{Hom}_{\mathcal{B}_S}(B,\,B') = \mathrm{Hom}_{\mathcal{A}}(R(B),\,R(B'))$, the naturality of $\varepsilon\colon RI \to 1_{\mathcal{A}}$ renders the following diagram commutative:

$$
\begin{array}{ccc}
RIR(B) & \xrightarrow{\;\varepsilon_{R(B)}\;} & R(B) \\
\downarrow{RI(f)} & & \downarrow{f} \\
RIR(B') & \xrightarrow{\;\varepsilon_{R(B')}\;} & R(B')
\end{array}
\qquad \text{i.e.,}\quad f \circ \varepsilon_{R(B)} = \varepsilon_{R(B')} \circ RI(f).
$$

$$(3.75)$$

Therefore, for any $f \in \mathrm{Hom}_{\mathcal{B}_S}(B,\,B') = \mathrm{Hom}_{\mathcal{A}}(R(B),\,R(B'))$ we have

$$
H(f) \overset{(3.15)}{=} H\big(\underline{f \circ \varepsilon_{R(B)}} \circ R(\eta_B)\big)
$$

$$
\overset{(3.75)}{=} H\big(\varepsilon_{R(B')} \circ RI(f) \circ R(\eta_B)\big)
$$

$$
= H\big(\underline{\varepsilon_{R(B')}}\big) \circ HRI(f) \circ HR(\eta_B)
$$

$$
\overset{(3.74)}{=} G(\eta_{B'})^{-1} \circ \underline{HFI}(f) \circ \underline{HF}(\eta_B)
$$

$$
\overset{(3.73)}{=} G(\eta_{B'})^{-1} \circ GI(f) \circ G(\eta_B).
$$

We define the functor $H\colon \mathcal{B}_S \to \mathcal{D}$ as follows:

$H(B) = G(B)$, for all $B \in \mathrm{Ob}\,\mathcal{B}_S$;
$H(f) = G(\eta_{B'})^{-1} \circ GI(f) \circ G(\eta_B)$, for all $f \in \mathrm{Hom}_{\mathcal{B}_S}(B,\,B')$.

The above discussion proves that H is the unique functor which might render diagram (3.73) commutative. We are left to prove that indeed H makes diagram (3.73) commute. To this end we will use the naturality of η, i.e., the commutativity of the above diagram for any $g \in \mathrm{Hom}_{\mathcal{B}}(B, B')$:

$$
\begin{array}{ccc}
B & \xrightarrow{\;\eta_B\;} & IR(B) \\
{\scriptstyle g}\big\downarrow & & \big\downarrow{\scriptstyle IR(g)} \\
B' & \xrightarrow[\;\eta_{B'}\;]{} & IR(B')
\end{array}
\qquad \text{i.e.,} \quad IR(g) \circ \eta_B = \eta_{B'} \circ g.
$$

$$(3.76)$$

Obviously, for any $B \in \mathrm{Ob}\,\mathcal{B}_S$ we have $H \circ F(B) = H(B) = G(B)$. Moreover, for any $g \in \mathrm{Hom}_{\mathcal{B}}(B, B')$ we have

$$
H \circ F(g) = H(R(g)) = G(\eta_{B'})^{-1} \circ GIR(g) \circ G(\eta_B)
$$

$$
= G(\eta_{B'})^{-1} \circ G(\underline{IR(g) \circ \eta_B})
$$

$$
\overset{(3.76)}{=} G(\eta_{B'})^{-1} \circ G(\eta_{B'} \circ g) = G(g).
$$

Next we show that the category $\mathcal{B}_S$ is equivalent to $\mathcal{A}$. Indeed, consider the functor $T \colon \mathcal{A} \to \mathcal{B}_S$ defined as follows:

$T(A) = I(A)$, for all $A \in \mathrm{Ob}\,\mathcal{A}$;
$T(f) = RI(f)$, for all $f \in \mathrm{Hom}_{\mathcal{A}}(A, A')$.

T is well-defined as for all $f \in \mathrm{Hom}_{\mathcal{A}}(A, A')$, we have

$$
RI(f) \in \mathrm{Hom}_{\mathcal{A}}\big(RI(A), RI(A')\big) = \mathrm{Hom}_{\mathcal{B}_S}\big(I(A), I(A')\big).
$$

Furthermore, T is fully faithful as RI is naturally isomorphic to $1_{\mathcal{A}}$ via ε. Indeed, let $h_1, h_2 \in \mathrm{Hom}_{\mathcal{A}}(A, A')$ such that $RI(h_1) = RI(h_2)$. The naturality of ε renders the following diagrams commutative for $i = 1, 2$:

$$
\begin{array}{ccc}
RI(A) & \xrightarrow{\;\varepsilon_A\;} & A \\
{\scriptstyle RI(h_i)}\big\downarrow & & \big\downarrow{\scriptstyle h_i} \\
RI(A') & \xrightarrow[\;\varepsilon_{A'}\;]{} & A'
\end{array}
\qquad \text{i.e.,} \quad h_i \circ \varepsilon_A = \varepsilon_{A'} \circ RI(h_i).
$$

$$(3.77)$$

Hence we obtain $h_1 \circ \varepsilon_A = h_2 \circ \varepsilon_A$ and since ε_A is an isomorphism we get $h_1 = h_2$ as desired. This shows that T is faithful.

Consider now $A, A' \in \mathrm{Ob}\,\mathcal{A}$, $v \in \mathrm{Hom}_{\mathcal{A}}(RI(A), RI(A'))$ and define

$$u = \varepsilon_{A'} \circ v \circ \varepsilon_A^{-1} \in \mathrm{Hom}_C(A, A'). \qquad (3.78)$$

We will prove that $RI(u) = v$. Indeed, using again the naturality of ε we obtain

$$\varepsilon_{A'} \circ RI(u) \circ \varepsilon_A^{-1} \overset{(3.77)}{=} u \overset{(3.78)}{=} \varepsilon_{A'} \circ v \circ \varepsilon_A^{-1}.$$

Since $\varepsilon_{A'}$ and ε_A are isomorphisms we get $RI(u) = v$ and we have proved that T is full.

Moreover, for any $B \in \mathrm{Ob}\,\mathcal{B}_S$ we have an isomorphism

$$R(\eta_B) \in \mathrm{Hom}_{\mathcal{A}}(R(B), RIR(B)) = \mathrm{Hom}_{\mathcal{B}_S}(B, IR(B)) = \mathrm{Hom}_{\mathcal{B}_S}(B, TR(B)),$$

and this shows that T is essentially surjective as well. Therefore, by Theorem 3.8.5, (2) T is an equivalence of categories and the proof is now finished. $\square$

Our next example connects ring and module localizations with the categorical notion introduced in Definition 3.9.1. We will show that the category of modules over the localized ring $S^{-1}R$ is equivalent to a localization, in the categorical sense, of the category of modules over R.

Example 3.9.5 Let R be a commutative ring with unity, S a multiplicative subset of R, and $(S^{-1}R, j)$ the corresponding localization ring. If $M \in \mathrm{Ob}\,_R\mathcal{M}$, we denote by $(S^{-1}M, \varphi_M)$ the corresponding localization module with respect to S, where $S^{-1}M \in \mathrm{Ob}_{S^{-1}R}\mathcal{M}$ and $\varphi_M : M \to S^{-1}M$ is the R-module homomorphism defined by $\varphi_M(m) = \frac{m}{1}$, for all $m \in M$. Throughout, $_R\mathcal{M}^{S-\mathrm{aut}}$ stands for the category of left R-modules on which S acts as an automorphism (see Example 1.2.2, (14)).

Consider now the inclusion functor $I : {_R\mathcal{M}^{S-\mathrm{aut}}} \to {_R\mathcal{M}}$ and the functor $L : {_R\mathcal{M}} \to {_R\mathcal{M}^{S-\mathrm{aut}}}$ defined as follows for all R-modules M and $f \in \mathrm{Hom}_{_R\mathcal{M}}(M, N)$:

$$L(M) = S^{-1}M, \qquad L(f) = \widetilde{f},$$

$$\widetilde{f} : S^{-1}M \to S^{-1}N, \quad \widetilde{f}\left(\frac{x}{s}\right) = \frac{f(x)}{s}, \quad \text{for all } x \in M, \; s \in S.$$

Note that we see $S^{-1}M$ as an R-module via j and by [3, Proposition 12.1], the multiplication map $\mu_s : S^{-1}M \to S^{-1}M$ is bijective for all $s \in S$ and therefore L is well-defined.

We will show that L is left adjoint to the inclusion functor I. To this end, let $\varphi : 1_{_R\mathcal{M}} \to IL$ be the natural transformation defined for all R-modules M by the R-module homomorphism $\varphi_M : M \to S^{-1}M$ associated with the localization

$S^{-1}M$. Indeed, if $f \in \mathrm{Hom}_{R\mathcal{M}}(M, N)$ and $m \in M$ we have

$$(\tilde{f} \circ \varphi_M)(m) = \tilde{f}\Big(\frac{m}{1}\Big) = \frac{f(m)}{1} = \varphi_N(f(m)) = (\varphi_N \circ f)(m),$$

which shows that φ is a natural transformation, as claimed.

Consider $u \in \mathrm{Hom}_{R\mathcal{M}}(M, I(N))$, with $N \in \mathrm{Ob}\, _R\mathcal{M}^{S-\mathrm{aut}}$. Now let $\bar{u}\colon S^{-1}M \to N$ be defined for all $x \in M$ and $s \in S$ by $\bar{u}\big(\frac{x}{s}\big) = y$, where y is the unique element of N such that $sy = u(x)$; note that since the multiplication by s is a bijection on N, we have a unique such y. It is straightforward to see that $\bar{u}$ is a well-defined R-module homomorphism. Furthermore, $\bar{u}\colon S^{-1}M \to N$ is the unique R-module homomorphism such that $\bar{u} \circ \varphi_M = u$. As both $S^{-1}M$ and N are objects in $_R\mathcal{M}^{S-\mathrm{aut}}$, which is a full subcategory of $_R\mathcal{M}$, we obtain that $\bar{u}$ is a morphism in $_R\mathcal{M}^{S-\mathrm{aut}}$ as well. To summarize, we have a unique morphism $\bar{u}\colon S^{-1}M \to N$ in $_R\mathcal{M}^{S-\mathrm{aut}}$ such that $I(\bar{u}) \circ \varphi_M = u$ and Theorem 3.6.1, (2) shows that L is left adjoint to the inclusion functor I. Therefore, $_R\mathcal{M}^{S-\mathrm{aut}}$ is a reflective subcategory of $_R\mathcal{M}$. Now Theorem 3.9.4 implies that $_R\mathcal{M}^{S-\mathrm{aut}}$ is equivalent to the localization (in the sense of Definition 3.9.1) of the category $_R\mathcal{M}$ with respect to the family of morphisms f in $_R\mathcal{M}$ for which $L(f)$ is an isomorphism. Furthermore, recall from Example 3.8.6 that $_R\mathcal{M}^{S-\mathrm{aut}}$ is also equivalent to $_{S^{-1}R}\mathcal{M}$. We can now conclude by Proposition 3.8.4, (3) that $_{S^{-1}R}\mathcal{M}$ is equivalent to a localization, in the sense of category theory, of the category $_R\mathcal{M}$. □

3.10   (Co)limits as Adjoint Functors

We start by recalling from Theorem 2.2.16 (resp. Theorem 2.2.17) that taking (co)limits yields a functor. It turns out that in certain conditions this limit (resp. colimit) functor has a left (resp. right) adjoint, namely the diagonal functor defined in Proposition 1.9.8.

Theorem 3.10.1 *Let I be a small category and C an arbitrary category.*

(1) The diagonal functor $\Delta\colon C \to \mathrm{Fun}(I, C)$ has a right adjoint if and only if C is complete. In this case, the right adjoint is the limit functor $\mathrm{lim}\colon \mathrm{Fun}(I, C) \to C$.

(2) The diagonal functor $\Delta\colon C \to \mathrm{Fun}(I, C)$ has a left adjoint if and only if C is cocomplete. In this case, the left adjoint is the colimit functor $\mathrm{colim}\colon \mathrm{Fun}(I, C) \to C$.

Proof

(1) Assume first that any functor $F\colon I \to C$, where I is a small category, has a limit. We will define a bijective map $\theta\colon \mathrm{Hom}_{\mathrm{Fun}(I,C)}(\Delta, -) \to \mathrm{Hom}_C(-, \mathrm{lim})$, natural in both variables. To this end, for any $X \in \mathrm{Ob}\,C$,

$F \in \mathrm{Ob}\big(\mathrm{Fun}(I,\,C)\big)$ and any natural transformation $\alpha \colon \Delta(X) \to F$, we define $\theta_{X,\,F}(\alpha) = f$, where $f \colon X \to \lim F$ is the unique morphism in C which makes the following diagram commutative for all $i \in \mathrm{Ob}\,I$:

$$\lim F \xrightarrow{\;p_i\;} F(i) \qquad \text{i.e.,} \quad p_i \circ f = \alpha_i$$

with $f \colon X \to \lim F$ and α_i. (3.79)

and $\big(\lim F,\, (p_i \colon \lim F \to F(i))_{i \in \mathrm{Ob}\,I}\big)$ denotes the limit of F.

First we prove that each map $\theta_{X,\,F}$ is bijective. Indeed, consider two natural transformations $\alpha, \beta \in \mathrm{Hom}_{\mathrm{Fun}(I,\,C)}(\Delta(X),\,F)$ such that $\theta_{X,\,F}(\alpha) = \theta_{X,\,F}(\beta) = f$. This implies that for all $i \in \mathrm{Ob}\,I$ we have $p_i \circ f = \alpha_i$ and $p_i \circ f = \beta_i$. Hence $\alpha_i = \beta_i$ for all $i \in \mathrm{Ob}\,I$, which implies that the two natural transformations α and β coincide. This shows that $\theta_{X,\,F}$ is injective.

Furthermore, consider $f \in \mathrm{Hom}_C(X,\,\lim F)$ and for all $i \in \mathrm{Ob}\,I$ define $\alpha_i \in \mathrm{Hom}_C(X,\,F(i))$ by $\alpha_i = p_i \circ f$. We will show that the family of morphisms $\big(\alpha_i \colon \Delta(X)(i) \to F(i)\big)_{i \in \mathrm{Ob}\,I}$ form a natural transformation $\alpha \colon \Delta(X) \to F$. To this end, let $u \in \mathrm{Hom}_I(i,\,j)$; we will show that the following diagram is commutative:

$$\begin{array}{ccc} \Delta(X)(i) & \xrightarrow{\;\alpha_i\;} & F(i) \\ {\scriptstyle 1_X}\big\downarrow & & \big\downarrow{\scriptstyle F(u)} \\ \Delta(X)(j) & \xrightarrow{\;\alpha_j\;} & F(j) \end{array} \qquad \text{i.e.,} \quad F(u) \circ \alpha_i = \alpha_j.$$

Indeed, recall that $\big(\lim F,\, (p_i \colon \lim F \to F(i))_{i \in \mathrm{Ob}\,I}\big)$ is in particular a cone on F and therefore we have $F(u) \circ p_i = p_j$. This yields $F(u) \circ \alpha_i = F(u) \circ p_i \circ f = p_j \circ f = \alpha_j$, which shows that the above diagram is indeed commutative.

Next we show that θ is natural in both variables. First, consider $f \in \mathrm{Hom}_C(X',\,X)$. We will prove the commutativity of the following diagram, which ensures the naturality in the first variable:

$$\begin{array}{ccc} \mathrm{Hom}_{\mathrm{Fun}(I,\,C)}(\Delta(X),\,F) & \xrightarrow{\;\theta_{X,\,F}\;} & \mathrm{Hom}_C(X,\,\lim F) \\ {\scriptstyle \mathrm{Hom}_{\mathrm{Fun}(I,\,C)}(\Delta(f),\,F)}\big\downarrow & & \big\downarrow{\scriptstyle \mathrm{Hom}_C(f,\,\lim F)} \\ \mathrm{Hom}_{\mathrm{Fun}(I,\,C)}(\Delta(X'),\,F) & \xrightarrow{\;\theta_{X',\,F}\;} & \mathrm{Hom}_C(X',\,\lim F) \end{array}$$

$$\text{i.e.,} \quad \mathrm{Hom}_C(f,\,\lim F) \circ \theta_{X,\,F} = \theta_{X',\,F} \circ \mathrm{Hom}_{\mathrm{Fun}(I,\,C)}(\Delta(f),\,F). \tag{3.80}$$

To this end, let $\alpha \in \mathrm{Hom}_{\mathrm{Fun}(I,\,C)}(\Delta(X),\,F)$, i.e., $\alpha\colon \Delta(X) \to F$ is a natural transformation. We obtain

$$\mathrm{Hom}_C(f,\,\lim F) \circ \theta_{X,\,F}(\alpha) = t \circ f,$$

$$\theta_{X',\,F} \circ \mathrm{Hom}_{\mathrm{Fun}(I,\,C)}(\Delta(f),\,F)(\alpha) = \theta_{X',\,F}\big(\alpha \circ \Delta(f)\big),$$

where $t\colon X \to \lim F$ is the unique morphism in C which makes the following diagram commutative for all $i \in \mathrm{Ob}\,I$:

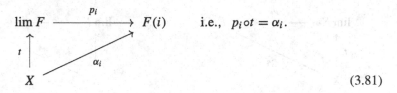

$$\text{i.e., } \quad p_i \circ t = \alpha_i.$$

$$(3.81)$$

Hence, we are left to show that $\theta_{X',\,F}\big(\alpha \circ \Delta(f)\big) = t \circ f$. Having in mind the way θ was defined, this comes down to proving that $t \circ f$ makes the following diagram commutative for all $i \in \mathrm{Ob}\,I$:

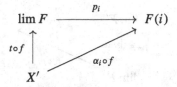

Indeed, we have $p_i \circ t \circ f \stackrel{(3.81)}{=} \alpha_i \circ f$ for all $i \in \mathrm{Ob}\,I$ and this shows that (3.80) holds.

Consider now two functors $F,\,F'\colon I \to C$ and $\beta \in \mathrm{Hom}_{\mathrm{Fun}(I,\,C)}(F,\,F')$, i.e., $\beta\colon F \to F'$ is a natural transformation. We denote by $\big(\lim F,\,(p_i\colon \lim F \to F(i))_{i \in \mathrm{Ob}\,I}\big)$ and $\big(\lim F',\,(s_i\colon \lim F' \to F'(i))_{i \in \mathrm{Ob}\,I}\big)$ the limit of F and F' respectively. The naturality of θ in the second variable comes down to proving the commutativity of the following diagram:

$$
\begin{array}{ccc}
\mathrm{Hom}_{\mathrm{Fun}(I,\,C)}(\Delta(X),\,F) & \xrightarrow{\;\;\theta_{X,\,F}\;\;} & \mathrm{Hom}_C(X,\,\lim F) \\[2mm]
{\scriptstyle \mathrm{Hom}_{\mathrm{Fun}(I,\,C)}(\Delta(X),\,\beta)}\Big\downarrow & & \Big\downarrow {\scriptstyle \mathrm{Hom}_C(X,\,\lim \beta)} \\[2mm]
\mathrm{Hom}_{\mathrm{Fun}(I,\,C)}(\Delta(X),\,F') & \xrightarrow[\;\;\theta_{X,\,F'}\;\;]{} & \mathrm{Hom}_C(X',\,\lim F')
\end{array}
$$

i.e., $\quad \mathrm{Hom}_C(X,\,\lim \beta) \circ \theta_{X,\,F} = \theta_{X,\,F'} \circ \mathrm{Hom}_{\mathrm{Fun}(I,\,C)}(\Delta(X),\,\beta).$

$$(3.82)$$

To this end, let $\gamma \in \mathrm{Hom}_{\mathrm{Fun}(I,C)}(\Delta(X), F)$, i.e., $\alpha: \Delta(X) \to F$ is a natural transformation. We obtain

$$\mathrm{Hom}_C(X, \lim \beta) \circ \theta_{X, F}(\gamma) = \lim \beta \circ t,$$

$$\theta_{X, F'} \circ \mathrm{Hom}_{\mathrm{Fun}(I,C)}(\Delta(X), \beta)(\gamma) = \theta_{X, F'}(\beta \circ \gamma) = r,$$

where $t: X \to \lim F$ and $r: X \to \lim F'$ are the unique morphisms in C which make the following diagrams commutative for all $i \in \mathrm{Ob}\, I$:

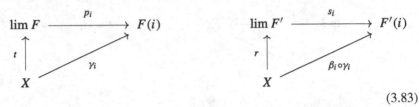

$$(3.83)$$

We are left to show that $\lim \beta \circ t = r$. To this end, recall that $\lim \beta \in \mathrm{Hom}_C(\lim F, \lim F')$ is the unique morphism in C which makes the following diagram commute for all $i \in \mathrm{Ob}\, I$:

$$\begin{array}{ccc}
\lim F' & \xrightarrow{\ s_i\ } & F'(i) \\
{\scriptstyle \lim \beta}\uparrow & \nearrow_{\beta_i \circ p_i} & \\
\lim F & &
\end{array}$$

$$(3.84)$$

By Proposition 2.2.14, (1) we only need to show that $s_i \circ \lim \beta \circ t = s_i \circ r$ for all $i \in \mathrm{Ob}\, I$. Indeed, we have $s_i \circ \lim \beta \circ t \overset{(3.84)}{=} \beta_i \circ p_i \circ t \overset{(3.83)}{=} \beta_i \circ \gamma_i \overset{(3.83)}{=} s_i \circ r$, as desired.

Assume now that the diagonal functor $\Delta: C \to \mathrm{Fun}(I, C)$ has a right adjoint, denoted by $R: \mathrm{Fun}(I, C) \to C$. We will show that any functor $F: I \to C$ has a limit. To this end, let $\varepsilon: \Delta R \to 1_{\mathrm{Fun}(I,C)}$ and $\theta: \mathrm{Hom}_{\mathrm{Fun}(I,C)}(\Delta, -) \to \mathrm{Hom}_C(-, R)$ be the counit and respectively the natural bijection induced by the adjunction $\Delta \dashv R$. In particular, $\varepsilon_F: \Delta_{R(F)} \to F$ is a natural transformation for any functor $F: I \to C$. Proposition 2.2.4, (1) implies that $\big(R(F), (\varepsilon_F)_i\big): R(F) \to F(i)_{i\in \mathrm{Ob}\, I}$ is a cone on F. We will show that $\big(R(F), (\varepsilon_F)_i\big): R(F) \to F(i)_{i\in \mathrm{Ob}\, I}$ is in fact the limit of F. Indeed, consider another cone $\big(X, (\alpha_i): X \to F(i)_{i\in \mathrm{Ob}\, I}\big)$ on F. Using again Proposition 2.2.4, (1) we obtain that $\alpha: \Delta_X \to F$, where $\alpha = (\alpha_i)_{i\in \mathrm{Ob}\, I}$, is a natural transformation, i.e., $\alpha \in \mathrm{Hom}_{\mathrm{Fun}(I,C)}(\Delta_X, F)$. Now Theorem 3.6.1, (3) yields a unique morphism $g \in \mathrm{Hom}_C(X, R(F))$ such that the following diagram is commutative:

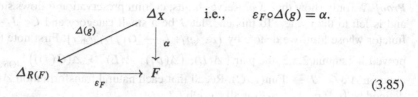

$$\varepsilon_F \circ \Delta(g) = \alpha.$$

(3.85)

The above equality between the natural transformations $\varepsilon_F \circ \Delta(g)$ and α comes down to identities between the corresponding morphisms associated to each $i \in \mathrm{Ob}\, I$. In light of Example 1.7.2, (5) we have $(\varepsilon_F)_i \circ g = \alpha_i$ for all $i \in \mathrm{Ob}\, I$ and therefore the following diagram is commutative:

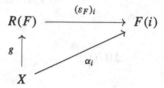

The proof will be finished once we show that g is the unique morphism in C which makes the above diagram commutative. To this end, assume that $h \in \mathrm{Hom}_C(X, R(F))$ such that $(\varepsilon_F)_i \circ h = \alpha_i$ for all $i \in \mathrm{Ob}\, I$. This leads to $\varepsilon_F \circ \Delta(h) = \alpha$ and the uniqueness of the morphism which makes diagram (3.85) commutative implies $g = h$, as desired.

$\square$

Remark 3.10.2 Functors having a left adjoint which is also a right adjoint are called *Frobenius functors* in the literature (see [17] for further details). The notion was first introduced in [16] and is motivated by the following example coming from ring theory: a ring extension $R \to S$ is Frobenius (in the sense of [41]) if and only if the corresponding restriction of scalars functor (Example 1.5.3, (32)) is Frobenius. In light of Theorem 3.10.1, since for complete and cocomplete categories C the diagonal functor $\Delta : C \to \mathrm{Fun}(I, C)$ has both a left and a right adjoint, it is natural to ask when the two adjoints are naturally isomorphic (or, equivalently, when the diagonal functor is Frobenius). This problem was considered in [20] and, given a complete and cocomplete category C, the small categories I for which the diagonal functor is Frobenius are characterized. $\square$

As a straightforward consequence of Theorem 3.10.1 we can easily conclude using Theorem 3.4.4 that if C is complete (resp. cocomplete) then the diagonal functor preserves colimits (resp. limits). However, we will see that even without the (co)completeness assumption on C, the diagonal functor still preserves all existing small (co)limits.

Proposition 3.10.3 *Let I be a small category and C an arbitrary category. Then the diagonal functor* $\Delta\colon C \to \text{Fun}(I, C)$ *preserves all existing small (co)limits.*

Proof We only show that Δ preserves limits; colimit preservation follows similarly and is left to the reader. To this end, let J be a small category and $G\colon J \to C$ a functor whose limit we denote by $\left(L, (p_j\colon L \to G(j))_{j\in\text{Ob}\,J}\right)$. First note that, as proved in Lemma 2.5.5, the pair $\left(\Delta(L), (\Delta(p_j)\colon \Delta(L) \to \Delta(G(j)))_{j\in\text{Ob}\,J}\right)$ is a cone on $\Delta \circ G\colon J \to \text{Fun}(I, C)$. Recall that each natural transformation $\Delta(p_j)$ is defined by $((\Delta(p_j))_i = p_j$ for all $i \in \text{Ob}\,I$.

Consider now another cone $\left(U, (q_j\colon U \to \Delta(G(j)))_{j\in\text{Ob}\,J}\right)$ on $\Delta \circ G$, where $U\colon I \to C$ is a functor and q_j is a natural transformation for all $j \in \text{Ob}\,J$. Hence, for all $u \in \text{Hom}_J(j, t)$, the following diagram is commutative:

$$
\begin{array}{ccc}
 & U & \\
\scriptstyle q_j \swarrow & & \searrow \scriptstyle q_t \\
\Delta(G(j)) & \xrightarrow{\;\Delta(G(u))\;} & \Delta(G(t))
\end{array}
\qquad\text{i.e.,}\quad \Delta(G(u))\circ q_j = q_t.
$$

$$(3.86)$$

Therefore, for all $i \in \text{Ob}\,I$ we have $(\Delta(G(u)))_i \circ (q_j)_i = (q_t)_i$, which comes down to $G(u) \circ (q_j)_i = (q_t)_i$, i.e., the following diagram is commutative:

$$
\begin{array}{ccc}
 & U(i) & \\
\scriptstyle (q_j)_i \swarrow & & \searrow \scriptstyle (q_t)_i \\
G(j) & \xrightarrow{\;G(u)\;} & G(t)
\end{array}
$$

$$(3.87)$$

Note that the commutativity of (3.87) implies that

$$
\left(U(i), ((q_j)_i\colon U(i) \to G(j))_{j\in\text{Ob}\,J}\right)
$$

is a cone on G. Since $\left(L, (p_j\colon L \to G(j))_{j\in\text{Ob}\,J}\right)$ is the limit of G, for any $i \in \text{Ob}\,I$, there exists a unique $g_i \in \text{Hom}_C(U(i), L)$ such that the following diagram is commutative for all $j \in \text{Ob}\,J$:

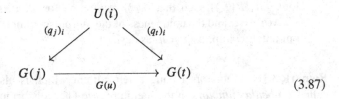

$$
\begin{array}{ccc}
U(i) & \xrightarrow{\;(q_j)_i\;} & G(j) \\
\scriptstyle g_i \downarrow & \nearrow \scriptstyle (\Delta(p_j))_i & \\
L & &
\end{array}
\qquad\text{i.e.,}\quad p_j\circ g_i = (q_j)_i.
$$

$$(3.88)$$

The proof will be finished once we show that $g = (g_i)_{i \in \mathrm{Ob}\, I} : U \to \Delta(L)$ is a natural transformation. To this end, let $v \in \mathrm{Hom}_I(i, s)$; we are left to prove the commutativity of the following diagram:

$$
\begin{array}{ccc}
U(i) & \xrightarrow{\ \ g_i\ \ } & \Delta(L)(i) \\
{\scriptstyle U(v)} \downarrow & & \downarrow {\scriptstyle \Delta(L)(v)=1_L} \\
U(s) & \xrightarrow[\ \ g_s\ \ ]{} & \Delta(L)(s)
\end{array}
\qquad \text{i.e.,} \quad g_i = g_s \circ U(v).
$$

$$(3.89)$$

In light of Proposition 2.2.14, (1) it is enough to prove that for all $j \in \mathrm{Ob}\, J$ we have $p_j \circ g_i = p_j \circ g_s \circ U(v)$. Indeed, we have

$$
\underline{p_j \circ g_s \circ U(v)} \overset{(3.88)}{=} \underline{(q_j)_s \circ U(v)} = \underline{(q_j)_i} \overset{(3.88)}{=} p_j \circ g_i,
$$

where the second equality holds because $q_j : U \to \Delta(G(j))$ is a natural transformation. $\qquad\square$

3.11 Freyd's Adjoint Functor Theorem

Theorem 3.4.4 shows that right (left) adjoints preserve all existing small limits (colimits). However, in general, small limit/colimit preservation alone does not guarantee the existence of a left/right adjoint. Indeed, consider the unique functor $T : \mathbf{Set}(\subseteq) \to \mathbf{1}$, where $\mathbf{1}$ is the discrete category with one object and $\mathbf{Set}(\subseteq)$ is the category defined in Example 1.2.2, (2). Note that the category $\mathbf{Set}(\subseteq)$ is cocomplete by Example 2.2.11 and does not posses a final object as shown in Example 1.3.10, (6). Therefore, T does not admit a right adjoint as can easily be seen from Example 3.4.1, (5) while it trivially preserves small colimits.

In this section we will prove that limit/colimit preservation is part of a necessary and sufficient condition which needs to be fulfilled by a functor in order to admit a left/right adjoint. Let $G : \mathcal{D} \to C$ be a functor, $X \in \mathrm{Ob}\, C$ and let $(X \downarrow G)$ be the comma category defined in Corollary 1.8.6, (1). We have an obvious forgetful functor $U : (X \downarrow G) \to \mathcal{D}$ defined for any $(f, Y) \in \mathrm{Ob}(X \downarrow G)$ and any morphism h in $(X \downarrow G)$ as follows:

$$
U(f, Y) = Y, \quad U(h) = h.
$$

Lemma 3.11.1 *Let $G : \mathcal{D} \to C$ be a functor.*

(1) The functor G admits a left adjoint if and only if for all $X \in \mathrm{Ob}\, C$ the comma category $(X \downarrow G)$ has an initial object.

(2) *The functor G admits a right adjoint if and only if for all $X \in \mathrm{Ob}\,C$ the comma category $(G \downarrow X)$ has a final object.*

Proof

(1) Suppose first that G has a left adjoint $F: C \to \mathcal{D}$ and let $\theta: \mathrm{Hom}_{\mathcal{D}}(F(-), -) \to \mathrm{Hom}_C(-, G(-))$ be the natural isomorphism corresponding to the adjunction $F \dashv G$. Now consider $X \in \mathrm{Ob}\,C$ and let $\eta: 1_C \to GF$ be the unit of the adjunction. We will prove that $(\eta_X, F(X))$ is the initial object of the category $(X \downarrow G)$. Let (v, W) be another object in $(X \downarrow G)$, i.e., $W \in \mathrm{Ob}\,\mathcal{D}$ and $v \in \mathrm{Hom}_C(X, G(W))$. To this end, we need to find a unique morphism $f: (\eta_X, F(X)) \to (v, W)$ in $(X \downarrow G)$, i.e., a morphism $f \in \mathrm{Hom}_{\mathcal{D}}(F(X), W)$ such that the following diagram is commutative:

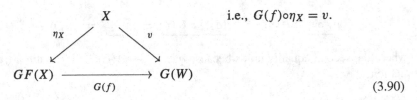

i.e., $G(f) \circ \eta_X = v$.

(3.90)

Recall from (the proof of) Theorem 3.5.1 that for all $u \in \mathrm{Hom}_{\mathcal{D}}(F(X), W)$ we have $G(u) \circ \eta_X = \theta_{X,W}(u)$. Now if we consider $f = \theta_{X,W}^{-1}(v)$ we obtain

$$v = \theta_{X,W}(f) = G(f) \circ \eta_X,$$

as desired. The uniqueness of f with this property follows from the bijectivity of $\theta_{X,W}$.

Conversely, assume now that for each $X \in \mathrm{Ob}\,C$ the comma category $(X \downarrow G)$ has an initial object, which we denote by (u_X, V_X), where $V_X \in \mathrm{Ob}\,\mathcal{D}$ and $u_X \in \mathrm{Hom}_C(X, G(V_X))$. Hence, for any $(f, Y) \in \mathrm{Ob}\,(X \downarrow G)$ there exists a unique morphism $h: (u_X, V_X) \to (f, Y)$ in $(X \downarrow G)$; in other words, for any $f \in \mathrm{Hom}_C(X, G(Y))$ there exists a unique morphism $h \in \mathrm{Hom}_{\mathcal{D}}(V_X, Y)$ making the following diagram commute:

$$
\begin{array}{ccc}
 & X & \\
u_X \swarrow & & \searrow f \\
G(V_X) & \xrightarrow{\;\;G(h)\;\;} & G(Y)
\end{array}
$$

i.e., $G(h) \circ u_X = f$.

(3.91)

We define a functor $F: C \to \mathcal{D}$ on objects by $F(X) = V_X$ for all $X \in \mathrm{Ob}\,C$. Consider now $f \in \mathrm{Hom}_C(X, X')$; then $u_{X'} \circ f \in \mathrm{Hom}_C(X, G(F(X')))$

and, using (3.91), we define $F(f) \in \mathrm{Hom}_{\mathcal{D}}(F(X), F(X'))$ to be the unique morphism such that

$$GF(f) \circ u_X = u_{X'} \circ f. \qquad (3.92)$$

Obviously $F(1_X) = 1_{F(X)}$ for all $X \in \mathrm{Ob}\,C$. Moreover, if $f \in \mathrm{Hom}_C(X, X')$ and $f' \in \mathrm{Hom}_C(X', X'')$ then $F(f' \circ f)$ and $F(f') \circ F(f)$ are both morphisms in the comma category $(X \downarrow G)$ from $(u_X, F(X))$ to $(u_{X''} \circ f' \circ f, F(X''))$ so they must be equal as $(u_X, F(X))$ is the initial object of $(X \downarrow G)$. Hence F is a functor and furthermore, according to (3.92), $u: 1_C \to GF$ is a natural transformation. To summarize, we have constructed a natural transformation $u: 1_C \to GF$ such that for any $f \in \mathrm{Hom}_C(X, G(Y))$ there exists a unique $h \in \mathrm{Hom}_{\mathcal{D}}(F(X), Y)$ satisfying $G(h) \circ u_X = f$. Now Theorem 3.6.1, (2) implies that F is left adjoint to G, as desired.

(2) Theorem 3.4.3 shows that $G: \mathcal{D} \to C$ admits a right adjoint if and only if $G^{op}: \mathcal{D}^{op} \to C^{op}$ admits a left adjoint. We have already proved in 1) that G^{op} has a left adjoint if and only if for all $X \in \mathrm{Ob}\,C^{op}$ the comma category $(X \downarrow G^{op})$ has an initial object. Furthermore, by Proposition 1.8.10 we have an isomorphism of categories between $(X \downarrow G^{op})^{op}$ and $(G \downarrow X)$. Now using Proposition 1.4.3, (2) we obtain that $(X \downarrow G^{op})$ has an initial object if and only if its opposite category, namely $(G \downarrow X)$, has a final object. By putting all the above together we obtain that G has a right adjoint if and only if for all $X \in \mathrm{Ob}\,C$ the comma category $(G \downarrow X)$ has a final object. $\qquad \square$

Definition 3.11.2 Let C be a category.

(1) A family $(K_i)_{i \in I}$ of objects of C, where I is a set, is called a *weakly initial set* if for any $C \in \mathrm{Ob}\,C$ there exists a morphism $t_C^j \in \mathrm{Hom}_C(K_j, C)$ for some $j \in I$.
(2) Dually, a family $(W_i)_{i \in I}$ of objects of C, where I is a set, is called a *weakly final set* if it is a weakly initial set in C^{op}; that is, if for any $C \in \mathrm{Ob}\,C$ there exists a morphism $l_C^j \in \mathrm{Hom}_C(C, W_j)$ for some $j \in I$.

Lemma 3.11.3 *Let C be a category. Then:*

(1) if C is complete then C has an initial object if and only if C has a weakly initial set;
(2) if C is cocomplete then C has a final object if and only if C has a weakly final set.

Proof

(1) Assume first that C has an initial object I; then $\{I\}$ is obviously a weakly initial set.

Conversely, let $(K_i)_{i \in I}$ be a weakly initial set. As C is complete and I is a set we can consider the product $\left(P, (\pi_i \colon P \to K_i)_{i \in I}\right)$ of the family of objects $(K_i)_{i \in I}$. Notice that for each $C \in \mathrm{Ob}\, C$ there exists at least one morphism $u_C \in \mathrm{Hom}_C(P, C)$ given by the composition $P \xrightarrow{\ \pi_j\ } K_j \xrightarrow{\ t_C^j\ } C$ for some $j \in I$.

Consider now the category J with $\mathrm{Ob}\, J = \{P\}$ and $\mathrm{Hom}_J(P, P) = \mathrm{Hom}_C(P, P)$ and let $(L, q \colon L \to P)$ be the limit of the inclusion functor $F \colon J \to C$.

We will prove that L is the initial object of the category C. Indeed, for any $C \in \mathrm{Ob}\, C$ there exists at least one morphism in $\mathrm{Hom}_C(L, C)$ given by the composition $L \xrightarrow{\ q\ } P \xrightarrow{\ u_C\ } C$. Suppose now that we have two such morphisms $f, g \in \mathrm{Hom}_C(L, C)$ and consider $(E, e \colon E \to L)$ to be the equalizer of (f, g). Since $E \in \mathrm{Ob}\, C$ there exists a morphism $u_E \in \mathrm{Hom}_C(P, E)$ given by the composition $P \xrightarrow{\ \pi_j\ } K_j \xrightarrow{\ t_E^j\ } E$ for some $j \in I$. Thus $q \circ e \circ u_E \in \mathrm{Hom}_C(P, P)$ and since $(L, q \colon L \to P)$ is in particular a cone on F, the following diagram is commutative:

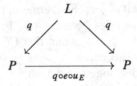

Thus we have $q \circ e \circ u_E \circ q = q = q \circ 1_L$ and by Proposition 2.2.14, (1) we get $e \circ u_E \circ q = 1_L$. This yields

$$f = f \circ 1_L = f \circ e \circ u_E \circ q = g \circ e \circ u_E \circ q = g \circ 1_L = g,$$

where in the third equality we used the fact that $(E, e \colon E \to L)$ is the equalizer of (f, g). We have obtained $f = g$ and hence L is an initial object of C.

(2) If C is cocomplete then C^{op} is complete and, as proved in 1), C^{op} has an initial object if and only if C^{op} has a weakly initial set. Equivalently, C has a final object if and only if C has a weakly final set.

$\square$

We are now ready to state the main result of this section:

Theorem 3.11.4 (Freyd's adjoint functor theorem)   *Let $G \colon \mathcal{D} \to C$ be a functor.*

(1) If $\mathcal{D}$ is a complete category then G has a left adjoint if and only if G preserves all small limits and for each $X \in \mathrm{Ob}\, C$ the comma category $(X \downarrow G)$ has a weakly initial set.

(2) *If $\mathcal{D}$ is a cocomplete category then G has a right adjoint if and only if G preserves all small colimits and for each $X \in \text{Ob}\,C$ the comma category $(G \downarrow X)$ has a weakly final set.*

Proof

(1) Suppose G has a left adjoint F. Then G is a right adjoint to F and by Theorem 3.4.4 it preserves limits. Moreover, by (the proof of) Lemma 3.11.1, (1) for any $X \in \text{Ob}\,C$, the pair $(\eta_X, F(X))$ is an initial object in $(X \downarrow G)$, where $\eta\colon 1_C \to GF$ is the unit of the adjunction (F, G).

Assume now that $G\colon \mathcal{D} \to C$ preserves small limits and for each $X \in \text{Ob}\,C$ the comma category $(X \downarrow G)$ has a weakly initial set. By Corollary 2.6.5, (1) the category $(X \downarrow G)$ is complete. Thus, from Lemma 3.11.3, (1) we obtain that $(X \downarrow G)$ has an initial object. The conclusion now follows by Lemma 3.11.1, (1).

(2) The category $\mathcal{D}^{op}$ is complete and by applying (1) for the functor $G^{op}\colon \mathcal{D}^{op} \to C^{op}$ it follows that G^{op} has a left adjoint if and only if G^{op} preserves small limits and for each $X \in \text{Ob}\,C$ the comma category $(X \downarrow G^{op})$ has a weakly initial set. Note also that by Theorem 3.4.3, G has a right adjoint if and only if G^{op} has a left adjoint. Furthermore, Proposition 1.8.10 shows that the comma categories $(X \downarrow G^{op})$ and $(G \downarrow X)^{op}$ are isomorphic and therefore a weakly initial set in $(X \downarrow G^{op})$ is a weakly final set in $(G \downarrow X)$. Putting all of the above together we can conclude that G has a right adjoint if and only if G preserves all small colimits and for each $X \in \text{Ob}\,C$ the comma category $(G \downarrow X)$ has a weakly final set, as desired.

$\square$

Theorem 3.11.4 can be stated in an equivalent form without the use of comma-categories. To this end, we introduce the following:

Definition 3.11.5 Let $F\colon C \to \mathcal{D}$ be a functor and $D \in \text{Ob}\,\mathcal{D}$. Then:

(1) F satisfies the *solution set condition* with respect to D if there exists a set U_D of objects of C such that for any $C \in \text{Ob}\,C$ and any $f \in \text{Hom}_{\mathcal{D}}(D, F(C))$, there exists an object $C' \in U_D$ and morphisms $u \in \text{Hom}_C(C', C)$, $g \in \text{Hom}_{\mathcal{D}}(D, F(C'))$ such that the following diagram is commutative:

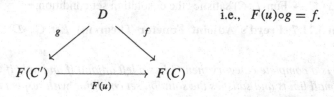

i.e., $F(u) \circ g = f$.

(2) F satisfies the *cosolution set condition* with respect to D if there exists a set W_D of objects of C such that for any $C \in \text{Ob}\,C$ and any $f \in \text{Hom}_{\mathcal{D}}(F(C), D)$, there exists an object $C' \in W_D$ and morphisms $u \in \text{Hom}_C(C, C')$, $g \in$

$\text{Hom}_{\mathcal{D}}(F(C'), D)$ such that the following diagram is commutative:

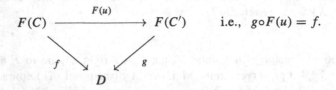

i.e., $g \circ F(u) = f$.

Example 3.11.6 If I is a small category and C is a cocomplete category then the diagonal functor $\Delta \colon C \to \text{Fun}(I, C)$ satisfies the solution set condition with respect to all $G \in \text{Ob}(\text{Fun}(I, C))$. Indeed, as $G \colon I \to C$ is a functor and C is cocomplete, G has a colimit, which we denote as usual by $(\text{colim } G, (q_i \colon G(i) \to \text{colim } G)_{i \in \text{Ob } I})$. Consider now the set $U_G = \{\text{colim } G\}$. If $C \in \text{Ob } C$ and $\alpha \in \text{Hom}_{\text{Fun}(I, C)}(G, \Delta_C)$ then by Proposition 2.2.4, (2) we obtain that $(C, (\alpha_i \colon G(i) \to C)_{i \in \text{Ob } I})$ is a cocone on G. As $(\text{colim } G, (q_i)_{i \in \text{Ob } I})$ is the colimit of G, there exists a unique $u \in \text{Hom}_C(\text{colim } G, C)$ such that for all $i \in \text{Ob } I$ we have

$$u \circ q_i = \alpha_i. \tag{3.93}$$

Furthermore, using again Proposition 2.2.4, (2) it follows that $\beta \colon G \to \Delta_{\text{colim } G}$ defined by $\beta_i = q_i$ for all $i \in \text{Ob } I$ is a natural transformation and, moreover, (3.93) implies

$$\Delta(u) \circ \beta = \alpha. \tag{3.94}$$

To summarize, given $C \in \text{Ob } C$ and $\alpha \in \text{Hom}_{\text{Fun}(I, C)}(G, \Delta_C)$, there exists $\text{colim } G \in U_G$ and two morphisms

$$u \in \text{Hom}_C(\text{colim } G, C), \qquad \beta \in \text{Hom}_{\text{Fun}(I, C)}(G, \Delta_{\text{colim } G})$$

such that (3.94) holds. This shows that $\Delta \colon C \to \text{Fun}(I, C)$ satisfies the solution set condition with respect to all $G \in \text{Ob}(\text{Fun}(I, C))$.

Similarly, it can be proved that if C is a complete category then the diagonal functor $\Delta \colon C \to \text{Fun}(I, C)$ satisfies the cosolution set condition. $\quad\square$

Theorem 3.11.7 (Freyd's Adjoint Functor Theorem) *Let $G \colon \mathcal{D} \to C$ be a functor.*

(1) If $\mathcal{D}$ is a complete category then G has a left adjoint if and only if G preserves all small limits and satisfies the solution set condition with respect to each $X \in \text{Ob } C$.

(2) If $\mathcal{D}$ is a cocomplete category then G has a right adjoint if and only if $G preserves all small colimits and satisfies the cosolution set condition with respect to each $X \in \text{Ob } C$.

Proof

(1) Note that G satisfies the solution set condition with respect to X if and only if the comma category $(X \downarrow G)$ has a weakly initial set and the conclusion follows by Theorem 3.11.4, (1). Indeed, assume first that G satisfies the solution set condition, let $X \in \mathrm{Ob}\, C$ and let U_X be the corresponding set as in Definition 3.11.5, (1). Then $\{(h, Y) \mid h \in \mathrm{Hom}_C(X, G(Y)), Y \in U_X\}$ is a weakly initial set in $(X \downarrow G)$. Conversely, if $K = \{(t, Z) \mid Z \in \mathrm{Ob}\, \mathcal{D}, t \in \mathrm{Hom}_C(X, G(Z))\}$ is a weakly initial set in $(X \downarrow G)$ then $U_X = \{D \in \mathrm{Ob}\, \mathcal{D} \mid$ there exists $t \in \mathrm{Hom}_C(X, G(Z))$ such that $(t, Z) \in K\}$ is a set which fulfills the condition in Definition 3.11.5, (1) and therefore G satisfies the solution set condition.

(2) Similarly, G satisfies the cosolution set condition with respect to X if and only if the comma category $(G \downarrow X)$ has a weakly final set and Theorem 3.11.4, (2) leads to the desired conclusion.

$\square$

We end this section with some applications of the adjoint functor theorem. As we will see, it can be used to show the existence of various free objects (such as free groups, free algebras, free modules etc.) without explicitly constructing them.

Examples 3.11.8

(1) Let $U \colon \mathbf{Grp} \to \mathbf{Set}$ be the forgetful functor. We will show that U admits a left adjoint by using Freyd's adjoint functor theorem. To start with, $\mathbf{Grp}$ is complete as shown in Example 2.4.3, (1) and U preserves small limits by Example 2.5.10, (1). In order to conclude that U admits a left adjoint we need to show that it satisfies the solution set condition with respect to each $X \in \mathrm{Ob}\,\mathbf{Set}$. To this end, for a given $X \in \mathrm{Ob}\,\mathbf{Set}$ we consider the class $\mathcal{U}_X$ of all isomorphism classes of groups of cardinality less than or equal to $\lambda = \max\{\aleph_0, |X|\}$, where $\aleph_0$ denotes the cardinality of the set of all natural numbers. First we show that $\mathcal{U}_X$ is in fact a set. Indeed, recall that there is only a set of composition laws (in particular, of composition laws which are group structures) on any given set. Thus, we can conclude that we have, up to isomorphism, only a set of group structures on any set whose cardinality is at most λ. This shows that $\mathcal{U}_X$, being a reunion of sets, is a set itself.

In order to show that $\mathcal{U}_X$ satisfies the conditions in Definition 3.11.5, (1) we start by proving that if $(g_x)_{x \in X}$ is a set of elements of a group G then the subgroup G_X of G generated by this set has cardinality at most λ. This can be easily seen by observing that the following map is surjective:

$$f \colon \coprod_{n \in \mathbb{N}} (X \times \mathbb{Z})^n \to G_X, \quad f\big((x_1, e_1), (x_2, e_2), \ldots (x_n, e_n)\big) = g_{x_1}^{e_1} g_{x_2}^{e_2} \cdots g_{x_n}^{e_n},$$

where the domain of f is the coproduct of the family $\{(X \times \mathbb{Z})^n\}_{n \in \mathbb{N}}$ in **Set**. We have

$$|G_X| \leqslant \left| \coprod_{n \in \mathbb{N}} (X \times \mathbb{Z})^n \right| = \sum_{n \in \mathbb{N}} |(X \times \mathbb{Z})^n|$$

$$= \begin{cases} \aleph_0, & \text{if } |X| \leqslant \aleph_0 \\ |X|, & \text{if } |X| \gtrless \aleph_0 \end{cases} \leqslant \max\{\aleph_0, |X|\} = \lambda.$$

Consider now $f \in \mathrm{Hom}_{\mathbf{Set}}(X, U(G))$ for some $G \in \mathrm{Ob}\,\mathbf{Grp}$ and let G' be the subgroup of G generated by the set $(f(x))_{x \in X}$. Then, according to the above discussion, G' is a group of cardinality at most λ and we can find a group $H \in \mathcal{U}_X$ and a group isomorphism $t \colon H \to G'$. Denote by $i \colon G' \to G$ the inclusion map and consider $i \circ t \in \mathrm{Hom}_{\mathbf{Grp}}(H, G)$ and $U(t^{-1}) \circ f \in \mathrm{Hom}_{\mathbf{Set}}(X, U(H))$ such that the following diagram is commutative:

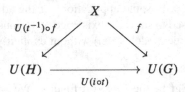

We can now conclude by Freyd's adjoint functor theorem that $U \colon \mathbf{Grp} \to \mathbf{Set}$ has a left adjoint.

(2) Let R be a ring, $M \in \mathrm{Ob}\,\mathcal{M}_R$, $N \in \mathrm{Ob}\,{}_R\mathcal{M}$ and $\mathrm{Bil}_{M,N} \colon \mathbf{Ab} \to \mathbf{Ab}$ the functor defined in Example 1.5.3, (28). We will show, using Freyd's adjoint functor theorem, that $\mathrm{Bil}_{M,N}$ admits a left adjoint. By Example 2.5.4, $\mathrm{Bil}_{M,N}$ preserves limits. We are left to show that $\mathrm{Bil}_{M,N}$ satisfies the solution set condition for each $A \in \mathrm{Ob}\,\mathbf{Ab}$. To this end, for a given $A \in \mathrm{Ob}\,\mathbf{Ab}$ we denote by $\mathcal{U}_A$ the class of all isomorphism classes of abelian groups of cardinality less than or equal to $\lambda = \max\{\aleph_0, |A| \cdot |M \times N|\}$. It follows easily, as in the previous example, that $\mathcal{U}_A$ is in fact a set.

We will show that $\mathcal{U}_A$ satisfies the conditions in Definition 3.11.5, (1). Indeed, let $B \in \mathrm{Ob}\,\mathbf{Ab}$ and $f \in \mathrm{Hom}_{\mathbf{Ab}}(A, \mathrm{Bil}_{M,N}(B))$. For all $a \in A$ let $P_a = \{f(a)(m, n) \mid m \in M, n \in N\}$ and let B' be the abelian subgroup of B generated by the set $P = \cup_{a \in A} P_a$. We start by proving that B' has cardinality at most λ. This can be easily seen by observing that the following map is surjective:

$$f \colon \coprod_{n \in \mathbb{N}} (P \times \mathbb{Z})^n \to B', \quad f\big((x_1, k_1), (x_2, k_2), \ldots (x_n, k_n)\big) = \sum_{i=1}^{n} k_i x_i,$$

where the domain of f is the coproduct of the family $\{(P \times \mathbb{Z})^n\}_{n \in \mathbb{N}}$ in **Set**. We have

$$|B'| \leq \left| \coprod_{n \in \mathbb{N}} (P \times \mathbb{Z})^n \right| = \left| \coprod_{n \in \mathbb{N}} \left((\cup_{a \in A} P_a) \times \mathbb{Z} \right)^n \right|$$

$$\leq \begin{cases} \aleph_0, & \text{if } |A| \cdot |M \times N| \leq \aleph_0 \\ |A| \cdot |M \times N|, & \text{if } |A| \cdot |M \times N| \gneq \aleph_0 \end{cases}$$

$$\leq \max\{\aleph_0, |A| \cdot |M \times N|\} = \lambda.$$

Hence, there exists an abelian group $H \in \mathcal{U}_A$ and a group isomorphism $t : H \to B'$. Now given the way we defined B', we obviously have $f \in \mathrm{Hom}_{\mathbf{Ab}}(A, \mathrm{Bil}_{M,N}(B'))$. Summarizing, if we denote by $i : B' \to B$ the inclusion map, we have found two maps $i \circ t \in \mathrm{Hom}_{\mathbf{Ab}}(H, B)$ and $\mathrm{Bil}_{M,N}(t^{-1}) \circ f \in \mathrm{Hom}_{\mathbf{Ab}}(A, \mathrm{Bil}_{M,N}(H))$ such that the following diagram is commutative:

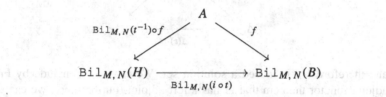

Now Freyd's adjoint functor theorem implies that $\mathrm{Bil}_{M,N} : \mathbf{Ab} \to \mathbf{Ab}$ has a left adjoint.

(3) Let I be a small category and $\Delta : \mathbf{Grp} \to \mathrm{Fun}(I, \mathbf{Grp})$ the diagonal functor. We show that Δ admits a left adjoint by using Freyd's adjoint functor theorem. To start with, **Grp** is a complete category by Example 2.4.3, (1) and the diagonal functor preserves limits as shown in Proposition 3.10.3. We are left to show that Δ also fulfills the solution set condition. To this end, let $F : I \to \mathbf{Grp}$ be a functor and let $X_i = F(i)$, for all $i \in \mathrm{Ob}\,I$. Consider $\lambda = |\coprod_{i \in \mathrm{Ob}\,I} X_j|$ and let U_F be the class of isomorphism classes of groups of cardinality less than or equal to λ, where $\coprod_{i \in \mathrm{Ob}\,I} X_j$ is the coproduct in **Set** of the underlying sets of the family of groups $(X_i)_{i \in I}$. As in the first example, U_F can be easily proved to be a set. Now let $C \in \mathrm{Ob}\,\mathbf{Grp}$ and $\psi : F \to \Delta_C$ be a natural transformation. Consider H to be the subgroup of C generated by $\bigcup_{j \in \mathrm{Ob}\,I} \psi_j(X_j)$, where $\psi_j : X_j \to C$, $j \in \mathrm{Ob}\,I$, are the group morphisms corresponding to the natural transformation ψ. In particular, we have $\psi : F \to \Delta_H$. Furthermore, the following holds:

$$|H| \leq \left| \bigcup_{j \in \mathrm{Ob}\,I} \psi_j(X_j) \right| \leq \left| \coprod_{i \in \mathrm{Ob}\,I} X_j \right| = \lambda.$$

Therefore we can find a group $G_\alpha \in U_F$ and a group isomorphism $u : H \to G_\alpha$. Consider now the natural transformation $\tau : F \to \Delta_{G_\alpha}$ defined for all $i \in \mathrm{Ob}\, I$ by $\tau_i(x) = u(\psi_i(x))$, $x \in X_i$. It can be easily seen that τ is a natural transformation; indeed, if $v \in \mathrm{Hom}_I(i, j)$ and $x \in X_i$ we have

$$\left(\tau_j \circ F(h)\right)(x) = u\left(\psi_j(F(h)(x))\right) = u\left(\psi_i(x)\right) = \tau_i(x),$$

where the second equality holds because ψ is a natural transformation. Now define $t : G_\alpha \to C$ by $t(y) = u^{-1}(y)$ for all $y \in G_\alpha$. Note that for all $x \in X_i$ we have

$$(t \circ \tau_i)(x) = t\left(u(\psi_i(x))\right) = u^{-1}\left(u(\psi_i(x))\right) = \psi_i(x).$$

This shows that the following diagram is commutative:

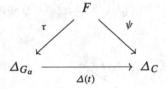

and therefore U_F is indeed a solution set. We can now conclude by Freyd's adjoint functor theorem that Δ has a left adjoint. Furthermore, we can derive the cocompleteness of **Grp** from Theorem 3.10.1, (2). $\square$

3.12 Special Adjoint Functor Theorem

In this section we show that under certain conditions on the domain category, a given functor admits a left (right) adjoint if and only if it preserves small limits (colimits). We start with some preparations.

Definition 3.12.1 Let C be a category.

(1) We say that C admits a *set of generators* (or *separators*) if there exists a set of objects $\{S_i \mid i \in I\}$ in C with the property that for any two morphisms u, $v \in \mathrm{Hom}_C(A, B)$ such that $u \neq v$ there exists a morphism $g \in \mathrm{Hom}_C(S_j, A)$, for some $j \in I$, satisfying $u \circ g \neq v \circ g$. An object S of C is called a *generator* (or *separator*) if $\{S\}$ is a set of generators.

(2) Dually, we say that C admits a *set of cogenerators* (or *coseparators*) if C^{op} admits a set of separators. More precisely, C admits a set of cogenerators if there exists a set of objects $\{U_i \mid i \in I\}$ in C with the property that for any two morphisms u, $v \in \mathrm{Hom}_C(A, B)$ such that $u \neq v$ there exists a morphism

$h \in \mathrm{Hom}_C(B, U_j)$, for some $j \in I$, satisfying $h \circ u \neq h \circ v$. An object U of C is called a *cogenerator* (or *coseparator*) if $\{U\}$ is a set of cogenerators.

Examples 3.12.2

(1) In **Set**, any set with only one element is a generator while any set with at least two elements set is a cogenerator. Indeed, let $u, v \colon X \to Y$ be two functions such that $u \neq v$. Hence there exists an $x_0 \in X$ such that $u(x_0) \neq v(x_0)$ and we can define a map $g \colon \{\star\} \to X$ by $g(\star) = x_0$. Therefore, we have a map g such that $u(g(\star)) = u(x_0) \neq v(x_0) = v(g(\star))$. This shows that $u \circ g \neq v \circ g$ and the singleton $\{\star\}$ is a generator in **Set**. Similarly, one can show that any singleton set is a generator in **Top**.

Next we look at cogenerators in **Set**. With the notations above, let $u(x_0) = y'$ and $v(x_0) = y''$ where $y, y' \in Y$ and $y' \neq y''$. We can now define a map $h \colon Y \to Z$ by

$$h(y) = \begin{cases} z', & \text{if } y = y', \\ z'', & \text{if } y \neq y', \end{cases}$$

where Z is a set with at least two elements and $z' \neq z''$. This leads to $h(u(x_0)) = h(y') = z' \neq z'' = h(y'') = h(v(x_0))$ and the desired conclusion follows.

(2) In **Grp**, the group of integers $(\mathbb{Z}, +)$ is a generator. To this end, let $u, v \in \mathrm{Hom}_{\mathbf{Grp}}(G, H)$ such that $u \neq v$ and consider $g_0 \in G$ such that $u(g_0) \neq v(g_0)$. Now define $g \colon \mathbb{Z} \to G$ by $g(k) = g_0^k$ for all $k \in \mathbb{Z}$, where the group structure on G is considered to be multiplicative. It can be easily seen that g is a morphism of groups and moreover $u(g(1)) \neq v(g(1))$, which leads to $u \circ g \neq v \circ g$, as desired.

On the other hand, **Grp** has no cogenerators. To this end, assume there exists a cogenerator U in **Grp** and let S be a simple group whose cardinality is larger than that of U.[12] Let $\mathrm{Id}_S, 0_S \in \mathrm{Hom}_{\mathbf{Grp}}(S, S)$, where Id_S denotes the identity morphism on S while 0_S is defined by $0_S(s) = 1_S$ for all $s \in S$. As S is a simple group we have $\mathrm{Id}_S \neq 0_S$ and since U is assumed to be a cogenerator in **Grp**, we have a group homomorphism $f \colon S \to U$ such that:

$$f \circ \mathrm{Id}_S \neq f \circ 0_S. \tag{3.95}$$

Now $\ker(f) \trianglelefteq S$ and since S is a simple group we have either $\ker(f) = S$ or $\ker(f) = \{1_S\}$. The first option is ruled out by (3.95) and therefore we obtain $\ker(f) = \{1_S\}$. This shows that f is injective, which implies that the cardinality of S is less than or equal to the

---

[12] Such a group is, for instance, the projective special linear group $\mathrm{PSL}(2, k)$, where k is the field of rational functions over the complex numbers in $|U|$ variables, i.e., $k = \mathbb{C}(X_u)_{u \in U}$ (see [50, Theorem 9.46]).

cardinality of U, contradicting our hypothesis. Therefore, we have reached a contradiction and we can conclude that **Grp** has no cogenerators.

(3) In **KHaus**, the category of compact Hausdorff spaces, the unit interval $[0, 1]$ is a cogenerator. Indeed, consider $u, v \in \mathrm{Hom}_{\mathbf{KHaus}}(H, K)$ such that $u \neq v$. Then, there exists an $h_0 \in h$ such that $u(h_0) \neq v(h_0)$. As K is in particular a Hausdorff space, there exist two disjoint neighborhoods U_0 and V_0 of $u(h_0)$ and $v(h_0)$, respectively. Now since any compact Hausdorff space is normal[13] ([39, Theorem 32.3]), we can apply Urysohn's lemma[14] to conclude that there exists a continuous map $f : K \rightarrow [0, 1]$ such that $f(x) = 0$ for all $x \in U_0$ and $f(x) = 1$ for all $x \in V_0$. In particular, we have $f\big(u(h_0)\big) = 0$ and $f\big(v(h_0)\big) = 1$, which shows that there exists an $f \in \mathrm{Hom}_{\mathbf{KHaus}}(K, [0, 1])$ such that $f \circ u \neq f \circ v$. This shows that $[0, 1]$ is a cogenerator in **KHaus**.

(4) If J is a small category, then the functor category $\mathrm{Fun}(J, \mathbf{Set})$ has a set of generators. Indeed, we will show that $\{\mathrm{Hom}_J(j, -) \mid j \in \mathrm{Ob}\, J\}$ is a set of generators. To this end, let $F, G \in \mathrm{Ob}\,\mathrm{Fun}(J, \mathbf{Set})$, i.e., F and G are functors, and consider two natural transformations $\alpha, \beta : F \rightarrow G$ such that $\alpha \neq \beta$. Since $\alpha \neq \beta$, there exists $j_0 \in \mathrm{Ob}\, J$ and some $x_0 \in F(j_0)$ such that $\alpha_{j_0}(x_0) \neq \beta_{j_0}(x_0)$. Consider now $h^{x_0} : \mathrm{Hom}_J(j_0, -) \rightarrow F$ defined for all $k \in \mathrm{Ob}\, J$ and $f \in \mathrm{Hom}_J(j_0, k)$ by $\big(h^{x_0}\big)_k(f) = F(f)(x_0)$. According to (the proof of) Yoneda's lemma we have $h^{x_0} \in \mathrm{Nat}\big(\mathrm{Hom}_J(j_0, -), F\big)$ and, consequently $\alpha \circ h^{x_0}, \beta \circ h^{x_0} \in \mathrm{Nat}\big(\mathrm{Hom}_J(j_0, -), G\big)$. Furthermore, we have

$$\big(\alpha \circ h^{x_0}\big)_{j_0}(1_{j_0}) = \alpha_{j_0} \circ (h^{x_0})_{j_0}(1_{j_0}) = \alpha_{j_0}\big(F(1_{j_0})(x_0)\big)$$

$$= \alpha_{j_0}\big(1_{F(j_0)}(x_0)\big) = \alpha_{j_0}(x_0).$$

A similar computation shows that $\big(\beta \circ h^{x_0}\big)_{j_0}(1_{j_0}) = \beta_{j_0}(x_0)$. As $\alpha_{j_0}(x_0) \neq \beta_{j_0}(x_0)$ we obtain $\big(\alpha \circ h^{x_0}\big)_{j_0}(1_{j_0}) \neq \big(\beta \circ h^{x_0}\big)_{j_0}(1_{j_0})$. Therefore, $\alpha \circ h^{x_0} \neq \beta \circ h^{x_0}$, as desired. $\square$

Proposition 3.12.3 *Let $S = \{X_i \mid i \in I\}$ be a set of objects of a category C.*

(1) Assume C has products, S is a set of cogenerators and consider the product $\big(P, (p_f)_{f \in \mathrm{Hom}_C(C, X_i)}\big)$ of the family S, where the product consists of as many copies of X_i as there are morphisms in $\mathrm{Hom}_C(C, X_i)$. Then, for any $C \in \mathrm{Ob}\, C$, the unique morphism $\gamma_C : C \rightarrow P$ which makes the following

---

[13] A topological space X is called *normal* if for each pair U, V of disjoint closed subsets of X there exist disjoint open subsets of X containing U and V ([39, Section 31]).

[14] Urysohn's lemma: Let X be a normal space and U and V two disjoint subsets of X. If $[a, b] \subset \mathbb{R}$ is a closed interval then there exists a continuous map $f : X \rightarrow [a, b]$ such that $f(x) = a$ for every $x \in U$ and $f(x) = b$ for every $x \in V$ ([39, Theorem 33.1]).

diagram commutative for all i ∈ I and all f ∈ Hom$_C$(C, X$_i$):

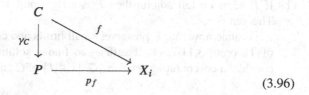

$$(3.96)$$

is a monomorphism.

(2) *Dually, assume C has coproducts, S is a set of generators and consider the coproduct $\left(Q, (q_f)_{f \in \text{Hom}_C(X_i, C)}\right)$ of the family S, where the coproduct consists of as many copies of X_i as there are morphisms in Hom$_C$(X$_i$, C). Then, for any C ∈ Ob C, the unique morphism ξ_C: P → C which makes the following diagram commutative for all i ∈ I and all f ∈ Hom$_C$(X$_i$, C):*

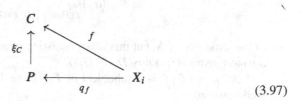

$$(3.97)$$

is an epimorphism.

Proof

(1) Consider $u, v \in \text{Hom}_C(A, C)$ such that $\gamma_C \circ u = \gamma_C \circ v$. This implies $p_f \circ \gamma_C \circ u = p_f \circ \gamma_C \circ v$ for all $i \in I$ and all $f \in \text{Hom}_C(C, X_i)$. The commutativity of (3.96) leads to $f \circ u = f \circ v$ for all $i \in I$ and all $f \in \text{Hom}_C(C, X_i)$. As S is a set of cogenerators we obtain $u = v$, as desired.

(2) Consider $u, v \in \text{Hom}_C(C, D)$ such that $u \circ \xi_C = v \circ \xi_C$. This implies $u \circ \xi_C \circ q_f = v \circ \xi_C \circ q_f$ for all $i \in I$ and all $f \in \text{Hom}_C(X_i, C)$. The commutativity of (3.97) leads to $u \circ f = v \circ f$ for all $i \in I$ and all $f \in \text{Hom}_C(X_i, C)$. As S is a set of generators we obtain $u = v$, as desired.

□

Having introduced the necessary concepts, we can now state the main result of this section.

Theorem 3.12.4 (Special Adjoint Functor Theorem) *Let F: C → D be a functor.*

(1) *Assume C is a complete, well-powered category which admits a cogenerating set. Then F admits a left adjoint if and only if F preserves small limits.*

(2) *Assume C is a cocomplete, co-well-powered category which admits a generating set. Then F admits a right adjoint if and only if F preserves small colimits.*

Proof

(1) If F admits a left adjoint then F is a right adjoint and it preserves limits by Theorem 3.4.4.

Assume now that F preserves small limits and consider $D \in \mathrm{Ob}\,\mathcal{D}$. In light of Theorem 3.11.7, (1) it suffices to find a solution set for D. To this end, consider a coseparating set $S = \{G_i \mid i \in I\}$ of C and denote by

$$\left(P,\ (p_f)_{\substack{i \in I \\ f \in \mathrm{Hom}_C(C, G_i)}}\right)$$

the product of the family S, where the product consists of as many copies of G_i as there are elements in $\mathrm{Hom}_C(C,\ G_i)$, where C is a fixed object. Similarly, we consider the product

$$\left(P',\ (q_f)_{\substack{i \in I \\ f \in \mathrm{Hom}_{\mathcal{D}}(D, F(G_i))}}\right)$$

of the same family S, but this time consisting of as many copies of G_i as there are morphisms in $\mathrm{Hom}_{\mathcal{D}}(D,\ F(G_i))$.

Let $U_D = \{T \mid T \text{ is a subobject of } P'\}$. Note that U_D is in fact a set as C is well-powered.

Consider now $g \in \mathrm{Hom}_{\mathcal{D}}(D,\ F(C))$. By Proposition 3.12.3, (1) the unique morphism $\alpha_C \in \mathrm{Hom}_C(C,\ P)$ such that the following holds for all $i \in I$ and all $f \in \mathrm{Hom}_C(C,\ G_i)$

$$\begin{array}{cc} & G_i \qquad\qquad \text{i.e.,}\quad p_f \circ \alpha_C = f, \\ \nearrow^{f} \quad \uparrow_{p_f} & \\ C \xrightarrow[\alpha_C]{} P & \end{array} \tag{3.98}$$

is a monomorphism.

As $\left(P,\ (p_f)_{\substack{i \in I \\ f \in \mathrm{Hom}_C(C, G_i)}}\right)$ is a product, there exists a unique morphism $\beta_C \colon P' \to P$ such that the following diagram is commutative for all $i \in I$ and all $f \in \mathrm{Hom}_C(C,\ G_i)$:

$$\begin{array}{cc} & G_i \qquad\qquad \text{i.e.,}\quad p_f \circ \beta_C = q_{F(f) \circ g}. \\ \nearrow^{q_{F(f) \circ g}} \quad \uparrow_{p_f} & \\ P' \xrightarrow[\beta_C]{} P & \end{array} \tag{3.99}$$

As C is a complete category, the pair of morphisms (α_C, β_C) admits a pullback, which we denote by (S, μ, γ)

$$
\begin{array}{ccc}
S & \xrightarrow{\ \gamma\ } & P' \\
{\scriptstyle \mu}\downarrow & & \downarrow{\scriptstyle \beta_C} \\
C & \xrightarrow[\alpha_C]{} & P
\end{array}
\qquad (3.100)
$$

Furthermore, since α_C is a monomorphism it follows by Proposition 2.1.17, (1) that γ is also a monomorphism and therefore we can assume without loss of generality that $S \in U_D$. Indeed, if $S \notin U_D$, then there exists some $S' \in U_D$ together with a monomorphism $\gamma': S' \to P'$ and an isomorphism $u \in \operatorname{Hom}_C(S', S)$ such that $\gamma' = \gamma \circ u$. Then, the triple $(S', \gamma \circ u, \mu \circ u)$ is also a pullback of the pair of morphisms (α_C, β_C) and $S' \in U_D$.

As F is assumed to be limit preserving, $\left(F(P'),\ \big(F(q_f)\big)_{\substack{i \in I \\ f \in \operatorname{Hom}_{\mathcal{D}}(D, F(G_i))}} \right)$ is the product of the family $\{F(G_i) \mid i \in I\}$, where the product consists of as many copies of G_i as there are morphisms in $\operatorname{Hom}_{\mathcal{D}}(D, F(G_i))$. Hence, we obtain a unique morphism $\lambda: D \to F(P')$ such that the following diagram is commutative for all $i \in I$ and all $h \in \operatorname{Hom}_{\mathcal{D}}(D, F(G_i))$:

$$
\begin{array}{ccc}
D & & \\
{\scriptstyle \lambda}\downarrow & \searrow^{h} & \\
F(P') & \xrightarrow[F(q_h)]{} & F(G_i)
\end{array}
\qquad \text{i.e.,}\quad F(q_h)\circ\lambda = h.
\qquad (3.101)
$$

Now using again the fact that F is limit preserving we obtain, in particular, that $\left(F(P),\ \big(F(p_f)\big)_{\substack{i \in I \\ f \in \operatorname{Hom}_C(C, G_i)}} \right)$ is the product of the family $\{F(G_i) \mid i \in I\}$, where the product consists of as many copies of G_i as there are morphisms in $\operatorname{Hom}_C(C, G_i)$. Therefore, for all $i \in I$ and all $f \in \operatorname{Hom}_C(C, G_i)$ we have

$$
F(p_f) \circ F(\alpha_C) \circ g = F(p_f \circ \alpha_C) \circ g \stackrel{(3.98)}{=} F(f) \circ g \stackrel{(3.101)}{=} F\big(q_{F(f)\circ g}\big) \circ \lambda
$$
$$
\stackrel{(3.99)}{=} F\big(p_f \circ \beta_C\big) \circ \lambda = F(p_f) \circ F(\beta_C) \circ \lambda.
$$

Proposition 2.2.14, (1) implies that $F(\alpha_C) \circ g = F(\beta_C) \circ \lambda$. As F is limit preserving it follows that $(F(S), F(\mu), F(\gamma))$ is the pullback of the pair of morphisms $(F(\alpha_C), F(\beta_C))$ and we obtain a unique morphism $g' \in \operatorname{Hom}_{\mathcal{D}}(D, F(S))$ such that $F(\mu) \circ g' = g$ and $F(\gamma) \circ g' = \lambda$. The complete picture is captured in the diagram below:

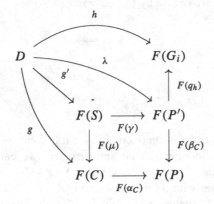

To conclude, we have proved that for any $C \in \mathrm{Ob}\,C$ and $g \in \mathrm{Hom}_{\mathcal{D}}(D, F(C))$ there exists some $S \in U_D$ together with morphisms $\mu \in \mathrm{Hom}_C(S, C)$ and $g' \in \mathrm{Hom}_{\mathcal{D}}(D, F(S))$ such that $F(\mu) \circ g' = g$. The desired conclusion now follows by Theorem 3.11.7, (1).

(2) To start with, note that the category C^{op} is complete, well-powered and admits a coseparating set. By applying 1) for the functor $F^{op} : C^{op} \to \mathcal{D}^{op}$, we obtain that F^{op} admits a left adjoint if and only if F^{op} preserves small limits. In light of Lemma 2.5.2 and Theorem 3.4.3 it follows that F admits a right adjoint if and only if F preserves small colimits, as desired.

□

Corollary 3.12.5 *Let C be a category.*

(1) If C is complete, well-powered and admits a coseparating set, then C is also cocomplete.

(2) If C is cocomplete, co-well-powered and admits a separating set, then C is also complete.

Proof

(1) Consider the diagonal functor $\Delta : C \to \mathrm{Fun}(I, C)$ which preserves small limits, as proved in Proposition 3.10.3. Since the conditions in Theorem 3.12.4, (1) are fulfilled it follows that Δ admits a left adjoint. Now Theorem 3.10.1, (2) implies that C is cocomplete.

(2) The diagonal functor $\Delta : C \to \mathrm{Fun}(I, C)$ also preserves small colimits, as proved in Proposition 3.10.3. Since the conditions in Theorem 3.12.4, (2) are fulfilled it follows that Δ admits a right adjoint. Now Theorem 3.10.1, (1) implies that C is complete.

□

Examples 3.12.6

(1) Let U: **KHaus** $\to$ **Top** be the forgetful functor. Recall that **KHaus**, the category **KHaus** of compact Hausdorff spaces, is well-powered, as shown in Example 1.3.16, and has a cogenerator by Example 3.12.2, (3). Furthermore, **KHaus** has products (Example 2.1.5, (4)) and equalizers (Example 2.1.10, (4)), which shows that is complete by Theorem 2.4.2. As both products and equalizers are constructed as in **Top** we can conclude by the Special Adjoint Functor Theorem that U has a left adjoint.

(2) Let F: $K \to J$ be a functor between small categories and consider the induced functor F^*: Fun$(J, \textbf{Set}) \to$ Fun$(K, \textbf{Set})$ defined in (1.37)

$$F^*(G) = GF, \quad F^*(\psi)_k = \psi_{F(k)}$$

for all functors G, H: $J \to$ **Set** and all natural transformations ψ: $G \to H$. We will use the Special Adjoint Functor Theorem to prove that F^* has a right adjoint. Indeed, note that the category Fun$(J, \textbf{Set})$ has a set of generators, as proved in Example 3.12.2, (4). Furthermore, **Set** is cocomplete by Example 2.4.4 while Example 2.7.4 shows that Fun$(J, \textbf{Set})$ is also cocomplete. As **Set** is cocomplete, it follows from Proposition 2.7.5 that a morphism ψ: $F \to G$ in Fun$(J, \textbf{Set})$ (i.e., a natural transformation) is an epimorphism if and only if each ψ_j: $F(j) \to G(j)$ is an epimorphism in **Set** for all $j \in$ Ob J. **Set** is co-well-powered by Example 1.3.16, which shows that each $F(j)$ has only a set of quotients and since J is small we can conclude that F has a set of quotients. Therefore, Fun$(J, \textbf{Set})$ is co-well-powered. We are left to show that F^* preserves colimits. To this end, let G: $I \to$ Fun$(J, \textbf{Set})$ and denote by $\left(H, (q_i: G(i) \to H)_{i \in \text{Ob} I}\right)$ its colimit, where H: $J \to$ **Set** is a functor and q_i is a natural transformation for all $i \in$ Ob I. The proof will be finished once we show that $\left(F^*(H), \left(F^*(q_i)\right)_{i \in \text{Ob} I}\right)$ is the colimit of the functor $F^* \circ G$: $I \to$ Fun$(K, \textbf{Set})$. In light of Theorem 2.7.2 it is enough to prove that for any $k \in$ Ob K the pair $\left(F^*(H)(k), \left(F^*(q_i)_k\right)_{i \in \text{Ob} I}\right)$ is the colimit of $(F^* \circ G)_k$: $I \to$ **Set**, where $(F^* \circ G)_k$ denotes the induced functor as defined in (2.40).

We start by showing that $\left(F^*(H)(k), \left(F^*(q_i)_k\right)_{i \in \text{Ob} I}\right)$ is a cocone on $(F^* \circ G)_k$: $I \to$ **Set**, i.e., for all $u \in$ Hom$_I(l, t)$ the following diagram is commutative:

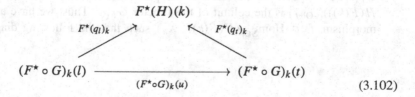

$$(3.102)$$

Indeed, recall from Theorem 2.7.2 that $\left(H(j), (q_i^j: G_j(i) \to H(j))_{i\in \mathrm{Ob}\, I}\right)$ is the colimit of the induced functor $G_j: I \to \mathbf{Set}$ and in particular a cocone on G_j, where $q_i^j = (q_i)_j$ for all $i \in \mathrm{Ob}\, I$ and $j \in \mathrm{Ob}\, J$. Therefore, the following diagram is commutative:

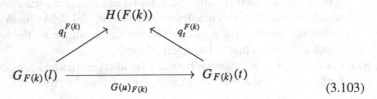

$$(3.103)$$

Now note that since we have

$$F^*(H)(k) = H(F(k)),$$

$$F^*(q_l)_k = (q_l)_{F(k)} = q_l^{F(k)},$$

$$(F^* \circ G)_k(u) = \left((F^* \circ G)(u)\right)_k = F^*(G(u))_k = G(u)_{F(k)},$$

$$(F^* \circ G)_k(l) = (F^* \circ G)(l)(k) = F^*(G(l))(k) = G(l)(F(k)) = G_{F(k)}(l),$$

it can be easily seen that the commutativity of (3.103) implies the commutativity of (3.102).

Consider now another cocone $\left(X_k, (s_i^k: (F^* \circ G)_k(i) \to X_k)_{i\in \mathrm{Ob}\, I}\right)$ on the functor $(F^* \circ G)_k: I \to \mathbf{Set}$. Hence, for all $u \in \mathrm{Hom}_I(l, t)$ the following diagram is commutative:

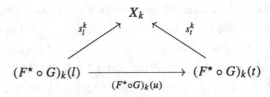

As $(F^* \circ G)_k(u) = G(u)_{F(k)}$ and $(F^* \circ G)_k(i) = G_{F(k)}(i)$, the commutativity of the above diagram comes down to $G(u)_{F(k)} \circ s_l^k = s_t^k$. Therefore, $\left(X_k, (s_i^k: G_{F(k)}(i) \to X_k)_{i\in \mathrm{Ob}\, I}\right)$ is a cocone on the induced functor $G_{F(k)}: I \to \mathbf{Set}$. Now recall that the pair $\left(H(F(k)), (q_i^{F(k)}: G_{F(k)}(i) \to H(F(k)))_{i\in \mathrm{Ob}\, I}\right)$ is the colimit of the functor $G_{F(k)}$. Thus, we have a unique morphism $f \in \mathrm{Hom}_{\mathbf{Set}}(H(F(k)), X_k)$ such that the following diagram is commutative for all $i \in \mathrm{Ob}\, I$:

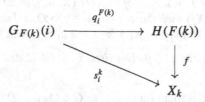

To conclude, there exists a unique morphism $f \in \mathrm{Hom}_{\mathbf{Set}}(H(F(k)), X_k)$ such that for all $i \in \mathrm{Ob}\, I$ we have

$$f \circ F^\star(q_i)_k = s_i^k.$$

This shows that $\left(F^\star(H)(k), \left(F^\star(q_i)_k\right)_{i \in \mathrm{Ob}\, I}\right)$ is the colimit of the functor $F^\star \circ G \colon I \to \mathrm{Fun}(K, \mathbf{Set})$, as desired. $\square$

3.13 Representable Functors Revisited

This section collects new representability criteria for certain classes of functors. The first one refers to limit preserving functors.

Theorem 3.13.1 (Representability criterion) *Let C be a complete category and $F \colon C \to \mathbf{Set}$ a functor such that*

(1) F preserves limits;
(2) there exists a set I, a family of objects $(X_i)_{i \in I}$ in C and for each $i \in I$ an element $f_i \in F(X_i)$ such that for any $Y \in \mathrm{Ob}\, C$ and any $g \in F(Y)$ there exists a morphism $\varphi \in \mathrm{Hom}_C(X_{i_0}, Y)$ for some $i_0 \in I$ such that $F(\varphi)(f_{i_0}) = g$.

Then F is representable, i.e., there exists $X \in \mathrm{Ob}\, C$ and a natural isomorphism $F \cong \mathrm{Hom}_C(X, -)$.

Proof To start with, note that condition 2) implies that $\{(f_i^*, X_i)_{i \in I}\}$ is a weakly initial set in the comma category $(\{\star\} \downarrow F)$, where $\{\star\}$ denotes a singleton set and $f_i^* \in \mathrm{Hom}_{\mathbf{Set}}(\{\star\}, F(X_i))$ is defined by $f_i^*(\star) = f_i \in F(X_i)$ for all $i \in I$. Since C is a complete category and F preserves limits, the comma category $(\{\star\} \downarrow F)$ is also complete by Corollary 2.6.5, (1). Now Lemma 3.11.3, (1) implies that $(\{\star\} \downarrow F)$ has an initial object and therefore F is representable by Proposition 1.8.8. $\square$

We state, for the sake of completeness, the contravariant version of the representability criterion:

Theorem 3.13.2 (Representability criterion for contravariant functors) *Let $\mathcal{D}$ be a cocomplete category and $G \colon \mathcal{D} \to \mathbf{Set}$ a contravariant functor such that*

(1) G turns colimits into limits;

(2) there exists a set I, a family of objects $(X_i)_{i \in I}$ in $\mathcal{D}$ and for each $i \in I$ an element $f_i \in G(X_i)$ such that for any $Y \in \text{Ob}\,\mathcal{D}$ and any $g \in G(Y)$ there exists a morphism $\varphi \in \text{Hom}_{\mathcal{D}}(Y, X_{i_0})$ for some $i_0 \in I$ such that $G(\varphi)(f_{i_0}) = g$.

Then F is representable, i.e., there exists $X \in \text{Ob}\,C$ and a natural isomorphism $F \cong \text{Hom}_C(-, X)$.

Proof Consider the covariant functor $F = G \circ O_{\mathcal{D}^{op}} : \mathcal{D}^{op} \to \textbf{Set}$. Note that $\mathcal{D}^{op}$ is a complete category and F preserves limits. Furthermore, condition (2) can be rephrased as follows: there exists a set I and a family of objects $(X_i)_{i \in I}$ in $\mathcal{D}^{op}$ and for each $i \in I$ an element $f_i \in F(X_i)$ such that for any $Y \in \text{Ob}\,\mathcal{D}^{op}$ and any $g \in F(Y)$ there exists a morphism $\varphi^{op} \in \text{Hom}_{\mathcal{D}^{op}}(X_i, Y)$ such that $F(\varphi)(f_i) = g$. Thus, F fulfills all conditions in Theorem 3.13.1 and therefore F is representable. This shows that G is a representable contravariant functor, as desired. □

Our next result relates representability to adjoint functors.

Theorem 3.13.3 *Let $F: C \to \mathcal{D}$ and $G: \mathcal{D} \to C$ be two functors. Then:*

(1) G has a left adjoint if and only if all functors

$$\text{Hom}_C(C, G(-)): \mathcal{D} \to \textbf{Set}$$

are representable for all $C \in \text{Ob}\,C$;

(2) F has a right adjoint if and only if all contravariant functors

$$\text{Hom}_{\mathcal{D}}(F(-), D): C \to \textbf{Set}$$

are representable for all $D \in \text{Ob}\,\mathcal{D}$.

Proof

(1) Assume first that $F: C \to \mathcal{D}$ is left adjoint of G. Then, for all $C \in \text{Ob}\,C$ and $D \in \text{Ob}\,\mathcal{D}$ we have a bijective map $\theta_{C,D}: \text{Hom}_{\mathcal{D}}(F(C), D) \to \text{Hom}_C(C, G(D))$ which is natural in both variables. In particular, naturality in the second variable implies that for all $C \in \text{Ob}\,C$, we have a natural isomorphism between the functors $\text{Hom}_{\mathcal{D}}(F(C), -)$ and $\text{Hom}_C(C, G(-))$. This shows precisely that the functor $\text{Hom}_C(C, G(-)): \mathcal{D} \to \textbf{Set}$ is representable and its representing object is $F(C)$.

Conversely, suppose that the functors $\text{Hom}_C(C, G(-)): \mathcal{D} \to \textbf{Set}$ are representable for all $C \in \text{Ob}\,C$. We will show that for all $C \in \text{Ob}\,C$, the comma category $(C \downarrow G)$ has an initial object. The conclusion will follow by Lemma 3.11.1, (1). Indeed, as $\text{Hom}_C(C, G(-)): \mathcal{D} \to \textbf{Set}$ is representable, there exists $X_C \in \text{Ob}\,\mathcal{D}$ and a natural isomorphism $\alpha: \text{Hom}_{\mathcal{D}}(X_C, -) \to \text{Hom}_C(C, G(-))$. We will prove that the pair $\big(\alpha_{X_C}(1_{X_C}), X_C\big)$ is the initial object of the comma category $(C \downarrow G)$. To this end, given $(g, D) \in \text{Ob}(C \downarrow G)$ we define $h \in \text{Hom}_{\mathcal{D}}(X_C, D)$ by $h = \alpha_D^{-1}(g)$. The proof will be finished once we show that h is the unique morphism such that $G(h) \circ \alpha_{X_C}(1_{X_C}) = g$. Recall that α is a natural transformation and therefore the following diagram is commutative:

$$
\begin{array}{ccc}
\mathrm{Hom}_{\mathcal{D}}(X_C, X_C) & \xrightarrow{\;\alpha_{X_C}\;} & \mathrm{Hom}_{\mathcal{C}}(C, G(X_C)) \\
{\scriptstyle \mathrm{Hom}_{\mathcal{D}}(X_C, h)}\Big\downarrow & & \Big\downarrow {\scriptstyle \mathrm{Hom}_{\mathcal{C}}(C, G(h))} \\
\mathrm{Hom}_{\mathcal{D}}(X_C, D) & \xrightarrow[\;\alpha_D\;]{} & \mathrm{Hom}_{\mathcal{C}}(C, G(D))
\end{array}
$$

The commutativity of the above diagram applied to the morphism $1_{X_C} \in \mathrm{Hom}_{\mathcal{D}}(X_C, X_C)$ yields $G(h) \circ \alpha_{X_C}(1_{X_C}) = \alpha_D(h)$, which comes down to

$$G(h) \circ \alpha_{X_C}(1_{X_C}) = g,$$

as desired. We are left to show that h is the unique morphism with this property. Indeed, assume there exists an $\overline{h} \in \mathrm{Hom}_{\mathcal{D}}(X_C, D)$ such that $G(\overline{h}) \circ \alpha_{X_C}(1_{X_C}) = g$. Using again the naturality of α this time for the morphism $\overline{h}$ yields $G(\overline{h}) \circ \alpha_{X_C}(1_{X_C}) = \alpha_D(\overline{h})$. It follows that $\alpha_D(\overline{h}) = g$ and since α_D is bijective we obtain $h = \overline{h}$, which finishes the proof.

(2) The second part follows in a similar manner. Indeed, if G is right adjoint to F then for all $C \in \mathrm{Ob}\,\mathcal{C}$ and $D \in \mathrm{Ob}\,\mathcal{D}$ we have a bijective map $\theta_{C,D} \colon \mathrm{Hom}_{\mathcal{D}}(F(C), D) \to \mathrm{Hom}_{\mathcal{C}}(C, G(D))$ which is natural in both variables. In particular, naturality in the first variable implies that for all $D \in \mathrm{Ob}\,\mathcal{D}$, we have a natural isomorphism between the functors $\mathrm{Hom}_{\mathcal{D}}(F(-), D)$ and $\mathrm{Hom}_{\mathcal{C}}(-, G(D))$. Hence the functor $\mathrm{Hom}_{\mathcal{D}}(F(-), D) \colon \mathcal{C} \to \mathbf{Set}$ is representable and its representing object is $G(D)$.

Conversely, as the functors $\mathrm{Hom}_{\mathcal{D}}(F(-), D) \colon \mathcal{C} \to \mathbf{Set}$ are representable for all $D \in \mathrm{Ob}\,\mathcal{D}$, there exists $Y_D \in \mathrm{Ob}\,\mathcal{C}$ and a natural isomorphism $\beta \colon \mathrm{Hom}_{\mathcal{C}}(-, Y_D) \to \mathrm{Hom}_{\mathcal{D}}(F(-), D)$. Then, it can be easily proved that $\big(Y_D, \beta_{Y_D}(1_{Y_D})\big)$ is the final object of the category $(F \downarrow D)$. The conclusion now follows by Lemma 3.11.1, (2).

$\square$

We end this section with an example which shows the existence of an algebraic object, namely the tensor product of modules, without explicitly constructing it.

Example 3.13.4 Let $\mathrm{Bil}_{M,N} \colon \mathbf{Ab} \to \mathbf{Ab}$ be the functor defined in Example 1.5.3, (28). It was proved in Example 3.11.8, (2) that $\mathrm{Bil}_{M,N}$ admits a left adjoint. Now Theorem 3.13.3, (1) shows that the functor $\mathrm{Hom}_{\mathbf{Ab}}(A, \mathrm{Bil}_{M,N}(-)) \colon \mathbf{Ab} \to \mathbf{Set}$ is representable for all abelian groups A. In particular, using Proposition 1.7.7, the representability of the functor $\mathrm{Hom}_{\mathbf{Ab}}(\{0\}, \mathrm{Bil}_{M,N}(-))$ implies the existence of a representing pair, denoted by $(M \otimes_R N, i)$. Hence, $M \otimes_R N$ is an abelian group and $i \in \mathrm{Bil}_{M,N}(M \otimes_R N)$, i.e., $i \colon M \times N \to M \otimes_R N$ is a bilinear map, such that for any other pair (A, f), where A is an abelian group and $f : M \times N \to A$ is a bilinear map, there exists a unique group homomorphism $g : M \otimes_R N \to A$ which makes the following diagram commutative:

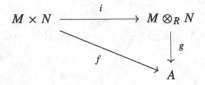

This means precisely that $M \otimes_R N$ is the tensor product of the R-modules M and N (see [45, Definition 1.5]).   □

It is worth to point out that the various adjoint functor theorems proved in this chapter have many notable applications, most of them exceeding the purpose of this introductory book. For instance, the following corollary to Freyd's theorem is presumedly *"more widely known than the theorem itself"* as stated in [7]: any functor between varieties of algebras which respects underlying sets has a left adjoint ([7, Corollary 8.17]). For precise definitions and more details we refer the reader to [7].

Furthermore, the adjoint functor theorems have been extended to various settings allowing for important applications. We only mention here the case of triangulated categories and an important consequence in algebraic geometry. A criterion for a functor between triangulated categories to admit a right adjoint was proved by building on a version of the representability theorem for triangulated categories. More precisely, a triangulated functor between triangulated categories satisfying certain technical conditions which commutes with arbitrary coproducts admits a right adjoint. A notable application of the aforementioned criterion on the existence of adjoints for triangulated functors is the Grothendieck duality theorem proved by A. Neeman (see [42] for further details).

3.14 Exercises

3.1 Prove that the forgetful functor $U :$ **Field** $\to$ **Ring** does not admit a right or a left adjoint.

3.2 Decide if the forgetful functor $U : \mathcal{A} \to$ Set admits a right adjoint, where $\mathcal{A}$ is **Grp**, **Ring** or $_R\mathcal{M}$.

3.3 Let R be a commutative ring. Show that the forgetful functor $U : {}_R\mathcal{M} \to$ **Ab** has both a left and a right adjoint.

3.4 If R is a commutative ring, show that the forgetful functor $F :$ **Alg**$_R \to {}_R\mathcal{M}$ (forgetting the multiplicative structure) has a left adjoint.

3.5 Decide if the inclusion functor $I :$ **Ring**$^c \to$ **Ring** has a left or a right adjoint.

3.6 Let $F : C \to \mathcal{D}$ and $G : \mathcal{D} \to C$ be functors such that $F \dashv G$. Prove that F preserves epimorphisms and G preserves monomorphisms.

3.7 Let $F: C \to \mathcal{D}$ and $G: \mathcal{D} \to C$ be functors such that $F \dashv G$. Show that if GF is fully faithful then F is fully faithful.

3.8 Let $F: C \to \mathcal{D}$ and $G: \mathcal{D} \to C$ be functors such that $F \dashv G$, and let $\eta: 1_C \to GF$ and $\varepsilon: FG \to 1_\mathcal{D}$ be the unit and respectively the counit of this adjunction. Then the following are equivalent:

(a) $F(\eta_C)$ is an isomorphism for all $C \in \mathrm{Ob}\, C$;
(b) $GF(\eta_C) = \eta_{GF(C)}$ for all $C \in \mathrm{Ob}\, C$;
(c) $\varepsilon_{F(C)}$ is an isomorphism for all $C \in \mathrm{Ob}\, C$;
(d) $G(\varepsilon_{F(C)})$ is an isomorphism for all $C \in \mathrm{Ob}\, C$.

3.9 Let $H: C \to \mathcal{D}$ and $F, G: \mathcal{D} \to C$ be functors such that $F \dashv H$ and $H \dashv G$. If $\eta: 1_\mathcal{D} \to HF$ is the unit of the adjunction $F \dashv H$ and $\varepsilon: HG \to 1_\mathcal{D}$ is the counit of the adjunction $H \dashv G$ then $\eta_D: D \to HF(D)$ is an epimorphism for every $D \in \mathrm{Ob}\, \mathcal{D}$ if and only if $\varepsilon_D: HG(D) \to D$ is a monomorphism for every $D \in \mathrm{Ob}\, \mathcal{D}$.

3.10 Let $F: C \to \mathcal{D}$ and $G: \mathcal{D} \to C$ be functors such that $F \dashv G$, and let $\eta: 1_C \to GF$ and $\varepsilon: FG \to 1_\mathcal{D}$ be the unit and respectively the counit of this adjunction. Show that the categories $\mathrm{Iso}(C)$ and $\mathrm{Iso}(\mathcal{D})$ are equivalent, where $\mathrm{Iso}(C)$ denotes the full subcategory of C consisting of those objects $C \in \mathrm{Ob}\, C$ for which η_C is an isomorphism and $\mathrm{Iso}(\mathcal{D})$ is the full subcategory of $\mathcal{D}$ consisting of those objects $D \in \mathrm{Ob}\, \mathcal{D}$ for which ε_D is an isomorphism.

3.11 Let $F, F': C \to \mathcal{D}, G, G': \mathcal{D} \to C$ be functors such that $F \dashv G$ and $F' \dashv G'$ and suppose that I is a small category.

(a) If $T: I \to \mathcal{D}$ is a functor which admits a limit and $\beta: G \to G'$ is a natural transformation such that $\beta_{T(i)}$ is an isomorphism in C for every $i \in \mathrm{Ob}\, I$, then $\beta_{\lim T}$ is also an isomorphism in C.
(b) If $H: I \to C$ is a functor which admits a colimit and $\alpha: F \to F'$ is a natural transformation such that $\alpha_{H(i)}$ is an isomorphism in $\mathcal{D}$ for every $i \in \mathrm{Ob}\, I$, then $\alpha_{\mathrm{colim}\, H}$ is also an isomorphism in $\mathcal{D}$.

3.12 Let $\mathrm{OB}: \mathbf{Cat} \to \mathbf{Set}$ be the objects functor defined as follows for all small categories $C, \mathcal{D}$ and all functors $F: C \to \mathcal{D}$:

(a) $\mathrm{OB}(C) = \mathrm{Ob}\, C$;
(b) $\mathrm{OB}(F) = F: \mathrm{Ob}\, C \to \mathrm{Ob}\, \mathcal{D}$.

Show that OB has both a left and a right adjoint.

3.13 Show that the inclusion functor $I: \mathbf{Poset} \to \mathbf{PreOrd}$ has a left adjoint.

3.14 Let X be a set and consider the cartesian product functor $X \times -: \mathbf{Set} \to \mathbf{Set}$. Find the sets X for which the functor $X \times -$ admits a left adjoint.

3.15 Show that the inclusion functor $I: \mathbf{Haus} \to \mathbf{Top}$ does not admit a right adjoint.

3.16 Let $F: C \to \mathcal{D}$ and $G: \mathcal{D} \to C$ be functors such that $F \dashv G$.

(a) If the functor $F': C \to \mathcal{D}$ is naturally isomorphic to F then $F' \dashv G$.
(b) If the functor $G': \mathcal{D} \to C$ is naturally isomorphic to G then $F \dashv G'$.

3.17 Let $F: C \to \mathcal{D}$ be an equivalence of categories. Prove that

(a) $f \in \mathrm{Hom}_C(C, C')$ is a monomorphism if and only if $F(f)$ is a monomorphism;

(b) $f \in \mathrm{Hom}_C(C, C')$ is an epimorphism if and only if $F(f)$ is an epimorphism;

(c) $f \in \mathrm{Hom}_C(C, C')$ is an isomorphism if and only if $F(f)$ is an isomorphism.

3.18 Let $F: C \to \mathcal{D}$ be a fully faithful functor. Show that C is equivalent to a full subcategory of $\mathcal{D}$.

3.19 Let R, S be two rings. Show that the product category $_RM \times _SM$ is equivalent to the category $_{R \times S}M$.

3.20 Show that the ring R is a generator in the category $_RM$ of left R-modules.

3.21 Let C be a pointed category which admits (co)products. Then C has a (co)generator if and only if it has a set of (co)generators.

3.22 Let $F: C \to \mathcal{D}$ and $G: \mathcal{D} \to C$ be functors such that $F \dashv G$. Prove that if G is faithful and S is a generator in C then $F(S)$ is a generator in $\mathcal{D}$.

3.23 Prove that an object S in a category C is a generator if and only if the functor $\mathrm{Hom}_C(S, -): C \to \mathbf{Set}$ is faithful. State and prove the dual.

3.24 Let C be a category and S a class of morphisms in C such that the localization category C_S exists. Then the localization of C^{op} by S^{op} exists too and we have an isomorphism of categories between $C^{op}_{S^{op}}$ and $(C_S)^{op}$, where S^{op} denotes the class of all opposites of morphisms in S.

3.25 Let $F: C \to \mathcal{D}$ and $G: \mathcal{D} \to C$ be two functors such that $F \dashv G$. Prove that

(a) if F preserves monomorphisms then G preserves injective objects;

(b) if G preserves epimorphisms then F preserves projective objects.

Chapter 4
Solutions to Selected Exercises

4.1 Chapter 1

1.4 (a) Let u denote the unique morphism in $\mathrm{Hom}_C(X, C)$. Then $m \circ u \in \mathrm{Hom}_C(X, X) = \{1_X\}$ and therefore we have

$$m \circ u = 1_X. \tag{4.1}$$

Furthermore, we have $\underline{m \circ u} \circ m \overset{(4.1)}{=} m$ and since m is a monomorphism we obtain $u \circ m = 1_C$.

(b) This claim follows by the duality principle; indeed, applying a) for the dual category C^{op} yields the desired claim. □

1.5 (b) Let $u, v \in \mathrm{Hom}_C(E, A)$ such that $f \circ u = f \circ v$. This implies $g \circ f \circ u = g \circ f \circ v$ and since $g \circ f$ is a monomorphism, we obtain $u = v$, as desired.

(c) Consider $t, w \in \mathrm{Hom}_C(C, D)$ such that $t \circ g = w \circ g$. This implies $t \circ g \circ f = w \circ g \circ f$ and since $g \circ f$ is an epimorphism, we obtain $t = w$. □

1.6 (a) Let $f \in \mathrm{Hom}_C(A, B)$ be a split monomorphism and denote by t its left inverse. If $g_1, g_2 \in \mathrm{Hom}_C(A', A)$ such that $f \circ g_1 = f \circ g_2$, then by composing on the left with t we obtain $g_1 = g_2$, which shows that f is a monomorphism.

For the converse consider the group morphism $f \colon \mathbb{Z}_2 \to \mathbb{Z}_4$ defined by $f(\overline{0}) = \hat{0}$ and $f(\overline{1}) = \hat{2}$, where $\overline{x}$ and $\hat{x}$ denote the residue classes modulo 2 and 4, respectively. It can be easily seen that f is a monomorphism; to this end, let $g_1, g_2 \in \mathrm{Hom}_{\mathbf{Grp}}(G, \mathbb{Z}_2)$ such that $f \circ g_1 = f \circ g_2$. If there exists some $x_0 \in G$ such that $g_1(x_0) \neq g_2(x_0)$ then we can assume without loss of generality that $g_1(x_0) = \overline{0}$ and $g_2(x_0) = \overline{1}$. This implies $\hat{0} = f(\overline{0}) = f(g_1(x_0)) = f(g_2(x_0)) = f(\overline{1}) = \hat{2}$, which is an obvious contradiction. Therefore, f is a monomorphism. We are left to show that f is not a split monomorphism. Indeed, assume that there exists a group morphism $g \colon \mathbb{Z}_4 \to \mathbb{Z}_2$ such that $g \circ f = 1_{\mathbb{Z}_2}$. This implies

$$g(\hat{0}) = g(f(\overline{0})) = \overline{0}, \qquad g(\hat{2}) = g(f(\overline{1})) = \overline{1}$$

and $g(\hat{1}) \in \{\overline{0}, \overline{1}\}$. If $g(\hat{1}) = \overline{0}$, it follows that $g(\hat{2}) = \overline{0}$, which is a contradiction. Thus, we must have $g(\hat{1}) = \overline{1}$, which leads to $g(\hat{2}) = \overline{2} = \overline{0}$, which is another contradiction. To conclude, we have proved that f is not a split monomorphism, as desired.

(b) Assume $f \in \mathrm{Hom}_C(A, B)$ is an epimorphism and split monomorphism. In particular, there exists a $t \in \mathrm{Hom}_C(B, A)$ such that $t \circ f = 1_A$. This implies that $f \circ t \circ f = f = 1_B \circ f$ and since f is an epimorphism we obtain $f \circ t = 1_B$. Hence, f is an isomorphism. The converse is obvious.
                                                                                          $\square$

1.7 Assume first that f is a strong epimorphism and a monomorphism. Then, the following commutative square

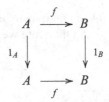

admits a unique $g \in \mathrm{Hom}_C(B, A)$ such that $g \circ f = 1_A$ and $f \circ g = 1_B$. This shows that f is an isomorphism.

Conversely, suppose now that f is an isomorphism and consider the following commutative square:

$$\begin{array}{ccc}
A & \xrightarrow{f} & B \\
\downarrow{\scriptstyle g} & & \downarrow{\scriptstyle h} \\
C & \xrightarrow{u} & D
\end{array} \qquad h \circ f = u \circ g,$$

where $u \in \mathrm{Hom}_C(C, D)$ is a monomorphism. It can be easily seen that $v = g \circ f^{-1} \colon B \to C$ is the unique morphism in C such that $v \circ f = g$ and $u \circ v = h$. We have proved that (a) is equivalent to (b). Furthermore, as f is an isomorphism in C if and only if f^{op} is an isomorphism in C^{op}, it follows that (a) is also equivalent to (c).
                                                                                          $\square$

1.8 Recall from Example 1.5.3, (8) that functors between categories associated to groups (in the sense of Example 1.2.2, (3)) are nothing but group homomorphisms between the given groups. Furthermore, the opposite of a category associated to a group is precisely the category associated to the opposite group[1]. Furthermore, for any group G, we have a group isomorphism $\theta: G \to G^{op}$ given by $\theta(g) = g^{-1}$ for all $g \in G$. This provides the desired isomorphism of categories between the category associated to the group G and its opposite.

For the case of monoids, it will suffice to consider the monoid described in Example 2.1.16, (10) as a counterexample. □

1.9 (a) The forgetful functor $U: \mathbf{Ring}^c \to \mathbf{Set}$ does not preserve epimorphisms. Indeed, recall from Example 1.3.2, (5) that the inclusion $i: \mathbb{Z} \to \mathbb{Q}$ is an epimorphism in $\mathbf{Ring}^c$ but not in $\mathbf{Set}$ as it is not a surjective map.

Consider now the abelianization functor $F: \mathbf{Grp} \to \mathbf{Ab}$ introduced in Example 1.5.3, (23). Let A_5 be the alternating group of degree 5 generated by the 3-cycles (123), (124), (125). Then the inclusion $i: C_3 \to A_5$ is an injective map and therefore a monomorphism in $\mathbf{Grp}$, where C_3 is the cyclic group generated by the 3-cycle (123). Now recall that $F(A_n) = \{1\}$ for all $n \geqslant 5$, while $F(C_n) = C_n$ for all $n \in \mathbb{N}^*$. Therefore $F(i): C_3 \to \{1\}$ is obviously not a monomorphism in $\mathbf{Ab}$.

(b) Let C be a category which is not a groupoid. Then, the inclusion functor $I: C^{\mathrm{grp}} \to C$ obviously reflects isomorphisms without being full, where C^{grp} is the core groupoid of the category C as defined in Example 1.3.6, (2). □

1.10 Consider the following two categories $\mathcal{B}$ and C:

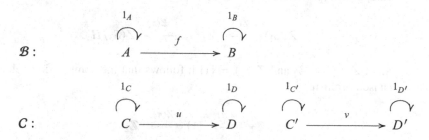

(a) The functor $F: \mathcal{B} \to C$ defined below is full while the morphism v and the object C' are not in its image:

$$F(A) = C, \quad F(B) = D,$$

$$F(1_A) = 1_C, \quad F(1_B) = 1_D, \quad F(f) = u.$$

(b) The functor $G: C \to \mathcal{B}$ defined as follows:

$$G(C) = G(C') = A, \quad G(D) = G(D') = B,$$

$$F(1_C) = F(1_{C'}) = 1_A, \quad F(1_D) = F(1_{D'}) = 1_B, \quad G(u) = G(v) = f$$

is faithful although we have $G(u) = G(v) = f$ and $G(C) = G(C')$. □

1.11 Assume there exists a functor $F: \mathbf{Grp} \to \mathbf{Grp}$ such that $F(G) = Z(G)$ for all groups G, where $Z(G)$ is the center of G. Consider the symmetric group S_3 on three letters and let H be the cyclic subgroup of S_3 generated by the 3-cycle $x = (123)$. It can be easily seen that H has order 3, H is a normal subgroup of S_3 and the quotient S_3/H has order 2 and is isomorphic to $\mathbb{Z}_2$. Furthermore, we consider $\mathbb{Z}_2$ to be the subgroup of S_3 generated by the transposition $y = (23)$. Consider now the following group morphisms:

$$\mathbb{Z}_2 \xrightarrow{\; i \;} S_3 \xrightarrow{\; \pi \;} S_3/H \ ,$$

where $i: \mathbb{Z}_2 \to S_3$ and $\pi: S_3 \to S_3/H$ are the inclusion map and respectively the quotient map. As $i(\mathbb{Z}_2) \cap H = \{1_{S_3}\}$ it follows that $\pi \circ i$ is a group isomorphism. In light of Proposition 1.6.9, (1) the following composition is a group isomorphism too:

$$Z(\mathbb{Z}_2) \xrightarrow{\; Z(i) \;} Z(S_3) \xrightarrow{\; Z(\pi) \;} Z(S_3/H).$$

Since $Z(\mathbb{Z}_2) = \mathbb{Z}_2$ and $Z(S_3) = \{1\}$ it follows that the composition below is an isomorphism

$$\mathbb{Z}_2 \xrightarrow{\; Z(i) \;} \{1\} \xrightarrow{\; Z(\pi) \;} \mathbb{Z}_2 \ ,$$

which is an obvious contradiction. Therefore, there is no functor $F: \mathbf{Grp} \to \mathbf{Grp}$ such that $F(G) = Z(G)$ for all groups G. □

1.12 Consider the functor $F: \mathbf{Top} \to \mathbf{Set}$ defined as follows for all topological spaces X, Y and $f \in \mathrm{Hom}_{\mathbf{Top}}(X, Y)$:

$$F(X) = \{X_i \mid X_i \text{ connected component of } X\},$$

$$F(f): F(X) \to F(Y), \quad F(f)(X_i) = \widetilde{X_i},$$

where $\widetilde{X_i}$ denotes the connected component of Y which contains $f(X_i)$. Since the image of a connected space under a continuous map is connected ([39, Theorem 23.5]), $\widetilde{X_i}$ is indeed a connected component of $F(Y)$ and therefore

$F(f)$ is well-defined. Furthermore, $F(1_X)$ is the identity on $F(X)$ and if $f \in$ $\mathrm{Hom}_{\mathbf{Top}}(X,\,Y)$, $g \in \mathrm{Hom}_{\mathbf{Top}}(Y,\,Z)$ and X_i is a connected component of X then $g\big(f(X_i)\big)$ is also connected. This shows that $F(g \circ f) = F(g) \circ F(f)$ and therefore F is indeed a functor. □

1.14 (a) Let $f,\,f' \in \mathrm{Hom}_C(A,\,B)$ such that $F(f) = F(f')$. Then we also have $GF(f) = GF(f')$ and since GF is faithful we obtain $f = f'$. This shows that F is faithful.

(b) Consider now $g \in \mathrm{Hom}_\mathcal{D}\big(F(A),\,F(B)\big)$. Then:

$$G(g) \in \mathrm{Hom}_\mathcal{D}\big(GF(A),\,GF(B)\big)$$

and since GF is full there exists some $f \in \mathrm{Hom}_C(A,\,B)$ such that $G(g) = GF(f)$. As G is faithful we obtain $g = F(f)$, which shows that F is full. □

1.15 (a) Let $G,\,G'\colon I \to C$ be two functors; we need to show that the following mapping is bijective:

$$\mathrm{Nat}(G,\,G') \to \mathrm{Nat}(FG,\,FG'), \quad \alpha \mapsto F\alpha,$$

where $\alpha\colon G \to G'$ is a natural transformation and $F\alpha$ is the whiskering of α on the right by F as defined in Example 1.7.2, (7). Suppose first that $\alpha,\,\alpha'\colon G \to G'$ are natural transformations such that $F\alpha = F\alpha'$. This implies that for all $i \in \mathrm{Ob}\,I$ we have $F(\alpha_i) = F(\alpha_i')$ and since F is faithful we obtain $\alpha_i = \alpha_i'$. This shows that $\alpha = \alpha'$ and therefore $F_\star$ is faithful.

Consider now $\gamma \in \mathrm{Nat}(FG,\,FG')$; for all $i \in \mathrm{Ob}\,I$ we have

$$\gamma_i \in \mathrm{Hom}_\mathcal{D}(FG(i),\,FG'(i))$$

and since F is fully faithful there exists a unique $\beta_i \in \mathrm{Hom}_\mathcal{D}(G(i),\,G'(i))$ such that $\gamma_i = F(\beta_i)$. We are left to show that the family of morphisms $\big(\beta_i\big)_{i \in \mathrm{Ob}\,I}$ form a natural transformation between the functors G and G'. To this end, consider $f \in \mathrm{Hom}_I(i,\,j)$. Since $\gamma\colon FG \to FG'$ is a natural transformation, the following diagram is commutative:

$$\begin{array}{ccc} FG(i) & \xrightarrow{\gamma_i} & FG'(j) \\ {\scriptstyle FG(f)}\downarrow & & \downarrow{\scriptstyle FG'(f)} \\ FG(j) & \xrightarrow[\gamma_j]{} & FG'(j) \end{array} \qquad \text{i.e., } FG'(f)\circ\gamma_i = \gamma_j\circ FG(f).$$

As $\gamma_i = F(\beta_i)$ for all $i \in \text{Ob}\,I$, the commutativity of the above diagram leads to $FG'(f) \circ F(\beta_i) = F(\beta_j) \circ FG(f)$ and since F is faithful we obtain $G'(f) \circ \beta_i = \beta_j \circ G(f)$, i.e., the following diagram is commutative:

$$
\begin{array}{ccc}
G(i) & \xrightarrow{\ \beta_i\ } & G'(j) \\
{\scriptstyle G(f)}\downarrow & & \downarrow{\scriptstyle G'(f)} \\
G(j) & \xrightarrow[\ \beta_j\ ]{} & G'(j)
\end{array}
$$

This shows that $\beta \colon G \to G'$ is indeed a natural transformation and, moreover, we have $F_\star(\beta) = \gamma$.

(b) As proved above, $F_\star$ is fully faithful and the desired conclusion now follows by Proposition 1.6.9, (2). $\qquad\square$

1.17 If $C \in \text{Ob}\,\mathcal{C}$, then the naturality of $\beta \colon H \to I$ applied to the morphism $\gamma_C \in \text{Hom}_{\mathcal{D}}(G(C), P(C))$ yields the following commutative diagram:

$$
\begin{array}{ccc}
H(G(C)) & \xrightarrow{\ \beta_{G(C)}\ } & I(G(C)) \\
{\scriptstyle H(\gamma_C)}\downarrow & & \downarrow{\scriptstyle I(\gamma_C)} \\
H(P(C)) & \xrightarrow[\ \beta_{P(C)}\ ]{} & I(P(C))
\end{array}
$$

i.e., $\quad \beta_{P(C)} \circ H(\gamma_C) = I(\gamma_C) \circ \beta_{G(C)}.$ $\qquad\qquad$ (4.2)

Therefore, for all $C \in \text{Ob}\,\mathcal{C}$, we have

$$
\begin{aligned}
\big((\delta \circ \beta) \ast (\gamma \circ \alpha)\big)_C \ &\overset{(1.18)}{=}\ (\delta \circ \beta)_{P(C)} \circ H\big((\gamma \circ \alpha)_C\big) \\
&=\ \delta_{P(C)} \circ \underline{\beta_{P(C)} \circ H(\gamma_C)} \circ H(\alpha_C) \\
&\overset{(4.2)}{=}\ \delta_{P(C)} \circ \underline{I(\gamma_C) \circ \beta_{G(C)}} \circ H(\alpha_C) \\
&\overset{(1.18)}{=}\ (\delta \ast \gamma)_C \circ (\beta \ast \alpha)_C \\
&=\ \big((\delta \ast \gamma) \circ (\beta \ast \alpha)\big)_C.
\end{aligned}
$$

$\hfill\square$

1.20 Let $\{v_i \mid i = \overline{1, n}\}$ be the basis of V over the field K and denote by $\{v_i^*\}_{i=\overline{1,n}}$ the dual basis of V^*. Define two natural transformations as follows:

$$\alpha: V^* \otimes - \to \mathrm{Hom}_K \mathcal{M}(V, -), \quad \alpha_U(f \otimes u)(v) = f(v)u,$$

$$\beta: \mathrm{Hom}_K \mathcal{M}(V, -) \to V^* \otimes -, \quad \beta_U(t) = \sum_{i=1}^n v_i^* \otimes t(v_i),$$

for all $U \in \mathrm{Ob}\,_K\mathcal{M}$, $u \in U$, $v \in V$, $f \in V^*$ and $t \in \mathrm{Hom}_K \mathcal{M}(V, U)$. It is straightforward to check that α and β are indeed natural transformations and that for all $U \in \mathrm{Ob}\,_K\mathcal{M}$ we have $\alpha_U \circ \beta_U = 1_{\mathrm{Hom}_K \mathcal{M}(V, U)}$ and $\beta_U \circ \alpha_U = 1_{V^* \otimes U}$. □

1.21 If C is the empty category then the empty functor from the empty category to **Set** is obviously not representable. For non-empty categories C, consider $A \in \mathrm{Ob}\,\mathbf{Set}$ such that $|A| = 2$ and define $F: C \to \mathbf{Set}$ as follows for all C, $D \in \mathrm{Ob}\,C$ and $f \in \mathrm{Hom}_C(C, D)$:

$$F(C) = A, \qquad F(f) = 1_A.$$

Assume F is representable and let (X, x) be the representing pair, where $X \in \mathrm{Ob}\,C$ and $x \in F(X) = A$. Consider now another pair (X', x') where $x' \in F(X') = A$ and $x' \neq x$. By Proposition 1.7.7 there exists a unique $f \in \mathrm{Hom}_C(X, X')$ such that $F(f)(x) = x'$. On the other hand, we have $F(f)(x) = 1_A(x) = x$ and $x \neq x'$. We have reached a contradiction and therefore F is not representable. □

1.25 Consider $\psi: \mathrm{Fun}\big(I, \mathrm{Fun}(J, C)\big) \to \mathrm{Fun}(I \times J, C)$ defined as follows for all functors $F, G: I \to \mathrm{Fun}(J, C)$ and all natural transformations $\eta: F \to G$:

$$\psi(F) = \mathcal{F}_F, \quad \psi(\eta)(i, j) = (\eta_i)_j, \quad \text{for all } (i, j) \in \mathrm{Ob}\,(I \times J),$$

where $\mathcal{F}_F: I \times J \to C$ is the functor defined by

$$\mathcal{F}_F(i, j) = F(i)(j), \quad \mathcal{F}_F(u, v) = F(k)(v) \circ F(u)_j,$$

for all $(i, j), (k, l) \in \mathrm{Ob}\,(I \times J)$ and $(u, v) \in \mathrm{Hom}_{I \times J}\big((i, j), (k, l)\big)$.

It can be easily checked that $\mathcal{F}_F$ is indeed a functor. We will show that ψ is an isomorphism of categories. To start with, we first show that ψ is indeed a functor. We have

$$\psi(1_F)(i, j) = \big((1_F)_i\big)_j = (1_{F(i)})_j = 1_{F(i)(j)},$$

which shows that ψ respects identities. Furthermore, for any two natural transformations $\eta: F \to G, \zeta: G \to H$, where $F, G, H: I \to \mathrm{Fun}(J, C)$ are functors, and any $(i, j) \in \mathrm{Ob}\,(I \times J)$, we have

$$\psi(\zeta \circ \eta)(i, j) = (\zeta \circ \eta_i)_j = (\zeta_i \circ \eta_i)_j = (\zeta_i)_j \circ (\eta_i)_j$$

$$= \psi(\zeta)(i, j) \circ \psi(\eta)(i, j)$$

$$= \big(\psi(\zeta) \circ \psi(\eta)\big)(i, j).$$

This shows that ψ respects compositions as well and is therefore a functor. Consider now $\varphi: \mathrm{Fun}(I \times J, C) \to \mathrm{Fun}\big(I, \mathrm{Fun}(J, C)\big)$ defined as follows for all functors $F, G: I \times J \to C$ and all natural transformations $\eta: F \to G$:

$$\varphi(F) = \mathcal{G}_F, \quad \big(\varphi(\eta)_i\big)_j = \eta_{(i, j)}, \quad \text{for all } i \in \mathrm{Ob}\,I, \ j \in \mathrm{Ob}\,J,$$

where $\mathcal{G}_F: I \to \mathrm{Fun}(J, C)$ is the functor defined by

$$\mathcal{G}_F(i) = F(i, -), \quad \mathcal{G}_F(f) = F(f, 1_j),$$

$$\text{for all } i, l \in \mathrm{Ob}\,I, \ j \in \mathrm{Ob}\,J \text{ and } f \in \mathrm{Hom}_I(i, l).$$

It can be easily checked by a straightforward computation that both $\mathcal{G}_F$ and φ are functors. We will show that ψ and φ are inverses to each other. Indeed, for all functors $F: I \to \mathrm{Fun}(J, C)$ and all $i \in \mathrm{Ob}\,I, j \in \mathrm{Ob}\,J$ we have

$$\varphi\big(\psi(F)\big)(i)(j) = \mathcal{F}_F(i, j) = F(i)(j).$$

Therefore, we have proved that $(\varphi \circ \psi)(F) = F$ for all functors $F: I \to \mathrm{Fun}(J, C)$. Consider now a natural transformation $\eta: F \to G$, where $F, G: I \to \mathrm{Fun}(J, C)$ are functors. Then, for all $i \in \mathrm{Ob}\,I, j \in \mathrm{Ob}\,J$ we have

$$\Big(\varphi\big(\psi(\eta)\big)_i\Big)_j = (\psi(\eta)_{(i, j)} = (\eta_i)_j.$$

Hence we obtain $\varphi\big(\psi(\eta)\big) = \eta$. To summarize, we have proved that $\varphi \circ \psi = 1_{\mathrm{Fun}\big(I, \mathrm{Fun}(J, C)\big)}$. The proof will be finished once we show that $\psi \circ \varphi = 1_{\mathrm{Fun}(I \times J, C)}$. Let $H: I \times J \to C$ be a functor. Then for all $(i, j) \in \mathrm{Ob}\,(I \times J)$ we have

$$\psi\big(\varphi(H)\big)(i, j) = \varphi(H)(i, j) = H(i, j),$$

as desired. Furthermore, if $\eta\colon H \to G$ is a natural transformation, where H, $G\colon I \times J \to C$ are functors, and $(i, j) \in \mathrm{Ob}\,(I \times J)$, we obtain

$$\psi\big(\varphi(\eta)\big)(i,\ j) = \big(\varphi(\eta)_i\big)_j = \eta_{(i,\ j)}.$$

Therefore $\psi \circ \varphi = 1_{\mathrm{Fun}(I \times J, C)}$ and the proof is finished. $\square$

4.2   Chapter 2

2.3 (a) $\Rightarrow$ (b) By assumption, (E, e) is the equalizer of (f, f), so there exists
a unique $w \in \mathrm{Hom}_C(A, E)$ such that the following diagram is
commutative:

$$\qquad\qquad E \xrightarrow{\ e\ } A \underset{f}{\overset{f}{\rightrightarrows}} B \qquad \text{i.e.,}\quad e \circ w = 1_A.$$

(4.3)

Consider $h_1, h_2 \in \mathrm{Hom}_C(A, B')$ such that $h_1 \circ e = h_2 \circ e$. Composing this equality with w on the right and using the commutativity of diagram (4.3) yields $h_1 = h_2$. This shows that e is an epimorphism.

(b) $\Rightarrow$ (c) In particular, we have $f \circ e = g \circ e$ and, since e is an epimorphism, we obtain $f = g$. Therefore, we have a unique $w \in \mathrm{Hom}_C(A, E)$ such that diagram (4.3) is commutative. Furthermore, we have $e \circ (w \circ e) = (e \circ w) \circ e \overset{(4.3)}{=} e$ and since e is an epimorphism we obtain $w \circ e = 1_E$. Putting everything together it follows that w is the inverse of e.

(c) $\Rightarrow$ (d) As e is an isomorphism and $f \circ e = g \circ e$ we obtain, after composing on the right with the inverse of e, that $f = g$. Now note that given $u \in \mathrm{Hom}_C(C, A)$ we have a unique morphism in $\mathrm{Hom}_C(C, A)$, namely u, which makes the following diagram commutative:

$$\qquad\qquad E \xrightarrow{\ 1_A\ } A \underset{f}{\overset{f}{\rightrightarrows}} B$$

(d) $\Rightarrow$ (a) Since $(A, 1_A)$ is the equalizer of (f, g) we have $f \circ 1_A = g \circ 1_A$ and therefore $f = g$, as desired. $\square$

2.6 Let $t \in \mathrm{Hom}_C(B, C')$ such that $t \circ f = t \circ g$. First we prove that $t \circ v$ makes the following diagram commutative:

$$
\begin{array}{ccc}
A \underset{g}{\overset{f}{\rightrightarrows}} & B & \xrightarrow{\ h\ } & C \\
 & & t\searrow & \downarrow{\scriptstyle t \circ v} \\
 & & & C'
\end{array}
$$

Indeed, we have $t \circ \underline{v \circ h} = t \circ f \circ u = t \circ g \circ u = t$. Assume now that there exists another morphism $w \in \mathrm{Hom}_C(C, C')$ which makes the above diagram commutative, i.e., $w \circ h = t$. By composing this last equality on the right with v and having in mind that $h \circ v = 1_C$, we obtain $w = t \circ v$. This shows that $t \circ v$ is the unique morphism which makes the above diagram commutative.   □

2.7 First, we have

$$
F(h) \circ F(f) = F(\underline{h \circ f}) = F(h \circ g) = F(h) \circ F(g).
$$

Consider now $e \in \mathrm{Hom}_C(F(B), E)$ such that $e \circ F(f) = e \circ F(g)$. We will show that $e \circ F(v)$ is the unique morphism which makes the following diagram commutative:

$$
\begin{array}{ccc}
F(A) \underset{F(g)}{\overset{F(f)}{\rightrightarrows}} & F(B) & \xrightarrow{\ F(h)\ } & F(C) \\
 & & e\searrow & \downarrow{\scriptstyle e \circ F(v)} \\
 & & & E
\end{array}
$$

To this end, we have

$$
e \circ F(v) \circ F(h) = e \circ F(\underline{v \circ h}) = e \circ F(f \circ u) = \underline{e \circ F(f)} \circ F(u)
$$

$$
= e \circ F(g) \circ F(u) = e \circ F(\underline{g \circ u}) = e.
$$

Furthermore, suppose there exists a $w \in \mathrm{Hom}_C(F(C), E)$ such that the above diagram is commutative, i.e., $w \circ F(h) = e$. By composing this equality on the right with $F(v)$ and having in mind that $h \circ v = 1_C$, we obtain $w = e \circ F(v)$.   □

2.8 Since (E, p) is the equalizer in C of (f, g) we have

$$
f \circ p = g \circ p. \tag{4.4}
$$

Therefore, we have

$$p_X \circ \overline{f} \circ p = 1_X \circ p = p_X \circ \overline{g} \circ p,$$

$$p_Y \circ \overline{f} \circ p = f \circ p \stackrel{(4.4)}{=} g \circ p = p_Y \circ \overline{g} \circ p.$$

Now using Proposition 2.2.14, (1) we obtain $\overline{f} \circ p = \overline{g} \circ p$.

Consider now $u, v \in \text{Hom}_C(E', X)$ such that $\overline{f} \circ u = \overline{g} \circ v$. Composing this identity on the left with p_X gives $v = u$. Similarly, by composing on the left with p_Y yields $f \circ u = g \circ v$. Putting everything together we obtain $f \circ u = g \circ u$. As (E, p) is the equalizer in C of (f, g) we obtain a unique morphism $w \in \text{Hom}_C(E', E)$ such that the following diagram is commutative:

$$
\begin{array}{ccc}
E & \xrightarrow{\ p\ } & X \rightrightarrows Y \qquad \text{i.e., } p.\circ w = u \\
w \uparrow \ \nearrow & & \\
E' & &
\end{array}
$$

with $\overset{f}{\underset{g}{\rightrightarrows}}$ and u labelling the slanted arrow.

This shows that w is the unique morphism which makes the following diagram commutative:

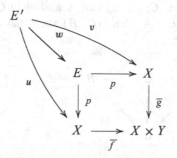

Therefore, (E, p, p) is the pullback of $(\overline{f}, \overline{g})$, as desired.   □

2.9 Proposition 1.7.7 shows that F is representable if and only if there exists a representing pair (A, a), where $A \in \text{Ob}\,C$ and $a = (\alpha_i)_{i \in \text{Ob}\,I} \in F(A) = \prod_{i \in I} \text{Hom}_C(A_i, A)$. Hence, F is representable if and only if for any $C \in \text{Ob}\,C$ and any $x = (f_i)_{i \in \text{Ob}\,I} \in F(C) = \prod_{i \in I} \text{Hom}_C(A_i, C)$, there exists a unique $f \in \text{Hom}_C(A, C)$ such that $F(f)\big((\alpha_i)_{i \in \text{Ob}\,I}\big) = (f_i)_{i \in \text{Ob}\,I}$. In other words, F is representable if and only if for any $C \in \text{Ob}\,C$ and any $x = (f_i)_{i \in \text{Ob}\,I}) \in F(C) = \prod_{i \in I} \text{Hom}_C(A_i, C)$, there exists a unique $f \in \text{Hom}_C(A, C)$ such that $f \circ \alpha_i = f_i$ for all $i \in \text{Ob}\,I$. Now observe that the last condition is equivalent to $\big(A, (\alpha_i)_{i \in \text{Ob}\,I}\big)$ being the coproduct of the family of objects $(A_i)_{i \in \text{Ob}\,I}$ in C.   □

2.10 (a) Let I be the initial object of C and consider $A, B \in \text{Ob}\,C$. We first construct the coproduct of A and B. As I is the initial object of C, we have unique morphisms $f \in \text{Hom}_C(I, A)$ and $g \in \text{Hom}_C(I, B)$. Consider now the pushout (C, q_A, q_B) of (f, g). In particular, we have $q_A \circ f = q_B \circ g$. We will prove that (C, q_A, q_B) is the coproduct of A and B. To this end, let $u \in \text{Hom}_C(A, D)$ and $v \in \text{Hom}_C(B, D)$. Since I is the initial object of C we have a unique morphism from I to D and therefore we obtain $u \circ f = v \circ g$. As (C, q_A, q_B) is the pushout of (f, g), there exists a unique morphism $\Theta \in \text{Hom}_C(C, D)$ such that $\Theta \circ q_A = u$ and $\Theta \circ q_B = v$. This shows that (C, q_A, q_B) is the coproduct of A and B.

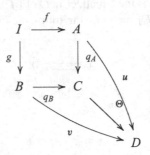

Next we construct coequalizers. Let $\alpha, \beta \in \text{Hom}_C(A, B)$ and consider the coproduct (C, q_A, q_B) of A and B in C. Thus, there exist unique morphisms $u, v \in \text{Hom}_C(C, B)$ such that the following diagrams are commutative:

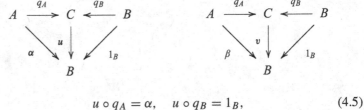

$$u \circ q_A = \alpha, \quad u \circ q_B = 1_B, \tag{4.5}$$

$$v \circ q_A = \beta, \quad v \circ q_B = 1_B. \tag{4.6}$$

Consider now (D, p, q) to be the pushout of (u, v). In particular, we have $q \circ u = p \circ v$ and by composing on the right with q_A and using (4.5) and (4.6) we obtain $q \circ \alpha = p \circ \beta$. On the other hand, composing the equality $q \circ u = p \circ v$ on the right with q_B and using again (4.5) and (4.6) yields $p = q$. Putting everything together we have $p \circ \alpha = p \circ \beta$. We will show that (D, p) is the coequalizer of (α, β). Indeed, consider

$t \in \mathrm{Hom}_C(B, D')$ such that $t \circ \alpha = t \circ \beta$.

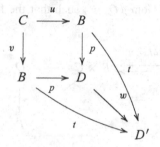

$$A \underset{\beta}{\overset{\alpha}{\rightrightarrows}} B \xrightarrow{p} D$$

(4.7)

We obtain

$$t \circ \underline{u} \circ q_A = t \circ \alpha = t \circ \underline{\beta} = t \circ v \circ q_A,$$

$$t \circ \underline{u} \circ q_B = t \circ 1_B = t \circ v \circ q_B.$$

Therefore, we have $(t \circ u) \circ q_A = (t \circ v) \circ q_A$ and $(t \circ u) \circ q_B = (t \circ v) \circ q_B$. Now Proposition 2.2.14, (2) implies that $t \circ u = t \circ v$.

Since (D, p, p) is the pushout of (u, v), we obtain a unique $w \in \mathrm{Hom}_C(D, D')$ such that $w \circ p = t$. This shows that w is the unique morphism which makes (4.7) commutative and (D, p) is the coequalizer of (α, β), as desired.

(b) Follows by duality.   □

2.11 (a) Let $f \in \mathrm{Hom}_C(A, B), g \in \mathrm{Hom}_C(A, C)$ and consider $\big(B \times C, (q_B, q_C)\big)$ to be the coproduct of B and C, where $q_B \in \mathrm{Hom}_C(B, B \times C)$ and $q_C \in \mathrm{Hom}_C(C, B \times C)$. Furthermore, let (Q, q) be the coequalizer of $(q_B \circ f, q_C \circ g)$. In particular, we have $q \circ q_B \circ f = q \circ q_C \circ g$.

We will prove that $(Q, q \circ q_C, q \circ q_B)$ is the pushout of (f, g). Indeed, let $f' \in \mathrm{Hom}_C(C, P)$ and $g' \in \mathrm{Hom}_C(B, P)$ such that

$$g' \circ f = f' \circ g.$$

(4.8)

Since $\big(B \times C, (q_B, q_C)\big)$ is the coproduct of B and C, there exists a unique $u \in \mathrm{Hom}_C(B \times C, P)$ such that the following diagram is commutative:

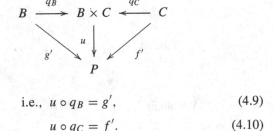

$$\text{i.e., } u \circ q_B = g', \tag{4.9}$$
$$u \circ q_C = f'. \tag{4.10}$$

We obtain

$$u \circ q_B \circ f \overset{(4.9)}{=} g' \circ f \overset{(4.8)}{=} f' \circ g \overset{(4.10)}{=} u \circ q_C \circ g.$$

Now, since (Q, q) is the coequalizer of $(q_B \circ f, q_C \circ g)$, there exists a unique $v \in \mathrm{Hom}_C(Q, P)$ such that the following diagram is commutative:

$$A \overset{q_B \circ f}{\underset{q_C \circ g}{\rightrightarrows}} B \times C \overset{q}{\longrightarrow} Q \qquad \text{i.e., } v \circ q = u.$$

(diagram with u, v, P)

$$\tag{4.11}$$

Moreover, we obtain

$$v \circ q \circ q_C \overset{(4.11)}{=} u \circ q_C \overset{(4.10)}{=} f',$$
$$v \circ q \circ q_B \overset{(4.11)}{=} u \circ q_B \overset{(4.9)}{=} g',$$

which shows that the following diagram is commutative:

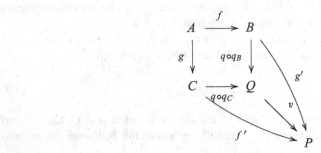

We are left to show that v is the unique morphism which makes the above diagram commutative. Indeed, suppose there exists a $t \in \text{Hom}_C(Q, P)$ such that $t \circ q \circ q_C = f'$ and $t \circ q \circ q_B = g'$. This yields

$$t \circ q \circ q_B = g' \overset{(4.9)}{=} u \circ q_B,$$

$$t \circ q \circ q_C = f' \overset{(4.10)}{=} u \circ q_C.$$

Now, Proposition 2.2.14, (2) implies $t \circ q = u$ and since v is the unique morphism which makes diagram (4.11) commutative, we obtain $t = v$, as desired.

   (b) Follows by duality.   □

2.12 (a) As shown in (the proof of) Proposition 2.1.4, any non-empty family of objects in C admits a coproduct. Furthermore, the existence of an initial object in C implies that an empty family of objects of C admits a coproduct as well and therefore, C has all finite coproducts. Now showing that a category with finite coproducts and coequalizers has all finite limits goes much in the same fashion as the proof of Theorem 2.4.2.

   (b) Follows by duality.   □

2.13 (a) By exercise 2.10, (a), C has binary coproducts and coequalizers. Thus, C has an initial object, binary coproducts and coequalizers and the conclusion now follows from Exercise 2.12, (a).

   (b) Follows by duality.   □

2.14 (a) Let $f \in \text{Hom}_C(A, B)$ be a split epimorphism and consider its right inverse $g \in \text{Hom}_C(B, A)$, i.e., $f \circ g = 1_B$. We will show that (B, f) is the coequalizer of the pair of morphisms $(g \circ f, 1_A)$. Indeed, we have $f \circ (g \circ f) = 1_B \circ f = f = f \circ 1_A$. Moreover, if $t \in \text{Hom}_C(A, B')$ such that $t \circ 1_A = t \circ (g \circ f)$, then $u = t \circ g \in \text{Hom}_C(B, B')$ is the unique morphism which makes the following diagram commutative:

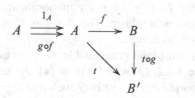

   (b) Let $f \in \text{Hom}_C(A, B)$ be a regular epimorphism, i.e., we can find two morphisms $u, v \in \text{Hom}_C(C, A)$ such that (B, f) is the coequalizer of the pair (u, v). Consider now the following commutative square:

$$A \xrightarrow{f} B \qquad \text{i.e., } h \circ f = m \circ g,$$

with $g: A \to C$, $h: B \to D$, $m: C \to D$

where $m \in \mathrm{Hom}_C(C, D)$ is a monomorphism. We have

$$m \circ g \circ u = h \circ f \circ u = h \circ f \circ v = m \circ g \circ v,$$

and since m is a monomorphism we obtain $g \circ u = g \circ v$. Since (B, f) is the coequalizer of the pair (u, v), there exists a unique $t \in \mathrm{Hom}_C(B, C)$ such that the following diagram is commutative:

$$C \underset{v}{\overset{u}{\rightrightarrows}} A \xrightarrow{f} B \qquad \text{i.e., } t \circ f = g.$$

with $g: A \to C$, $t: B \to C$

Furthermore, we have $m \circ t \circ f = m \circ g = h \circ f$ and since f is an epimorphism we obtain $m \circ t = h$. To conclude, we have a unique morphism $t \in \mathrm{Hom}_C(B, C)$ such that $t \circ f = g$ and $m \circ t = h$; this shows that f is a strong epimorphism. □

2.15 Assume f is an epimorphism such that $f = g \circ h$, where g is a monomorphism. Since $f = g \circ h$ is an epimorphism, Exercise 1.5, (c), shows that g is also an epimorphism. Therefore, as C is a balanced category, f is an isomorphism, as desired. □

2.16 Let G be a non-trivial group and let $\mathcal{G}$ be its associated category in the sense of Example 1.2.2, (3) with $\mathrm{Ob}\,\mathcal{G} = \{\bullet\}$. As noticed in Example 2.1.10, (10) a pair of morphisms (x, y) in $\mathcal{G}$ such that $x \neq y$ does not have an equalizer. However, the category $\mathcal{G}$ admits pullbacks for any pair of morphisms (x, y). To this end, we will show that the triple $(\bullet, y^{-1}, x^{-1})$ is the pullback of (x, y), where x^{-1} and y^{-1} denote the inverse of the elements x and y, respectively, in the group G.

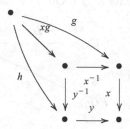

Indeed, it is straightforward to see that ($\bullet$, y^{-1}, x^{-1}) makes the above square commutative. Furthermore, for any other g, $h \in G$ such that $xg = yh$, the unique morphism in $\mathcal{G}$ which makes the two triangles above commutative is xg. $\qquad\square$

2.20 Let A, $B \in \mathrm{Ob}\,C$ and f, $g \in \mathrm{Hom}_C(A, B)$ such that $\mathcal{F}_{A, B}(f) = \mathcal{F}_{A, B}(g)$, where $\mathcal{F}_{A, B} \colon \mathrm{Hom}_C(A, B) \to \mathrm{Hom}_{\mathcal{D}}\big(F(A), F(B)\big)$ is the induced map defined in (1.12). Then, we have $F(f) = F(g)$. As C has equalizers we can consider the equalizer (E, e) of (f, g). Since F preserves equalizers we obtain that $(F(E), F(e))$ is the equalizer of $(F(f), F(f))$. Exercise 2.3 implies that $F(e)$ is an isomorphism. Furthermore, F reflects isomorphisms, which implies that e is also an isomorphism. Hence, by Exercise 2.3, we obtain $f = g$, and therefore $\mathcal{F}_{A, B}$ is injective, as desired. $\qquad\square$

2.21 Let $i_o \in \mathrm{Ob}\,I$ be the initial object of I and for any $j \in \mathrm{Ob}\,I$ denote by u_j the unique morphism in I between i_0 and j. Then the pair $\big(F(i_0), (p_j)_{j \in \mathrm{Ob}\,I}\big)$ is a cone on F, where $p_j = F(u_j) \in \mathrm{Hom}_C(F(i_0), F(j))$ for all $j \in \mathrm{Ob}\,I$. Indeed, if $d \in \mathrm{Hom}_I(i, j)$ then we have u_j, $d \circ u_i \in \mathrm{Hom}_I(i_0, j)$ and since i_0 is the initial object of I we obtain $u_j = d \circ u_i$. Applying F yields $p_j = F(d) \circ p_i$, as desired. Consider now another cone $\big(C, (f_j \in \mathrm{Hom}_C(C, F(j)))_{j \in \mathrm{Ob}\,I}\big)$ on F. In particular, this implies that the following diagram is commutative for all $j \in \mathrm{Ob}\,I$:

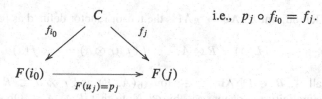 i.e., $p_j \circ f_{i_0} = f_j$.

In other words, the morphism $f_{i_0} \in \mathrm{Hom}_C(C, F(i_0))$ makes the following diagram commutative for all $j \in \mathrm{Ob}\, I$:

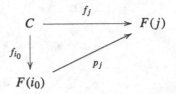

The proof will be finished once we show that f_{i_0} is the unique morphism with this property. To this end, let $g \in \mathrm{Hom}_C(C, F(i_0))$ such that $p_j \circ g = f_j$ for all $j \in \mathrm{Ob}\, I$. We obtain

$$g = F(1_{i_0}) \circ g = F(u_{i_0}) \circ g = p_{i_0} \circ g = f_{i_0},$$

as desired.   □

2.23 See Exercise 2.16.   □

4.3   Chapter 3

3.1 Assume $U\colon \mathbf{Field} \to \mathbf{Ring}$ has a left adjoint $L\colon \mathbf{Ring} \to \mathbf{Field}$. As noted in Example 1.3.10, (3) $\mathbb{Z}$ is the initial object of $\mathbf{Ring}$ and Theorem 3.4.4 implies that $L(\mathbb{Z})$ is the initial object of $\mathbf{Field}$. This contradicts Example 1.3.10, (4); therefore U does not admit a left adjoint.

Assume now that $U\colon \mathbf{Field} \to \mathbf{Ring}$ has a right adjoint $R\colon \mathbf{Ring} \to \mathbf{Field}$. As noted in Example 1.3.10, (3) the zero ring is the final object of $\mathbf{Ring}$ and Theorem 3.4.4 implies that $R(\{0\})$ is the final object of $\mathbf{Field}$. Again, this contradicts Example 1.3.10, (4); therefore U does not admit a right adjoint.

□

3.3 The left adjoint $L\colon \mathbf{Ab} \to {}_R\mathcal{M}$ is the tensor functor defined as follows:

$$L(A) = R \otimes A, \qquad L(f)(r \otimes a) = r \otimes f(a)$$

for all $A, B \in \mathrm{Ob}\,\mathbf{Ab}$, $f \in \mathrm{Hom}_{\mathbf{Ab}}(A, B)$ and $r \otimes a \in R \otimes A$, where for simplicity we denote $\otimes_{\mathbb{Z}}$ by $\otimes$. Note that $R \otimes A \in \mathrm{Ob}\, {}_R\mathcal{M}$ with the left R-module structure given by $r(r' \otimes a) = rr' \otimes a$ for all $r, r' \in R$ and $a \in A$. We use Theorem 3.5.1 to show that $L \dashv U$. Indeed, consider the natural transformations $\eta\colon 1_{\mathbf{Ab}} \to UL$ and $\varepsilon\colon LU \to 1_{{}_R\mathcal{M}}$ defined as follows for all $A \in \mathrm{Ob}\,\mathbf{Ab}$ and $M \in \mathrm{Ob}\,_R\mathcal{M}$:

$$\eta_A: A \to R \otimes A, \quad \eta_A(a) = 1_R \otimes a,$$

$$\varepsilon_M: R \otimes M \to M, \quad \varepsilon_M(r \otimes m) = rm.$$

It will suffice to show that (3.15) and (3.16) hold. To this end, for all $r \otimes a \in R \otimes A$ we have

$$\varepsilon_{R \otimes A} \circ L(\eta_A)(r \otimes a) = \varepsilon_{R \otimes A}(r \otimes \eta_A(a)) = \varepsilon_{R \otimes A}(r \otimes 1_R \otimes a)$$

$$= r(1_R \otimes a) = r \otimes a = 1_{R \otimes A}(r \otimes a),$$

i.e., (3.15) holds. Furthermore, for all $m \in M$ we have

$$U(\varepsilon_M) \circ \eta_M(m) = U(\varepsilon_M)(1_R \otimes m) = 1_R m = m = 1_M(m),$$

which shows that (3.16) also holds. Thus, in light of Theorem 3.5.1, we obtain that $L \dashv U$, as desired.

Next, the right adjoint of U is the hom functor $T: \mathbf{Ab} \to \,_R\mathcal{M}$ defined as follows:

$$T(A) = \mathrm{Hom}_{\mathbb{Z}}(R,\, A), \qquad T(f)(g) = f \circ g,$$

for all $A \in \mathrm{Ob}\,\mathbf{Ab}$, $f \in \mathrm{Hom}_{\mathbf{Ab}}(A,\, B)$ and $g \in \mathrm{Hom}_{\mathbb{Z}}(R,\, A)$. Note that $\mathrm{Hom}_{\mathbb{Z}}(R,\, A) \in \mathrm{Ob}\,_R\mathcal{M}$ with the left R-module structure given by $(rf)(t) = f(rt)$ for all $r, t \in R$ and $f \in \mathrm{Hom}_{\mathbb{Z}}(R,\, A)$. Again, we use Theorem 3.5.1 to show that $U \dashv R$. Indeed, consider the natural transformations $\eta: 1_{_R\mathcal{M}} \to TU$ and $\varepsilon: UT \to 1_{\mathbf{Ab}}$ defined as follows for all $A, B \in \mathrm{Ob}\,\mathbf{Ab}$, $M \in \mathrm{Ob}\,_R\mathcal{M}$, $g \in \mathrm{Hom}_{\mathbb{Z}}(R,\, A)$ and $m \in M$:

$$\eta_M: M \to \mathrm{Hom}_{\mathbb{Z}}(R,\, U(M)), \quad \eta_M(m) = \psi_m: R \to U(M),$$

$$\psi_m(r) = rm,$$

$$\varepsilon_A: \mathrm{Hom}_{\mathbb{Z}}(R,\, A) \to A, \qquad \varepsilon_A(g) = g(1_R).$$

It will suffice to show that (3.15) and (3.16) hold. To this end, for all $m \in M$ we have

$$\left(\varepsilon_{U(M)} \circ U(\eta_M)\right)(m) = \varepsilon_{U(M)}(\psi_m) = \psi_m(1_R) = 1_R m = m = 1_{U(M)}(m),$$

and therefore (3.15) holds. Furthermore, for all $g \in \mathrm{Hom}_{\mathbb{Z}}(R,\, A)$ and $r \in R$ we have $\varepsilon_A \circ \psi_g(r) = \varepsilon_A(rg) = (rg)(1_R) = g(r1_R) = g(r)$. Hence, we obtain

$$\varepsilon_A \circ \psi_g = g. \tag{4.12}$$

This leads to the following:

$$\big(T(\varepsilon_A) \circ \eta_{T(A)}\big)(g) = T(\varepsilon_A)(\psi_g) = \varepsilon_A \circ \psi_g \overset{(4.12)}{=} g,$$

i.e., $T(\varepsilon_A) \circ \eta_{T(A)} = 1_{T(A)}$, which shows that (3.16) holds as well and we obtain the desired adjunction. $\square$

3.8 The equivalence (a) $\Leftrightarrow$ (c) follows easily from (3.15).

(b) $\Rightarrow$ (d) For all $C \in \mathrm{Ob}\,C$ we have

$$1_{GF(C)} \overset{(3.16)}{=} G(\varepsilon_{F(C)}) \circ \underline{\eta_{GF(C)}} \overset{b)}{=} G(\varepsilon_{F(C)}) \circ GF(\eta_C).$$

Thus, for all $C \in \mathrm{Ob}\,C$ we have

$$1_{GF(C)} = G(\varepsilon_{F(C)}) \circ GF(\eta_C). \tag{4.13}$$

On the other hand, we also have

$$G\big(F(\eta_C) \circ \varepsilon_{F(C)}\big) \circ \eta_{GF(C)} = GF(\eta_C) \circ \underline{G(\varepsilon_{F(C)}) \circ \eta_{GF(C)}}$$

$$\overset{(3.16)}{=} \underline{GF(\eta_C)} \overset{b)}{=} \eta_{GF(C)}.$$

Now Corollary 3.6.2, (1) implies that for all $C \in \mathrm{Ob}\,C$ we have $F(\eta_C) \circ \varepsilon_{F(C)} = 1_{FGF(C)}$, and therefore

$$GF(\eta_C) \circ G(\varepsilon_{F(C)}) = 1_{GFGF(C)}. \tag{4.14}$$

(4.13) together with (4.14) imply that $G(\varepsilon_{F(C)})$ is an isomorphism.

(d) $\Rightarrow$ (b) If we assume $G(\varepsilon_{F(C)})$ is an isomorphism, (3.15) implies that its inverse is $GF(\eta_C)$. Furthermore, from (3.16) we obtain $1_{GF(C)} = G(\varepsilon_{F(C)}) \circ \eta_{GF(C)}$ and therefore the inverse of $G(\varepsilon_{F(C)})$ is $\eta_{GF(C)}$. This shows that $GF(\eta_C) = \eta_{GF(C)}$.

(a) $\Rightarrow$ (d) Using (3.15), for all $C \in \mathrm{Ob}\,C$ we have

$$1_{GF(C)} = G(\varepsilon_{F(C)}) \circ GF(\eta_C). \tag{4.15}$$

Since by Proposition 1.6.9, (1) any functor preserves isomorphisms, it follows that $GF(\eta_C)$ is an isomorphism. Now (4.15) implies that $G(\varepsilon_{F(C)})$ is an isomorphism as well.

(d) $\Rightarrow$ (a) We have already proved that d) implies the following for all $C \in \mathrm{Ob}\,C$:

$$GF(\eta_C) = \eta_{GF(C)}. \tag{4.16}$$

Thus for all $C \in \mathrm{Ob}\,C$ we have

$$G\big(F(\eta_C) \circ \varepsilon_{F(C)}\big) \circ \eta_{GF(C)} = GF(\eta_C) \circ \underline{G(\varepsilon_{F(C)}) \circ \eta_{GF(C)}} \overset{(3.16)}{=} GF(\eta_C)$$

$$\overset{(4.16)}{=} \eta_{GF(C)} = G(1_{FGF(C)}) \circ \eta_{GF(C)}.$$

Now Corollary 3.6.2, (1) implies $F(\eta_C) \circ \varepsilon_{F(C)} = 1_{FGF(C)}$. Using (3.15) we obtain that $F(\eta_C)$ is indeed an isomorphism. □

3.9 Suppose first that $\eta_D \in \mathrm{Hom}_{\mathcal{D}}(D,\,HF(D))$ is an epimorphism for any $d \in \mathrm{Ob}\,\mathcal{D}$ and let $f, g \in \mathrm{Hom}_{\mathcal{D}}(D',\,HG(D))$ such that

$$\varepsilon_D \circ f = \varepsilon_D \circ g. \tag{4.17}$$

By Theorem 3.6.1, (1) there exists a unique $f' \in \mathrm{Hom}_C(F(D'),\,G(D))$ such that $H(f') \circ \eta_{D'} = f$. Similarly, we have a unique $g' \in \mathrm{Hom}_C(F(D'),\,G(D'))$ such that $H(g') \circ \eta_{D'} = g$. Then (4.17) comes down to $\varepsilon_D \circ H(f') \circ \eta_{D'} = \varepsilon_D \circ H(g') \circ \eta_{D'}$ and since $\eta_{D'}$ is an epimorphism we obtain $\varepsilon_D \circ H(f') = \varepsilon_D \circ H(g')$. Now Corollary 3.6.2, (2) implies $f' = g'$ and therefore $f = g$.

The converse follows by duality. Indeed, by Corollary 3.5.4 we have $G^{op} \dashv H^{op}$ with unit ε^{op} and $H^{op} \dashv F^{op}$ with counit η^{op}. □

3.10 For all objects C in $\mathrm{Iso}(C)$, $\eta_C\colon C \to GF(C)$ is an isomorphism and by Proposition 1.6.9, (1) it follows that $F(\eta_C)$ is also an isomorphism. Furthermore, (3.15) now implies that $\varepsilon_{F(C)}$ is an isomorphism and therefore the restriction of F to the subcategory $\mathrm{Iso}(C)$ of C, denoted by $\overline{F}$, is a functor with codomain $\mathrm{Iso}(\mathcal{D})$, i.e., $\overline{F}\colon \mathrm{Iso}(C) \to \mathrm{Iso}(\mathcal{D})$. Similarly, using (3.16) this time, it can be easily seen that the restriction of G to the subcategory $\mathrm{Iso}(\mathcal{D})$ of $\mathcal{D}$, denoted by $\overline{G}$, is a functor with codomain $\mathrm{Iso}(C)$, i.e., $\overline{G}\colon \mathrm{Iso}(\mathcal{D}) \to \mathrm{Iso}(C)$.

Furthermore, we can consider the natural transformations $\overline{\eta}\colon 1_{\mathrm{Iso}(C)} \to \overline{GF}$ and $\overline{\varepsilon}\colon \overline{FG} \to 1_{\mathrm{Iso}(\mathcal{D})}$ defined by $\overline{\eta}_C = \eta_C$ and $\overline{\varepsilon}_D = \varepsilon_D$ for all $C \in \mathrm{Ob}\big(\mathrm{Iso}(C)\big)$ and $D \in \mathrm{Ob}\big(\mathrm{Iso}(\mathcal{D})\big)$. Then $\overline{F}$ and $\overline{G}$ form a pair of adjoint functors with unit and counit given by $\overline{\eta}$ and $\overline{\varepsilon}$ respectively. As $\overline{\eta}$ and $\overline{\varepsilon}$ are natural isomorphisms it follows, using Theorem 3.8.5, that the categories $\mathrm{Iso}(C)$ and $\mathrm{Iso}(\mathcal{D})$ are equivalent. □

3.11 (a) Let $\big(\lim T,\,(q_i\colon \lim T \to T(i))_{i \in \mathrm{Ob}\,I}\big)$ be the limit of T. The naturality of β renders the following diagram commutative for all $i \in \mathrm{Ob}\,I$:

$$\begin{array}{ccc}
G(T(i)) & \xrightarrow{\ \beta_{T(i)}\ } & G'(T(i)) \\
{\scriptstyle G(q_i)}\big\uparrow & & \big\uparrow{\scriptstyle G'(q_i)} \\
G(\lim T) & \xrightarrow[\ \beta_{\lim T}\ ]{} & G'(\lim T)
\end{array}
\qquad \text{i.e.,}\ \ \beta_{T(i)} \circ G(q_i) = G'(q_i) \circ \beta_{\lim T}.$$

$$\tag{4.18}$$

As G is the right adjoint of F, it preserves limits by Theorem 3.4.4. Therefore, the pair $\left(G(\lim T), (G(q_i))_{i \in \mathrm{Ob}\, I}\right)$ is the limit of $GT: I \to C$. Hence, there exists a unique $\psi: G'(\lim T) \to G(\lim T)$ such that the following diagram is commutative for all $i \in \mathrm{Ob}\, I$:

$$
\begin{array}{ccc}
G(\lim T) & \xrightarrow{\ G(q_i)\ } & G(T(i)) \\
\psi \uparrow & \nearrow{\scriptstyle \beta_{T(i)}^{-1} \circ G'(q_i)} & \\
G'(\lim T) & &
\end{array}
\qquad \text{i.e.,} \quad G(q_i) \circ \psi = \beta_{T(i)}^{-1} \circ G'(q_i).
$$

$$\tag{4.19}$$

Therefore, for all $i \in \mathrm{Ob}\, I$ we have

$$
G'(q_i) \overset{(4.19)}{=} \beta_{T(i)} \circ G(q_i) \circ \psi \overset{(4.18)}{=} G'(q_i) \circ \beta_{\lim T} \circ \psi.
$$

Now since $\left(G'(\lim T), (G'(q_i))_{i \in \mathrm{Ob}\, I}\right)$ is the limit of $G'T: I \to C$, Proposition 2.2.14, (1) implies that $1_{G'(\lim T)} = \beta_{\lim T} \circ \psi$. Furthermore, for all $i \in \mathrm{Ob}\, I$ we have

$$
G(q_i) \circ \psi \circ \beta_{\lim T} \overset{(4.19)}{=} \beta_{T(i)}^{-1} \circ G'(q_i) \circ \beta_{\lim T} \overset{(4.18)}{=} \beta_{T(i)}^{-1} \circ \beta_{T(i)} \circ G(q_i)
$$
$$
= G(q_i).
$$

Proposition 2.2.14, (1) implies $\psi \circ \beta_{\lim T} = 1_{G(\lim T)}$. Therefore, $\beta_{\lim T}$ is invertible and its inverse is equal to ψ.

(b) Follows by duality. $\square$

3.13 First we construct a functor $F: \mathbf{PreOrd} \to \mathbf{Poset}$ which will turn out to be the left adjoint of the inclusion functor. To start with, given a pre-ordered set $(P, \leqslant)$, we consider on P the relation '$\sim$' defined as follows for all $x, y \in P$: $x \sim y$ if and only if $x \leqslant y$ and $y \leqslant x$. A straightforward computation shows that '$\sim$' is in fact an equivalence relation on P and we denote by $\overline{P}$ the set of equivalence classes with respect to '$\sim$' and by $\overline{x}$ the equivalence class in $\overline{P}$ of some element $x \in P$. Moreover, $(\overline{P}, \widetilde{\leqslant})$ is a partially-ordered set, where $\widetilde{\leqslant}$ is defined as follows for all $\overline{x}, \overline{y} \in \overline{P}$: $\overline{x} \widetilde{\leqslant} \overline{y}$ if and only if $x \leqslant y$. Note that $\widetilde{\leqslant}$ is a well-defined relation on $\overline{P}$; indeed, if $\overline{x} = \overline{x'}$ and $\overline{y} = \overline{y'}$ we have $x \leqslant x'$, $x' \leqslant x$, $y \leqslant y'$, $y' \leqslant y$ and if $x \leqslant y$ then $x' \leqslant x \leqslant y \leqslant y'$, which implies $x' \leqslant y'$, as desired.

Furthermore, if $f: (P, \leqslant) \to (Q, \ll)$ is a morphism in $\mathbf{PreOrd}$ (i.e., an order preserving map $f: P \to Q$), then we can define an order preserving map $f_\star: (\overline{P}, \widetilde{\leqslant}) \to (\overline{Q}, \widetilde{\ll})$ by $f_\star(\overline{x}) = \overline{f(x)}$ for all $\overline{x} \in \overline{P}$. We only show that $f_\star$ is well-defined: if $\overline{x}, \overline{y} \in \overline{P}$ such that $\overline{x} = \overline{y}$ then $x \leqslant y$ and $y \leqslant x$ and since f is order preserving we obtain $\overline{f(x)} = \overline{f(y)}$ or, equivalently, that $f_\star(\overline{x}) = f_\star(\overline{y})$.

We can now define a functor $F\colon \mathbf{PreOrd} \to \mathbf{Poset}$ as follows for all pre-ordered sets $(P, \leqslant)$, $(Q, \ll)$ and order-preserving maps $f\colon (P, \leqslant) \to (Q, \ll)$:

$$F(P, \leqslant) = (\overline{P}, \widetilde{\leqslant}), \qquad F(f) = f_\star.$$

Next we use Theorem 3.6.1, (1) in order to show that $F \dashv I$. To this end, note that for all pre-ordered sets $(P, \leqslant)$ we have an order-preserving map $\pi_P\colon (P, \leqslant) \to (\overline{P}, \widetilde{\leqslant})$ defined by $\pi_P(x) = \overline{x}$ for all $x \in P$. Moreover, $\pi\colon 1_{\mathbf{PreOrd}} \to IF$ defined for any pre-ordered set $(P, \leqslant)$ by π_P is a natural transformation. Consider now two pre-ordered sets $(P, \leqslant)$, $(Q, \ll)$ and a morphism $f\colon (P, \leqslant) \to (\overline{Q}, \widetilde{\ll})$ in $\mathbf{PreOrd}$; the proof will be finished once we show that there exists a unique morphism $g\colon (\overline{P}, \widetilde{\leqslant}), \to (\overline{Q}, \widetilde{\ll})$ in $\mathbf{Poset}$ such that $I(g) \circ \pi_P = f$. Define $g\colon (\overline{P}, \widetilde{\leqslant}), \to (\overline{Q}, \widetilde{\ll})$ by $g(\overline{x}) = f(x)$ for all $\overline{x} \in \overline{P}$. The only thing left to prove is that g is well-defined. Indeed, if $\overline{x} = \overline{y}$ then $x \leqslant y$ and $y \leqslant x$ and since f is order-preserving we obtain $f(x)\widetilde{\ll}f(y)$ and $f(y)\widetilde{\ll}f(x)$. Now recall that $\widetilde{\ll}$ is a partial order on $\overline{Q}$ and the anti-symmetry implies $f(x) = f(y)$, as desired.   □

3.14 We will show that, unless X is a singleton set, the cartesian product functor $X \times -$ does not preserve products and, therefore, by virtue of Theorem 3.4.4, it does not admit a left adjoint. Note that if X is a singleton set then the identity functor is obviously the left adjoint of the corresponding cartesian product functor.

Hereafter, let X be a set such that $|X| \neq 1$. Let $Y, Z \in \mathrm{Ob}\,\mathbf{Set}$ and consider their product $(Y \times Z, (p_1, p_2))$ in $\mathbf{Set}$, as constructed in Example 2.1.5, (1) i.e., $Y \times Z$ is the cartesian product of the two sets while $p_1\colon Y \times Z \to Y$ and $p_2\colon Y \times Z \to Z$ denote the projections on the first and second component, respectively. Furthermore, let $(X \times Y \times X \times Z, (\pi_1, \pi_2))$ be the product of $X \times Y$ and $X \times Z$ in $\mathbf{Set}$, as constructed in Example 2.1.5, (1) where $\pi_1\colon X \times Y \times X \times Z \to X \times Y$, $\pi_2\colon X \times Y \times X \times Z \to X \times Z$ denote the projections on $X \times Y$ and $X \times Z$, respectively. Assume now that the cartesian product functor $X \times -$ preserves products. Then, in light of Proposition 2.1.3, there exists a unique isomorphism $f\colon X \times Y \times Z \to X \times Y \times X \times Z$ in $\mathbf{Set}$ (set bijection) such that the following diagram is commutative:

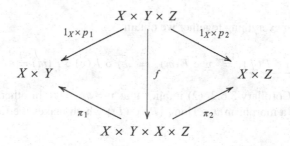

It can be easily seen that any map $f : X \times Y \times Z \to X \times Y \times X \times Z$ which makes the above diagram commutative is given by $f(x, y, z) = (x, y, x, z)$ for all $x \in X$, $y \in Y$ and $z \in Z$. The map f is obviously not surjective whenever X has more than one element and we have reached a contradiction. Hence, $X \times -$ does not preserve products. $\square$

3.22 Let $u, v \in \mathrm{Hom}_{\mathcal{D}}(D, D')$ such that $u \neq v$. As G is a faithful functor we have $G(u) \neq G(v)$ and since S is a generator in C, there exists a morphism $h \in \mathrm{Hom}_C(S, G(D))$ such that

$$G(u) \circ h \neq G(v) \circ h. \tag{4.20}$$

By Theorem 3.6.1, (2) there exists a unique morphism $w \in \mathrm{Hom}_{\mathcal{D}}(F(S), D)$ such that $G(w) \circ \eta_S = h$, where $\eta : 1_C \to GF$ is the unit of the adjunction $F \dashv G$. Then (4.20) becomes

$$G(u \circ w) \circ \eta_S \neq G(v \circ w) \circ \eta_S. \tag{4.21}$$

This shows that $u \circ w \neq v \circ w$ and therefore $F(S)$ is a generator in $\mathcal{D}$. Indeed, if $u \circ w = v \circ w$ then we would have $G(u \circ w) \circ \eta_S = G(v \circ w) \circ \eta_S$, which contradicts (4.21). $\square$

3.25 Let I be an injective object in $\mathcal{D}$ and consider $f \in \mathrm{Hom}_C(A, G(I))$ and $m \in \mathrm{Hom}_C(A, B)$ a monomorphism. As I is an injective object in $\mathcal{D}$ and $F(m)$ is a monomorphism, there exists a morphism $w \in \mathrm{Hom}_{\mathcal{D}}(F(B), I)$ which makes the following diagram commutative:

$$
\begin{array}{lll}
F(A) & \qquad\qquad & \text{i.e.,} \quad \varepsilon_I \circ F(f) = w \circ F(m), \\
\Big\downarrow{\scriptstyle F(m)} \searrow{\scriptstyle \varepsilon_I \circ F(f)} & & \\
F(B) \xrightarrow{w} I & &
\end{array}
\tag{4.22}
$$

where $\varepsilon : FG \to 1_{\mathcal{D}}$ is the counit of the adjunction $F \dashv G$. Furthermore, by Theorem 3.6.1, (3) there exists a unique $v \in \mathrm{Hom}_C(B, G(I))$ such that

$$\varepsilon_I \circ F(v) = w. \tag{4.23}$$

Putting everything together we obtain

$$\underline{\varepsilon_I \circ F(f)} \overset{(4.22)}{=} \underline{w} \circ F(m) \overset{(4.23)}{=} \varepsilon_I \circ F(v) \circ F(m) = \varepsilon_I \circ F(v \circ m).$$

Now Corollary 3.6.2, (2) implies that $v \circ m = f$. In other words, we have found a morphism $v \in \mathrm{Hom}_C(B, G(I))$ which makes the following diagram commutative:

This shows that $G(I)$ is an injective object in C. The second claim follows by duality. □

References

1. Adamek, J., Rosicky, J.: Locally Presentable and Accessible Categories. London Mathematical Society Lecture Note Series, vol. 189. Cambridge University Press, Cambridge (1994)
2. Adamek, J., Herrlich, H., Strecker, G.E.: Abstract and concrete categories. Pure and Applied Mathematics. John Wiley & Sons, New York (1990)
3. Altman, A., Kleiman, S.: A Term of Commutative Algebra. Worldwide Center of Mathematics, Cambridge (2013)
4. Arkowitz, M.: Introduction to Homotopy Theory. Universitext, Springer, New York (2011)
5. Awodey, S.: Category Theory. Oxford Logic Guides, vol. 52. Oxford University Press, Oxford (2010)
6. Bergman, G.M.: An Invitation to General Algebra and Universal Constructions. Universitext. Springer, Cham (2015)
7. Bergman, G.M., Hausknecht, A.O.: Co-Groups and Co-Rings in Categories of Associative Rings. Mathematical Surveys and Monographs, vol. 45. American Mathematical Society, Providence (1996)
8. Borceux, F.: Handbook of Categorical Algebra. 1. Encyclopedia of Mathematics and Its Applications, vol. 50. Cambridge University Press, Cambridge (1994)
9. Borceux, F.: Handbook of Categorical Algebra. 2. Encyclopedia of Mathematics and Its Applications, vol. 51. Cambridge University Press, Cambridge (1994)
10. Borceux, F.: Handbook of Categorical Algebra. 3. Encyclopedia of Mathematics and Its Applications, vol. 52. Cambridge University Press, Cambridge (1994)
11. Bosch, S.: Algebraic Geometry and Commutative Algebra. Universitext. Springer, London (2022)
12. Bradley, T.-D., Bryson, T., Terilla, J.: Topology: A Categorical Approach. MIT Press, Cambridge (2020)
13. Bucur, I., Deleanu, A.: Introduction to the Theory of Categories and Functors. Pure and Applied Mathematics, vol. XIX. John Wiley & Sons, London-New York-Sydney (1968)
14. Burris, S., Sankappanavar, H.P.: A Course in Universal Algebra. Graduate Texts in Mathematics, vol. 78. Springer-Verlag, New York-Berlin (1981)
15. Burton, D.M.: A First Course in Rings and Ideals. Addison-Wesley Series in Mathematics. Addison-Wesley Publishing Company, Boston (1970)
16. Caenepeel, S., Militaru, G., Zhu, S.: Doi-Hopf modules, Yetter-Drinfel'd modules and Frobenius type properties. Trans. Am. Math. Soc. **349**, 4311–4342 (1997)
17. Caenepeel, S., Militaru, G., Zhu, S.: Frobenius and Separable Functors for Generalized Module Categories and Nonlinear Equations. Lecture Notes in Mathematics, vol. 1787. Springer-Verlag, Berlin (2002)

18. Card, E.F.: Ring Extensions With Identity Element. https://dalspace.library.dal.ca//handle/10222/79584 (1994). Accessed 15 Aug 2023
19. Cartan, H., Eilenberg, S.: Homological Algebra. Princeton University Press, Princeton (1956)
20. Chirvăsitu, A.: When is the diagonal functor Frobenius? Commun. Algebra **39**, 1208–1225 (2011)
21. Eilenberg, S., Mac Lane, S.: General theory of natural equivalences. Trans. Am. Math. Soc. **58**, 231–294 (1945)
22. Faith, C.: Algebra: Rings, Modules and Categories I. Grundlehren der mathematischen Wissenschaften, vol. 190. Springer, Berlin/Heidelberg (1973)
23. Freyd, P.: Abelian categories. An introduction to the theory of functors. Harper's Series in Modern Mathematics. Harper & Row, Publishers, New York (1964)
24. Freyd, P.: Homotopy is not concrete. Reprints in Theory Appl. Categ. **6**, 1–10 (2004)
25. Freyd, P., Scedrov, A.: Categories, Allegories. North-Holland Mathematical Library, vol. 39. North-Holland Publishing, Amsterdam (1990)
26. Gabriel, P., Zisman, M.: Calculus of fractions and homotopy theory, Springer-Verlag, Berlin (1967)
27. Gallian, J. A., Van Buskirk, J.: The number of homomorphisms from $\mathbb{Z}_n$ to $\mathbb{Z}_m$. Am. Math. Monthly **91**, 196–197 (1984)
28. Gelfand, S.I., Manin, Y.I.: Methods of Homological Algebra. Springer Monographs in Mathematics. Springer-Verlag, Berlin (2003)
29. Gomez-Ramirez, J.: A New Foundation for Representation in Cognitive and Brain Science. Springer Series in Cognitive and Neural Systems, vol. 7. Springer Dordrecht (2014)
30. Herrlich, H., Strecker, G. E.: Category Theory. Sigma Series in Pure Mathematics, vol. 1. Heldermann Verlag, Lemgo (2007)
31. Kan, D.M.: Adjoint functors. Trans. Am. Math. Soc. **87**, 294–329 (1958)
32. Kashiwara, M., Schapira, P.: Categories and sheaves. Grundlehren der mathematischen Wissenschaften, vol. 332. Springer, Berlin/Heidelberg (2006)
33. Kelly, G.M.: Basic Concepts of Enriched Category Theory. London Mathematical Society Lecture Note Series, vol. 64. Cambridge University Press, Cambridge-New York (1982)
34. Leinster, T.: Basic Category Theory. Cambridge Studies in Advanced Mathematics, vol. 143. Cambridge University Press, Cambridge (2014)
35. Mac Lane, S.: Categories for the Working Mathematician. Graduate Texts in Mathematics, vol. 5. Springer-Verlag, New York (1998)
36. Mac Lane, S., Moerdijk, I.: Sheaves in Geometry and Logic. Universitext. Springer-Verlag, New York (1994)
37. Moschovakis, Y.: Notes on Set Theory. Undergraduate Texts in Mathematics. Springer, New York (2006)
38. Müger, M.: Topology for the Working Mathematician. https://www.math.ru.nl/~mueger/topology.pdf (2022). Accessed 15 Aug 2023
39. Munkres, J.R.: Topology. Prentice Hall, Upper Saddle River (2000)
40. van Munster, B.: Hausdorffization and Homotopy. Am. Math. Mon. **124**, 81–82 (2017)
41. Nakayama, T., Tsuzuku, T.: On Frobenius extensions. I. Nagoya Math. J. **17**, 89-110 (1960)
42. Neeman, A.: Triangulated categories. Annals of Mathematics Studies, vol. 148. Princeton University Press, Princeton (2001)
43. Negrepontis, J.W.: Duality in analysis from the point of view of triples. J. Algebra **19**, 228–253 (1971)
44. Osborne, M.S.: Hausdorffization and Such. Am. Math. Mon. **121**, 727–733 (2014)
45. Pareigis, B.: Advanced Algebra. https://www.mathematik.uni-muenchen.de/~pareigis/Vorlesungen/01WS/advalg.pdf (2002). Accessed 15 Aug 2023
46. Pareigis, B.: Categories and Functors. Pure and Applied Mathematics, vol. 39. Academic Press, New York-London (1970)
47. Riehl, E.: Categorical Homotopy Theory. New Mathematical Monographs, vol. 24. Cambridge University Press, Cambridge (2014)

48. Riehl, E.: Category Theory in Context. Aurora: Dover Modern Math Originals. Dover Publications, Mineola (2016)
49. Rotman, J.J.: Advanced Modern Algebra. Prentice Hall, Upper Saddle River (2002)
50. Rotman, J.J.: An introduction to the Theory of Groups. Graduate Texts in Mathematics, vol. 148. Springer-Verlag, New York (1995)
51. Shulman, M.A.: Set Theory for Category Theory. https://arxiv.org/abs/0810.1279v2 (2008). Accessed 15 Aug 2023
52. Spivak, D. I.: Category Theory for the Sciences. MIT Press, Cambridge (2014)
53. Stenström, B.: Rings of Quotients. Grundlehren der mathematischen Wissenschaften, vol. 217. Springer-Verlag, New York-Heidelberg (1975)

Index

A
Abelianization, 30
  functor, 30
Adjunction/adjoint functors, 153
Alexandroff topology, 34
Anti-isomorphism of categories, 38
Associative law, 2
Axiom of choice, 1

B
Balanced category, 10
Bifunctor, 24
Bimorphism, 10

C
Cartesian product
  bifunctor, 28
  functor, 29
Category, 2
  of abelian groups, 4
  of (commutative) algebras, 5
  of cocones, 108
  with coequalizers, 92
  of commutative unitary rings, 4
  of compact Hausdorff topological spaces, 5
  of cones, 108
  with (finite) coproducts, 84
  defined by a pre-ordered set, 3
  of divisible groups, 4
  with equalizers, 92
  of fields, 4
  with finite colimits, 110
  with finite limits, 110

  of finite sets, 3
  of groups, 4
  of Hausdorff topological spaces, 5
  of left R-modules, 5
  of monoids, 4
  of objects over C (slice category), 59
  of objects under C (coslice category), 58
  of partially ordered sets, 5
  of pointed topological spaces, 5
  of pre-ordered sets, 5
  of presheaves, 63
  with (finite) products, 84
  with pullbacks, 97
  with pushouts, 98
  of relations, 4
  of right R-modules, 4
  of rings, 4
  of sets, 3
  of simple groups, 4
  of small categories, 35
  of topological spaces, 5
  of unitary rings, 4
Cayley's theorem, 76
Cocomplete category, 110
Cocone, 106
Codomain of a morphism, 2
Coequalizer, 92
Cogenerator (coseparator), 237
Cokernel of a morphism, 97
Colimit
  functor, 110, 118
  preserving functor, 127
  reflecting functor, 131
Comma-category, 55
Commutative diagram, 19

Commutator subgroup, 30
Compact-open topology, 27
Complete category, 110
Composition law, 2
Concrete category, 38
Cone, 106
Congruence (on a category), 20
Constant functor, 25
Continuity at a point, 162
Coproduct, 84
Coreflective subcategory, 194
Coreflector, 194
Core groupoid of a category, 10
Cosolution set condition, 231
Counit of an adjunction, 170
Covariant/contravariant functor, 24
Co-well-powered category, 16

D
Diagonal functor, 69
Diagram, 19
Discrete
    category on a set, 2
    topology, 9
Divisible group, 4
Domain of a morphism, 2
Dorroh extension
    functor, 34
    of a ring, 33
Double dual space functor, 28
Dual
    (opposite) category, 16
    functor, 52
    of a natural transformation, 54
    space (contravariant) functor, 28
Duality
    of categories, 199
    principle, 52

E
Edges of a graph, 18
Empty
    category, 3
    functor, 26
Epimorphism, 6
Equalizer, 91
Equivalence
    of categories, 199
    relation generated by a binary relation,
        94

Essentially
    small category, 209
    surjective functor, 36
Evaluation bifunctor, 68
Extremal
    epimorphism, 150
    monomorphism, 150

F
Faithful functor, 36
Final object, 11
Finite category, 2
Forgetful functor, 30
Free
    category on a graph, 18
    group functor, 155
    group on a set, 155
    module functor, 176
    module on a set, 176
    product of groups, 90
Freyd's adjoint functor theorem, 230
Frobenius functor, 225
Full
    functor, 36
    subcategory, 6
Fully faithful functor, 36
Functor
    category, 63
    preserving a property, 40
    reflecting a property, 40

G
Galois connection, 157
Gelfand-Naimark duality, 214
Generator (separator), 236
Godement product, 44
Graph, 18
Groupoid, 10

H
Hausdorff quotient functor, 33
Hom
    bifunctor, 26
    (contravariant) functor, 27
Homeomorphism, 9
Homotopy, 22
    category, 23
Horizontal composition of natural
        transformations (Godement
        product), 44

I

Identity
    law, 2
    morphism, 2
Image of a functor, 25
Inclusion functor, 25
Indiscrete topology, 7
Initial object, 11
Injective object, 80
Interchange law between vertical and
        horizontal composition of natural
        transformations, 80
Interior of a subset of a topological space, 94
Inverse
    of a functor, 38
    natural transformation, 43
Isomorphic
    categories, 38
    objects, 6
Isomorphism of categories, 38

K

Kernel of a morphism, 97

L

Left adjoint, 153
Left/right associated functor of a bifunctor,
        35
Limit
    functor, 110, 117
    preserving functor, 127
    reflecting functor, 131
Limitation of size axiom, 1
Localization
    of a category by a class of morphisms, 214
    functor with respect to a multiplicative set,
        33
Locally
    compact topological space, 162
    small category, 2

M

Monomorphism, 6
Morphisms of a category, 2
Multiplicative subset of a ring, 5

N

Natural isomorphism, 42
Natural transformation/isomorphism, 41
Neumann–Bernays–Gödel set theory, 1

O

Objects of a category, 2
Order preserving map, 5

P

Partially ordered set, 5
Pasting lemma, 23
Path (in a graph), 18
Pointed category, 11
Pointwise composition of functors, 25
Pontryagin duality, 214
Power set (contravariant) functor, 29
Precomposition functor, 68
Pre-ordered set, 3
Presheaf, 63
Product, 83
    category, 17
    topology, 29
Projection functor, 26
Projective object, 81
Pullback, 97
Pushout, 97

Q

Quotient
    category, 21
    functor, 25
    object, 14
    topology, 95

R

R-bilinear map, 32
Reflective subcategory, 194
Reflector, 194
Regular
    epimorphism, 150
    monomorphism, 150
Representability criterion, 245
Representable functor, 48
Representing
    object, 48
    pair of a functor, 50
Restriction of scalars functor, 34
Right adjoint, 153

S

Set
    of cogenerators (coseparators), 236
    of generators (separators), 236
Simple group, 4

Skeleton of a category, 208
Small
    category, 2
    colimit of a functor, 110
    graph, 18
    limit of a functor, 110
Solution set condition, 231
Special adjoint functor theorem, 239
Split
    epimorphism, 78
    (co)equalizer, 148
    monomorphism, 78
Stone duality, 214
Stone space, 214
Stone–Čech compactification functor, 31
Strong
    epimorphism, 78
    monomorphism, 78
Subcategory, 6
Subobject, 12
Subspace topology, 94

T
Tensor product functor, 29
Trivial path in a graph, 18
Tube lemma, 162

U
Unit of an adjunction, 170
Universal enveloping group functor, 31

V
Vertex
    of a cocone, 107
    of a cone, 106
Vertical composition of natural
            transformations, 43
Vertices of a graph, 18

W
Weakly
    final set, 229
    initial set, 229
Well-powered category, 16
Whiskering of a natural transformation by a
            functor, 44

Y
Yoneda's embedding, 76
Yoneda's lemma, 70

Z
Zermelo–Fraenkel set theory, 1
Zero-morphism, 11
Zero object, 11

Printed in the United States
by Baker & Taylor Publisher Services